Supplements to the 2nd Edition of

RODD'S CHEMISTRY OF CARBON COMPOUNDS

ELSEVIER SCIENCE PUBLISHERS B.V.
Sara Burgerhartstraat 25
P.O. Box 211, 1000 AE Amsterdam, The Netherlands

Distributors for the United States and Canada:

ELSEVIER SCIENCE PUBLISHING COMPANY INC.
52, Vanderbilt Avenue
New York, NY 10017

Library of Congress Card Number: 64-4605

ISBN 0-444-42897-6

© Elsevier Science Publishers B.V., 1987

Printed In The Netherlands

Supplements to the 2nd Edition of

RODD'S CHEMISTRY OF CARBON COMPOUNDS

VOLUME I

ALIPHATIC COMPOUNDS

★

VOLUME II

ALICYCLIC COMPOUNDS

★

VOLUME III

AROMATIC COMPOUNDS

★

VOLUME IV

HETEROCYCLIC COMPOUNDS

★

VOLUME V

MISCELLANEOUS
GENERAL INDEX

★

Supplements to the 2nd Edition (Editor S. Coffey) of

RODD'S CHEMISTRY OF CARBON COMPOUNDS

A modern comprehensive treatise

Edited by
MARTIN F. ANSELL
Ph.D., D.Sc. (London) F.R.S.C. C. Chem.
Reader Emeritus, Department of Chemistry,
Queen Mary College, University of London, Great Britain

Supplement to

VOLUME IV HETEROCYCLIC COMPOUNDS

Part G:
Six-Membered Heterocyclic Compounds with a Single Nitrogen Atom in the Ring to which are Fused Two or More Carbocyclic Ring Systems, and Six-Membered Ring Compounds where the Hetero-Atom is Phosphorus, Arsenic, Antimony or Bismuth. Alkaloids Containing a Six-Membered Heterocyclic Ring System

ELSEVIER
Amsterdam – Oxford – New York – Tokyo   1987

CONTRIBUTORS TO THIS VOLUME

Kenneth W. Bentley, M.A., D.Sc., D.Phil., F.R.S.E.
Department of Chemistry, Loughborough University,
Loughborough, Leicestershire LE11 3TU

John D. Hepworth, B.Sc., Ph.D., C.Chem., F.R.S.C.
Department of Chemistry, Lancashire Polytechnic,
Preston, Lancashire, PR1 2TQ

Robert Livingstone, B.Sc., Ph.D., F.R.S.C.
Department of Pure and Applied Chemistry, The Polytechnic,
Queensgate, Huddersfield, HD1 3DH

A. Reginald Pinder, D.Sc., Ph.D., D.Phil.
Department of Chemistry, The University,
Clemson, South Carolina, U.S.A.

Malcolm Sainsbury, D.Sc., Ph.D., C.Chem., F.R.S.C.
Department of Chemistry, The University,
Bath, BA2 7AY

Raymond E. Fairbairn, B.Sc., Ph.D., F.R.S.C.
Formerly of Research Department,
Dyestuffs Division,
I.C.I. (INDEX)

PREFACE TO SUPPLEMENT IVG

The publication of this volume continues the supplementation of the second edition of Rodd's Chemistry of Carbon Compounds, thus keeping this major work of reference up-to-date. In this volume Chapters 28 to 35 of the second edition are brought up-to-date and the supplement covers the advances that have occurred in the decade since the publication of Volume IVG in 1978.

I have been fortunate in that three of the contributors to the second edition, namely Professor Bentley, Professor Pinder and Dr Sainsbury have again provided valuable contributions, as have Dr Hepworth and Professor Livingstone who have previously contributed to other supplements to Rodd. To each of these authors I express my thanks and appreciation for providing clear, concise and interesting chapters. I also wish to thank Dr Fairbairn, who indexed the second edition, for again providing an extremely detailed index which greatly facilitates the use of this book.

At a time when there are many specialist reviews, monographs and reports available, there is still in my view an important place for a book such as Rodd, which gives a broader coverage of organic chemistry. One aspect of the value of this work is that it allows the expert in one field to quickly find out what is happening in other fields of chemistry. On the other hand a chemist looking for the way into a field of study will find in Rodd an outline of the important aspects of that area in chemistry together with leading references to other works to provide more detailed information.

This volume has been produced by direct reproduction of the manuscripts. I am most grateful to the contributors for all the care and effort both they and their secretaries have put into the production of the manuscripts, including the diagrams. I also wish to thank the staff at Elsevier for all the help they have given me and for seeing the transformation of authors' manuscripts to published work.

September 1987   Martin Ansell

CONTENTS

VOLUME IV G

Heterocyclic Compounds: Six-Membered Heterocyclic Compounds with a Single Nitrogen Atom in the Ring to which are Fused Two or More Carbocyclic Ring Systems, and Six-Membered Ring Compounds where the Hetero-Atom is Phosphorus, Arsenic, Antimony or Bismuth. Alkaloids Containing a Six-Membered Heterocyclic Ring System

*Chapter 28. Polycyclic Compounds Comprising a Pyridine and Two or More
Carbocyclic Rings*
by J.D. HEPWORTH

Chapter 35. Steroidal Alkaloids
by A.R. PINDER

OFFICIAL PUBLICATIONS

B.P.	British (United Kingdom) Patent
F.P.	French Patent
G.P.	German Patent
Sw.P.	Swiss Patent
U.S.P.	United States Patent
U.S.S.R.P.	Russian Patent
B.I.O.S.	British Intelligence Objectives Sub-Committee Reports
F.I.A.T.	Field Information Agency, Technical Reports of U.S. Group Control Council for Germany
B.S.	British Standards Specification
A.S.T.M.	American Society for Testing and Materials
A.P.I.	American Petroleum Institute Projects
C.I.	Colour Index Number of Dyestuffs and Pigments

SCIENTIFIC JOURNALS AND PERIODICALS

With few obvious and self-explanatory modifications the abbreviations used in references to journals and periodicals comprising the extensive literature on organic chemistry, are those used in the World List of Scientific Periodicals.

LIST OF COMMON ABBREVIATIONS AND
SYMBOLS USED

A	acid
Å	Ångström units
Ac	acetyl
a	axial; antarafacial
as, $asymm.$	asymmetrical
at	atmosphere
B	base
Bu	butyl
b.p.	boiling point
C, mC and μC	curie, millicurie and microcurie
c, C	concentration
C.D.	circular dichroism
conc.	concentrated
crit.	critical
D	Debye unit, 1×10^{-18} e.s.u.
D	dissociation energy
D	dextro-rotatory; dextro configuration
DL	optically inactive (externally compensated)
d	density
dec. or decomp.	with decomposition
deriv.	derivative
E	energy; extinction; electromeric effect; Entgegen (opposite) configuration
E1, E2	uni- and bi-molecular elimination mechanisms
E1cB	unimolecular elimination in conjugate base
e.s.r.	electron spin resonance
Et	ethyl
e	nuclear charge; equatorial
f	oscillator strength
f.p.	freezing point
G	free energy
g.l.c.	gas liquid chromatography
g	spectroscopic splitting factor, 2.0023
H	applied magnetic field; heat content
h	Planck's constant
Hz	hertz
I	spin quantum number; intensity; inductive effect
i.r.	infrared
J	coupling constant in n.m.r. spectra; joule
K	dissociation constant
kJ	kilojoule

LIST OF COMMON ABBREVIATIONS

k	Boltzmann constant; velocity constant
kcal	kilocalories
L	laevorotatory; laevo configuration
M	molecular weight; molar; mesomeric effect
Me	methyl
m	mass; mole; molecule; *meta-*
ml	millilitre
m.p.	melting point
Ms	mesyl (methanesulphonyl)
[M]	molecular rotation
N	Avogadro number; normal
nm	nanometre (10^{-9} metre)
n.m.r.	nuclear magnetic resonance
n	normal; refractive index; principal quantum number
o	*ortho-*
o.r.d.	optical rotatory dispersion
P	polarisation, probability; orbital state
Pr	propyl
Ph	phenyl
p	*para-*; orbital
p.m.r.	proton magnetic resonance
R	clockwise configuration
S	counterclockwise config.; entropy; net spin of incompleted electronic shells; orbital state
S_N1, S_N2	uni- and bi-molecular nucleophilic substitution mechanisms
$S_N i$	internal nucleophilic substitution mechanisms
s	symmetrical; orbital; suprafacial
sec	secondary
soln.	solution
symm.	symmetrical
T	absolute temperature
Tosyl	*p*-toluenesulphonyl
Trityl	triphenylmethyl
t	time
temp.	temperature (in degrees centigrade)
tert.	tertiary
U	potential energy
u.v.	ultraviolet
v	velocity
Z	zusammen (together) configuration

LIST OF COMMON ABBREVIATIONS

α	optical rotation (in water unless otherwise stated)
$[\alpha]$	specific optical rotation
αA	atomic susceptibility
αE	electronic susceptibility
ε	dielectric constant; extinction coefficient
μ	microns (10^{-4} cm); dipole moment; magnetic moment
μB	Bohr magneton
μg	microgram (10^{-6} g)
λ	wavelength
ν	frequency; wave number
$\chi, \chi d, \chi \mu$	magnetic, diamagnetic and paramagnetic susceptibilities
$\sim$	about
$(+)$	dextrorotatory
$(-)$	laevorotatory
$(\pm)$	racemic
$\ominus$	negative charge
$\oplus$	positive charge

Chapter 28

POLYCYCLIC COMPOUNDS COMPRISING A PYRIDINE AND TWO OR
MORE CARBOCYCLIC RINGS

J.D. HEPWORTH

1. *Acridine and its derivatives*

The widespread interest in acridine and its derivatives
is reflected in the reviews which relate to their
chemistry (S. Skonieczny, Heterocycles, 1977, 6, 987;
1980, 14, 985), biological activity (A. Nasim and
T. Brychcy, Mutat. Res., 1979, 65, 261 and M.R. Melamed
and Z. Darzynkiewicz in 'Histochemistry: Widening
Horizons in Applied Biomedical Science' ed. P.J. Stoward
and J.M. Polak, Wiley, Chichester, 1981, p. 237) and
their occurrence in alkaloids (M.F. Grundon, Nat. Prod.
Rep., 1985, 2, 393).

(i) *Acridines*

The principal synthetic route to acridines involves
formation of the C-9 – C-9a bond γ to the heteroatom,
although the exact nature and source of the immediate
acridine precursor varies quite appreciably.

A range of derivatives of 2-nitroacridine has been obtained by the acid catalysed cyclisation of 2-arylamino-5-nitrobenzaldehydes which probably proceeds through protonation of the carbonyl group (J. Rosevear and J.F.K. Wilshire, Austral. J. Chem., 1981, 34, 839). It is of interest to note that the presence of a *meta*-substituent in the arylamino moiety leads predominantly to the 6-substituted 2-nitroacridine; only a small amount of the 8-isomer is produced. Electron withdrawing substituents markedly decrease the rate of cyclisation in trifluoroacetic acid. The effect of electron releasing groups is not as simple, for whilst a methyl group accelerates cyclisation, a dimethylamino or a methoxy group has a retarding effect. It seems certain that under the strongly acidic conditions used to effect cyclisation the amine function is protonated and perhaps the methoxy group is similarly affected.

(i) $ArNH_2$, DMSO, Et_3N; (ii) CF_3COOH

Electron rich 2-arylaminoacetophenones cyclise readily on alumina (B. Kasum and R.H. Prager, Austral. J. Chem., 1983, 36, 1455) and polyphosphoric ethyl ester effects the cyclodehydration of amides (1) to 9-aminoacridines (D. Chambers and W.A. Denny, J. chem. Soc. Perkin I, 1986, 1055).

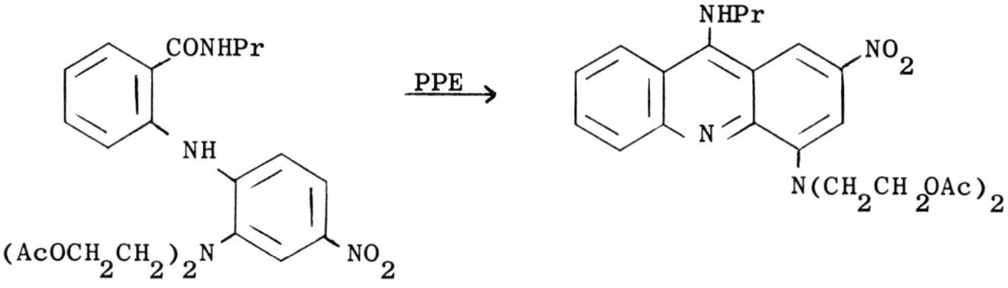

(1)

3-Acyl-2,5-bis(arylamino)-1,4-benzoquinones (2) undergo cyclisation in either sulphuric acid or methanolic hydrogen chloride to yield acridinequinones (3) (K. Joos, M. Pardo and W. Schafer, J. chem. Research (M), 1978, 4901). 2-Methoxyacridine is the starting point for the synthesis of both acridine-1,4-diones and the corresponding 1,2-quinones (J. Renault *et al.*, Eur. J. Med. Chem.-Chim. Ther., 1981, __16__, 24).

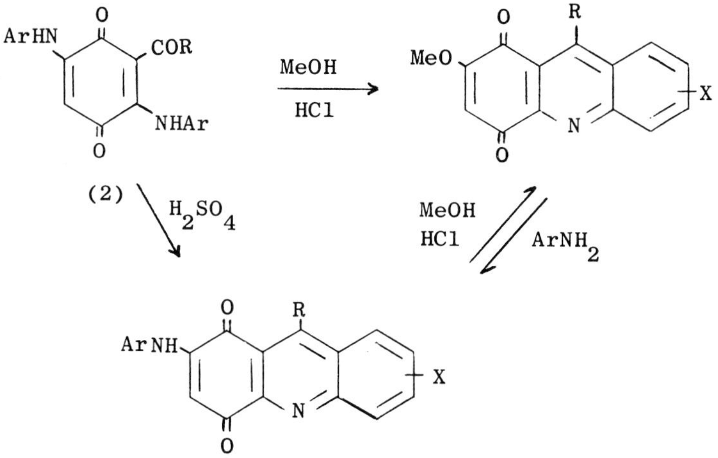

(3)

Cyclisation of 4-chloro-N-(3-nitrophenyl)anthranilic acid (4) by means of phosphorus oxychloride gives a mixture of 1- and 3-nitro-6,9-dichloroacridine. The 1-isomer reacts selectively with pyridine to give the 9-pyridinium salt and hence separation of the isomers is possible (B. Wysocka-Skrzela, K. Biskup and A. Ledochowski, Rocz. Chem., 1977, 51, 2411). The 9-chlorine atom in each isomer is exclusively displaced by phenol; the resulting 9-phenoxy substituent is also labile. For example, reaction with mono-Boc-protected amines enables mono-, di- and tri- 9-acridyl derivatives (5, 6 and 7) of polyamines to be prepared (J.B. Hansen and O. Buchardt, Chem. Comm., 1983, 162) and several 9-acridylamino acids have been obtained from 1-nitro-9-phenoxyacridine (B. Wysocka-Skrzela, G. Weltrowska and A. Ledochowski, Pol. J. Chem., 1980, 54, 619).

(4)   (5)

(6)   (7)

2,9-Dimethylacridine results from the reaction of 4-methyldiphenylamine with acetic acid in the presence of zinc chloride; the route involves decarboxylation of N-p-tolylanthranilic acid, itself a precursor of the acridine ring system (J.R. Patton and K.H. Dudley. J. heterocyclic Chem., 1979, 16, 257). It is also of interest to note that C-9 of acridine can be introduced directly by the vapour phase reaction of chloroform and diphenylamine (R.E. Busby *et al.*, J. chem. Research (M), 1980, 4935).

Flash vacuum pyrolysis of 2-azidodiphenylmethane (8) affords a mixture of acridine and its 9,10-dihydro derivative, the composition of the product varying with the reaction temperature (M.G. Hicks and G. Jones, Chem. Comm., 1983, 1277).

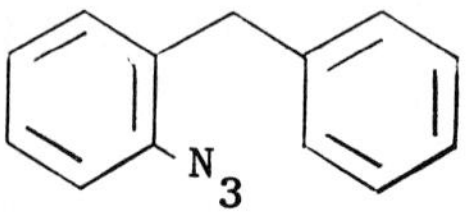

Temp	% composition	
(°C)	acridine	9,10-dihydro
350	10	90
500	33	66
700	95	5

(8)

Both a dihydroacridine and an acridine result from the thermolysis of 2-azidotriphenylmethanes (9) (R.N. Carde *et al.*, J. chem. Soc. Perkin I, 1978, 1211; 1981, 1132). In an analogous manner, 2-(phenylamino)phenylcarbenes (10), generated in the vapour phase from the tosylhydrazone sodium salts, insert into the adjacent *ortho*-position although giving only the dihydro derivative (W. D. Crow and H. McNab, Austral. J. Chem., 1981, 34, 1037).

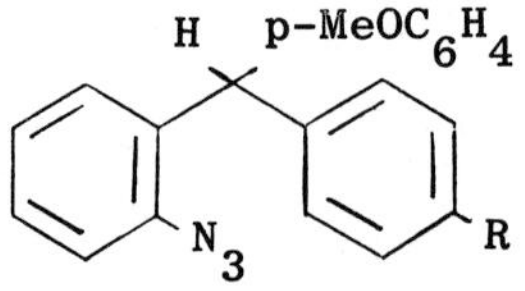

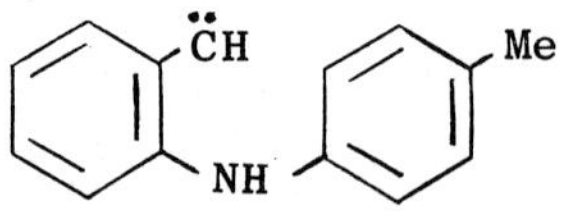

(9)

(10)

The quinone methide (11) derived from flavan by a retro-Diels-Alder reaction gives only a 4% yield of acridine when heated with aniline. However, the other products include the diphenylmethane (12) and the Mannich base (13) both of which yield the tricyclic compound upon pyrolysis (J.L. Asherson, O. Bilgic and D.W. Young, J. chem. Soc. Perkin I, 1980, 522).

Both 2-N-phenylaminobenzyl alcohol and 1-phenylbenzoxazine (14) break down to the azaxylylene (15) at high temperatures; in the presence of silica or alumina the reaction temperature is reduced from 650°C to 400°C. Electrocyclisation of the azaxylylene gives dihydroacridine with some acridine, the extent of dehydrogenation increasing at temperatures over 650°C (I. Hodgetts, S.J. Noyce and R.C. Storr, Tetrahedron Letters, 1984, 5435).

Alkylation of acridine at C-9 occurs on reaction with ω-alkoxyalkyl lithium compounds and subsequent oxidation of the resulting 9,10-dihydroacridine derivative (16). Various functional group interconversions are possible leading notably to acridines bearing phosphorus containing substituents at the 9-position (L. Horner and W. Hallenbach, Phosphorus and Sulphur, 1984, 20, 173).

(16)

Cyclisation of 2-carboxytriphenylamines by phosphorus oxychloride provides access to 9-amino-10-arylacridinium salts *via* the 9-chloro derivatives (R.M. Acheson, D.H. Birtwistle and P.B. Wyatt, J. chem. Research (M), 1986, 2762).

An alternative approach to 9-substituted acridinium salts and thence the corresponding acridines involves the conversion of 9-acridones into the 9-trifluoromethanesulphonyloxy acridinium salts (17), which react readily with halides, pseudohalides such as azide and isothiocyanate, and sulphur nucleophiles (B. Singer and G. Maas, Z. Naturforsch., 1984, 39b, 1399). The free base results on reaction with diisopropylethylamine. 9,9'-Bisacridine ethers are also available by this methodology.

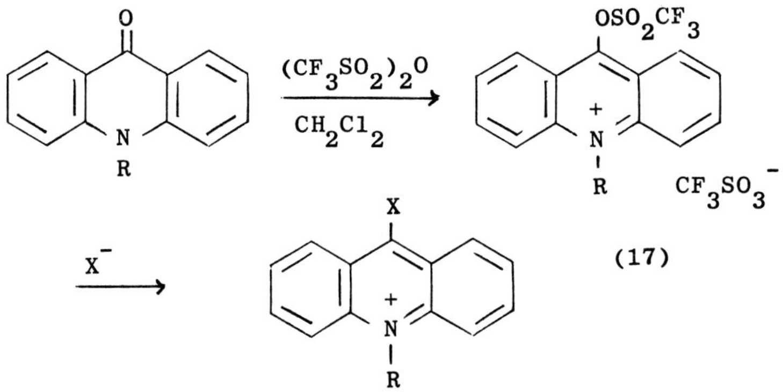

(17)

A range of acridine derivatives are available from 3,6-diaminoacridine, proflavine, through either diazotisation which yields mainly the mono-diazonium salt (W. Firth III and L.W. Yielding, J. org. Chem., 1982, 47, 3002) or direct iodination by iodide ion in the presence of chloramine-T (R.F. Martin and D.P. Kelly, Austral. J. Chem., 1979, 32, 2637).

It is well known that acridine reacts with dienophiles to yield the unbridged dihydroacridines rather than the bridged adducts. However, N-methyl- N -propargyl-9-acridinecarboxamide (18) undergoes an intramolecular Diels-Alder reaction, providing the first example of a

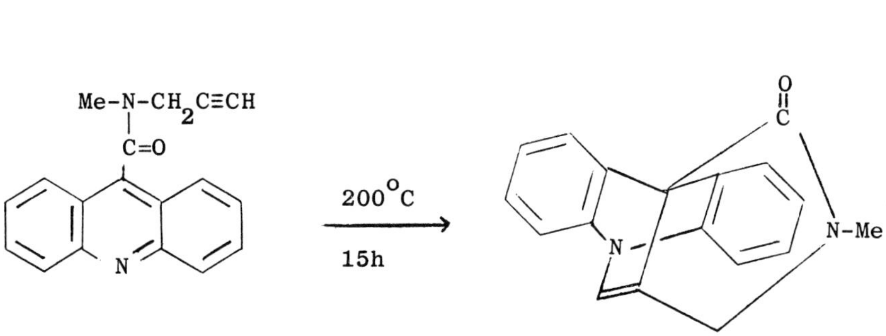

(18)

thermal [4+2] cycloaddition of an acridine (E. Ciganek, J. org. Chem., 1980, 45, 1497).

The [1]H- and [13]C- nmr data for acridine are shown below. A study of the nmr spectra of some substituted aminoacridines has shown that the electron density is significantly higher at the sites adjacent to the amino substituent in agreement with the observed pattern of electrophilic substitution (R.F. Martin and D.P. Kelly, Austral. J. Chem., 1979, 32, 2637). The [1]H- and [13]C- nmr spectra for some 9-substituted acridines (R. Faure *et al.*, Farmaco Ed. Sci., 1980, 35, 779; Chem. Scripta, 1980, 15, 62) and pmr spectra for alkoxy derivatives of 9-chloro-6-nitroacridines (S. Mager, I. Hopartean and D. Binisor, Monatsch., 1978, 109, 1393) have been reported.

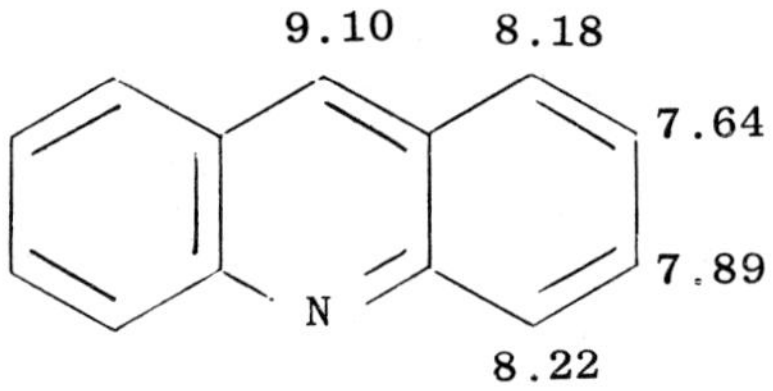

Coupling	1,2	1,4	1,9	2,3	3,4	4,9
J(Hz)	8.2	0.6	0.4	6.6	9.0	0.9

Proton chemical shifts (δ) and coupling constants for acridine.

^{13}C chemical shifts (δ) for acridine

Acridine coordinates with the shift reagent Eu(thd)$_3$ although the binding constant is much smaller than for pyridine or quinoline even though the pK$_a$ values are quite similar (D.M. Rackham, Spectros. Letters, 1980, 13, 517).

The radicals arising by X-ray irradiation of a single crystal of acridine are formed by the addition of a hydrogen atom to the heteroatom. Spin densities have been deduced from the proton hyperfine tensors determined by the ENDOR technique. The unpaired electron is extensively delocalised but there is a large spin density on C-9 (V.P. Chacko, C.A. McDowell and B.C. Singh, Molecular Physics, 1979, 38, 321).

Various electronic indices have been obtained for the five aminoacridines, several aminobenzacridines and the corresponding salts from molecular orbital calculations and these have been related to the pK$_a$ values and infrared, ultraviolet and visible spectra of the molecules (N.F. Ellerton and D.O. Jordan, Austral. J. Chem., 1978, 31, 1463).

(ii) 9,9'-Biacridines

The redox reaction of benzaldehyde and acridine in the presence of 3-benzylthiazolium salts gives high yields of 9,9',10,10'-tetrahydro-9,9'-biacridine, (biacridan) (19) by electron transfer from the activated aldehyde (H. Inoue and K. Higashiura, Chem. Comm., 1980, 549).

(19)

Oxidation of 10,10'-dimethyl-9,9'-dihydro-9,9'-biacridine by a range of π-acceptors leads to the 10-methylacridinium ion, presumably *via* the 10-methylacridanyl radical, (A.K. Colter *et al.*, Canad. J. Chem., 1985, 63, 445). This radical has been trapped by 2-methyl-2-nitrosopropane during the oxidation of 9,10-dihydro-10-methylacridine by 2,3-dicyano-1,4-benzoquinone (C.C. Lai and A.K. Colter, Chem. Comm., 1980, 1115).

Lucigenin, a charge transfer complex between 10,10'-dimethyl-9,9'-biacridinium and two nitrate anions (20) affords 10,10'-dimethyl-9,9'-biacridylidene (21) on irradiation at the charge transfer band (> 510 nm) in deaerated solution. However, irradiation at > 420 nm gives the red 7,16-dimethylbenzo[1,2,3-k1:6,5,4-

k'1']diacridine (22) (K. Moeda *et al.*, J. chem. Soc. Perkin II, 1984, 441). Reduction of lucigenin proceeds *via* a cation radical to the acridylidene (21) (E. Ahlberg, O. Hammerich and V.D. Parker, J. Amer. chem. Soc., 1981, 103, 844).

The two dihydroacridine systems have a dihedral conformation and are joined by an elongated C-C single bond of 1.58 Å in 10,10'-dimethyl-9,9',10,10'-tetrahydro-9,9'-biacridine (J. Preuss, V. Zanker and A. Gieren, Acta. Cryst., 1977, B33, 2317).

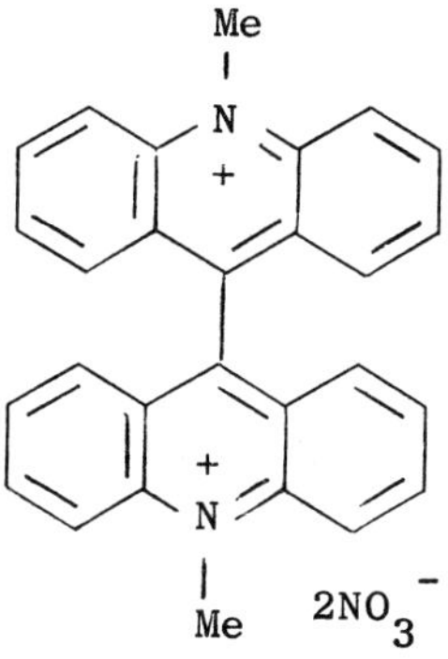

(20)

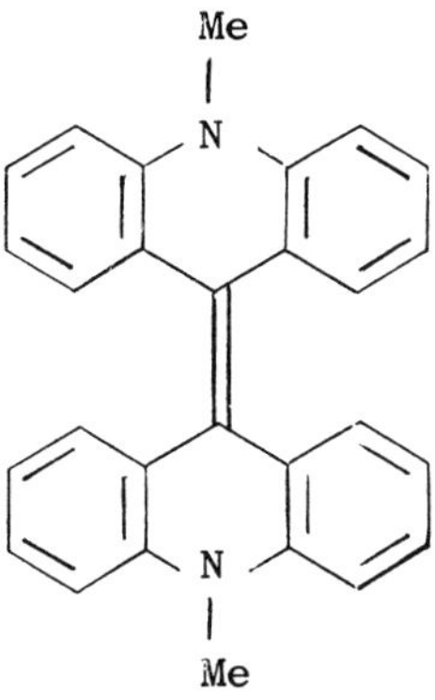

(21)

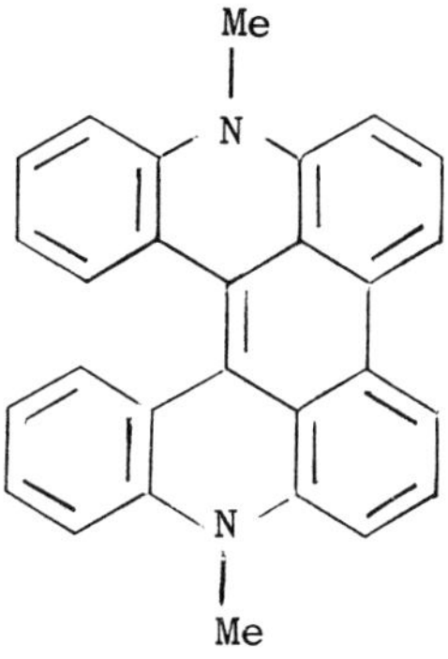

(22)

(iii)   Reduced acridines

The reduction of acridine by water gas   (CO + H_2O),
synthesis gas (CO + H_2) or hydrogen alone is catalysed by
transition metal carbonyls.   The reaction is highly
regioselective,   only   the   heterocyclic   ring   being
hydrogenated (R.H. Fish, A.D. Thormodsen and G.A. Cremer,
J.   Amer.   chem.   Soc.,   1982,   104,   5234).   A similar
selectivity is not observed using $(Ph_3P)_3RhCl$ as catalyst
and 1,2,3,4-tetrahydroacridine is formed in addition to
the   9,10-dihydroacridine   (R.H.   Fish.   J.L.   Tan   and
A.D. Thormodsen, J. org. Chem.,1984, 49, 4500).

9,10-Dihydroacridines arise from the reaction of ketones
with diarylamines at elevated temperatures and pressures.
Cyclic ketones lead to spirodihydroacridines (23).   The
9-aryl   spiro   derivatives   exist   in   a   twisted   boat
conformation,   with   the   aryl   substituent   pseudo
equatorial,   as   do   the   analogous   symmetrical   9,9-diaryl
dihydroacridines (W. Tritschler *et al.,* Ber., 1984, 117.
2703).

(23)

Acridone is reduced by sodium in deuteriated butanol to the 9,9-di-deuteriated acridan, which on oxidation affords 9-deuterioacridine. Acridinium salts are readily reduced by hydride ion donors, providing support for the suggestion that the disproportionation of 10-methylacridinium involves hydride ion transfer from the 9-position of the pseudo-base to the corresponding site of the acridinium salt (J. Clark and M. Bakavoli, J. chem. Soc. Perkin I, 1977, 1966). The kinetics of the cation→pseudo-base equilibrium for 9-substituted acridinium ions have been studied (J.W. Bunting *et al.*, Canad. J. Chem., 1984, **62**, 351) and the rate of oxidation of the 2- and the 3- methoxy derivative of *N*-methylacridan has also been investigated (A.K. Colter *et al.*, Canad. J. Chem., 1984, **62**, 1781).

The dehydration of formate to carbon dioxide by 10-methylacridinium mimics the behaviour of formate dehydrogenase (J.E.C. Hutchins, D. A. Binder and M. M.

Kreevoy. Tetrahedron, 1986, <u>42</u>, 993). Whilst 10–
methylacridinium iodide does not oxidise alcohols, the 3–
hydroxy derivative does so in the presence of
potassium t–butoxide, thereby behaving as an NAD^{+} model
oxidising agent (S. Shinkai et $al.$, Chem. Letters, 1980,
1235; J. org. Chem., 1981, <u>46</u>, 2333).

9,10–Dihydroacridine is efficiently oxidised to acridine
under phase transfer conditions by oxygen (E. Alneri,
G. Bottaccio and V. Carletti, Tetrahedron Letters, 1977,
2117).

5,10–Dihydroacridines are formed along with 5,6–diphenyl–
5,6–dihydrophenanthridines in the reaction of benzyne
with imines such as PhCH=NPh, which confirms the
intermediacy of benzazetidines (C.W.G. Fishwick, R.C.
Gupta and R.C. Storr, J. chem. Soc. Perkin I, 1984,
2827).

A variation on the Pfitzinger acridine synthesis utilises
the reaction between isatin, cyclohexanone and ammonia to
yield 1,2,3,4–tetrahydroacridine–9–carboxamides (24)
directly (J. Bielavsky, Coll. Czech. chem. Comm., 1977,
<u>42</u>, 2802). Hofmann degradation of the amides yields the
partially reduced 9–aminoacridines.

(24)

A number of 9-dimethylamino-1,2,3,4-tetrahydroacridines have been obtained by the reaction of methyl anthranilate and cyclohexanones in hexamethylphosphoric triamide (A. Osbirk and E.B. Pedersen, Acta Chem. Scand., 1979, 33B, 313). At lower temperatures, the corresponding acridone can be isolated and it seems likely that the dimethylamino function is introduced by the reaction of HMPT with the acridone at reflux temperature. The use of phosphorus pentoxide and diethylamine and its hydrochloride in place of HMPT yields reduced 9-aminoacridines.

In an extension of this method, the corresponding 9-arylamino derivatives result directly when the ester, cyclohexanone and an arylamine hydrochloride are heated with phosphorus pentoxide and NN-dimethylcyclohexylamine. Under these conditions, anthranilamide affords the unsubstituted 9-aminoacridine, presumably via the nitrile (N.S. Girgis and E.B. Pedersen, Synthesis, 1985, 547).

1,2,3,4,5,6,7,8-Octahydroacridine results from the reaction between cyclohexanone and 2-aminomethylene-cyclohexanone (R.P. Thummel and D.K. Kohli, J. org. Chem., 1977, 42, 2742).

1,2,3,4,5,6,7,8-Octahydroacridine is reduced by sodium in ethanol to the *trans-syn-trans* perhydroacridine (25), the $^{13}C-$ nmr spectrum of which has been fully assigned. All three rings exist in the chair form in this rigid molecule (R.W. Vierhapper and E. Eliel, J. org. Chem., 1975, 40, 2734; 1976, 41, 199).

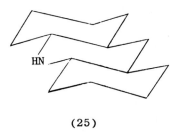

(25)

Treatment of 3-aminocyclohex-2-enone (26; R=H) with formaldehyde and acid yields 3,4,5,6,9,10-hexahydroacridine-1(2*H*),8(7*H*)-dione; the spiran (28), which is formed as a side-product, becomes the major product when trichloroacetic acid is used as the cyclising medium, and the exclusive product from dimedone enaminone (26; R=Me). The methylenebisenaminone (27) is also readily converted into a mixture of acridinedione and spiran (I. Chaaban, J.V. Greenhill and P. Akhtar, J. chem. Soc. Perkin I, 1979, 1593).

The use of acetaldehyde and benzaldehyde in place of the formaldehyde leads to the 9-methyl and 9-phenyl acridinedione, respectively, whilst 3-methylamino-cyclohex-2-enone gives the 10-methyl derivative.

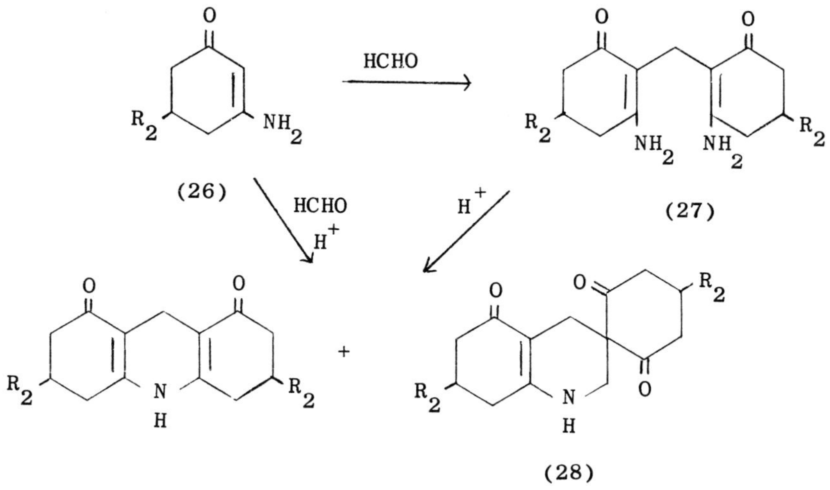

(26)   (27)

(28)

The reduced acridinediones are also produced when cyclohexan-1,3-diones react with a 3-amino-2-alkyl-acrolein, involving the loss of a carbon atom, possibly as formic acid. 7,8-Dihydroquinolin-5(6H)-ones are also formed (J.V. Greenhill *et al.*, J. chem. Research (M), 1981, 0821).

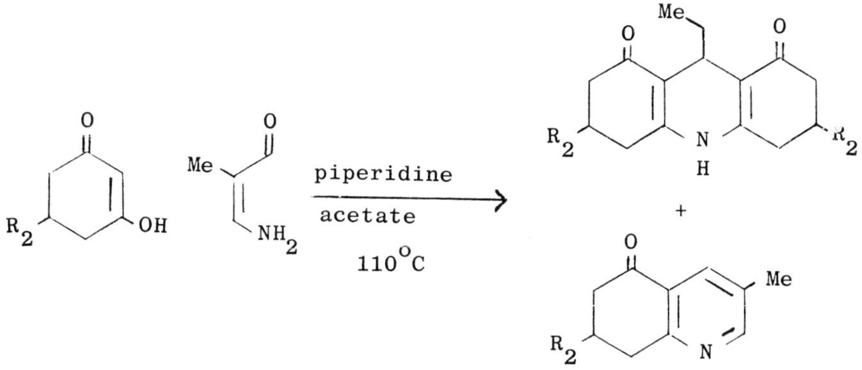

A ^{13}C- nmr study of 9,10-dihydroacridines indicates that there is very little delocalisation of the nitrogen lone pair of electrons into the aromatic rings in the case of the 10-acetyl derivative (E. Ragg *et al.*, J. chem. Soc. Perkin II, 1983, 1289). Although 9-hydroxy-1,2,3,4-tetrahydroacridine exists predominantly as the acridone, there is no evidence from ^{13}C- nmr spectral data for the presence of the tautomeric imine of 9-(*N*-methylamino)-1,2,3,4 -tetrahydroacridine (R. Faure *et al.*, J. Chim-phys., 1981, 78, 527).

(iv) Acridones

The cyclodehydration of 2-arylaminobenzoic acids, which are readily available from anilines and 2-halobenzoic acids by the Jourdan-Ullmann reaction, involving formation of the bond γ to the hetero-atom (C-9 – C-9a), is the most versatile and important route to 9-acridones and it continues to attract attention.

The effectiveness of a number of dehydrating agents for the cyclisation of some *N*-arylanthranilic acids has been investigated; polyphosphoric acid appears to be the reagent of choice (J.M. Kauffman and I.B. Taraporewala, J. heterocyclic Chem., 1982, 19, 1557).

Whilst electron-withdrawing groups on the amine component are considered to inhibit the Jourdan-Ullmann condensation (see J.M.F. Gagan in 'Chemistry of Heterocyclic Compounds, vol. 9, Acridines' ed. R.M. Acheson, Wiley-Interscience, New York, 1973), anthranilic acid reacts quantitatively with 2-halobenzoic acids to

give *N*-(2-carboxyphenylamino)benzoic acid and thence
9,10-dihydro-9-oxoacridine-4-carboxylic acid.

Of course, cyclisation of substituted examples of the
intermediate benzoic acids can lead to two different
products. An investigation of the factors affecting this
cyclisation recognised that the direction of ring closure
could be explained in terms of electronic and steric
influences of the substituent on the intermediate
carbocation (G.M. Stewart, G.W. Rewcastle and W.A. Denny
Austral. J. Chem., 1984, 37, 1939).   Thus, using
concentrated sulphuric acid as the cyclising medium,
precursors having electron withdrawing substituents yield
the acridone (29) in which the substituent and carboxyl
function are present in the same ring.   Electron
releasing substituents favour formation of the isomer
(30).   Similar results obtain with either phosphorus
oxychloride   or   ethyl   pyrophosphate   as   the
cyclodehydrating agent. Steric effects dominate when the
substituent is adjacent to the carboxyl group and often
result in almost exclusive formation of (30).

Despite these guidelines, the problem of isomer formation remains. This difficulty has been overcome by using a 2-haloisophthalic acid (31) as the acid component (G.W. Rewcastle and W.A. Denny, Synthesis, 1985, 217), although isomer formation still occurs when a *m*-substituted aniline is used in the reaction.

(31)

A further solution to the problem of isomer formation in the synthesis of oxoacridine carboxylic acids is to mask one of the two carboxyl groups. The requisite 2-(2-methoxycarbonylphenylamino)benzoic acids are available from methyl anthranilates by reaction with diphenyliodonium carboxylates. The esters which result after cyclisation with polyphosphoric ethyl ester undergo ready alkaline hydrolysis to the acid (G.W. Rewcastle and W.A. Denny, Synthesis, 1985, 220).

When the synthesis of the appropriately substituted anthranilic acid cannot be achieved by the usual Jourdan-Ullmann method, the reaction of an aniline with diphenyliodonium carboxylate in the presence of copper(II) acetate offers an alternative approach (D. Chambers and W.A. Denny, J. chem. Soc. Perkin I, 1986, 1055).

Diphenylamines containing substituents other than carboxyl can act as precursors of acridones, though not necessarily *via* a carbocation intermediate and novel examples include the thiobenzoate (32) (J. Martens, K. Praefcke and U. Schulze, Synthesis, 1976, 532) and the Schiff's base (33).

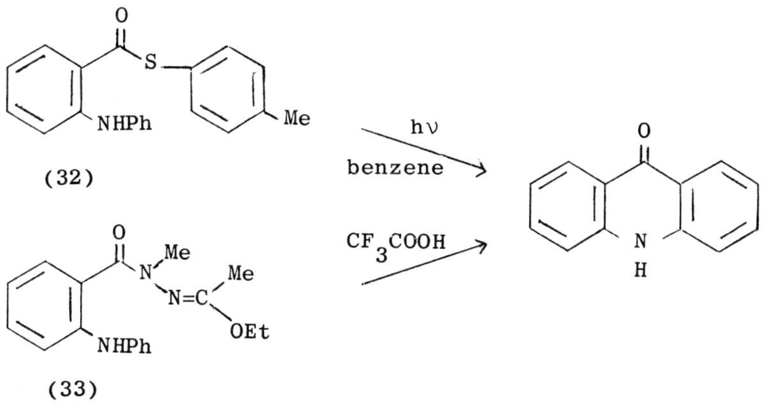

(32)

(33)

The cyclisation of *N*-arylanthranilamides into acridones is effected by prolonged boiling with heptafluorobutanoic acid (M. Iwao, J.N. Reed and V. Snieckus, J. Amer. chem. Soc., 1982, 104, 5531). The particular significance of this work lies in the regiospecific *ortho*-arylamination

of benzamides by directed metallation. The synthesis of
evoxanthine (34) in 13% overall yield from 3,4–
methylenedioxy–*NN*–dimethylbenzamide is illustrative of
the reaction sequence. The approach is not too
dissimilar from the formation of 4–methoxy–10–
methylacridone from the reaction between the lithio salt
of *N*–methylaniline and 2–(2,3–dimethoxyphenyl)oxazoline
(A.I. Meyers and R. Gabel, J. org. Chem., 1977, <u>42</u>,
2654).

A totally different approach to acridones involves
formation of the bond adjacent to the hetero–atom during
cyclisation. Thus, 2,4,6–trihydroxy–2'–nitrobenzophenone
yields 1,3–dihydroxyacridone after prior reduction (I.H.
Bowen, P. Gutpa and J.R. Lewis, Chem. Comm., 1970, 1625).
In a similar approach, the cyclisation of 2'–amino–2–
methoxybenzophenones and related compounds occurs on
treatment with sodium hydride in dimethylsulphoxide,
providing a route to acridone alkaloids
(J.H. Adams *et al.*, J. chem. Soc. Perkin I, 1977, 2173).

Treatment of 2,2',4,4'-tetranitrobenzophenone with an aromatic amine in DMSO leads to the displacement of both *ortho*-nitro groups and the formation of an *N*-aryl-3,6-dinitroacridone in high yield (J.H. Gorvin and D.P. Whalley, J. chem. Soc. Perkin I, 1979, 1364).

Formation of the C-N bond is also involved in several acridone syntheses based on thermal decompositions. The conversion of a 3-phenylbenz[3,4]isoxazole (35) into acridone has long been known (A. Kliegl, Ber., 1909, 42, 591; E. Bamberger, Ber., 1909, 42, 1707). The thermal transformation involves generation of a nitrene and leads to two products arising from either direct substitution

(35)

or *via* a rearrangement. The reaction is very sensitive to temperature, solvent, metal catalysts and to the nature of the substituents on the 3-phenyl ring (D.G. Hawkins and O. Meth-Cohn, J. chem. Soc. Perkin I, 1983, 2077).

Varying amounts of acridone are produced along with carbazoles when salicyl 2-azidobenzoates (36) are pyrolysed at 400°C. The acridone predominates with the phenyl ester (M.G. Clancy, M.M. Hesabi and O. Meth-Cohn, Chem. Comm., 1980, 1112) and later work indicates that the amount of acridone increases as the leaving group efficiency of OR increases (*idem*, J. chem. Soc. Perkin I, 1984, 429). Spray pyrolysis of 2,6-disubstituted phenyl 2-azidobenzoates leads to a mixture of 4-substituted 9,10-dihydroacridine and the corresponding acridine, possibly involving *ipso* attack by the nitrene.

(36)

Although the thermolysis of 3-aryl-1,2,3-benzotriazin-4(3H)-one (37) in paraffin oil (250°C) gives mainly the corresponding benzanilides (D.H. Hey, C.W. Rees and A.R. Todd, J. chem. Soc. C, 1968, 1028), in 1-methylnaphthalene a more rapid decomposition leads to 2-substituted 9-acridones. Lower yields of the acridones result in the absence of solvent (A.J. Barker *et al.*, J. chem. Soc. Perkin I, 1979, 2203).

(37)

The pyrolysis of 2-aminobenzophenone in the presence of silica affords acridone (I. Hodgetts, S.J. Noyce and R.C. Storr, Tetrahedron Letters, 1984, 5435).

The susceptibility of the chlorine in 9-chloroacridine towards nucleophilic displacement is well known and provides access to acridones *via* the phosphorus oxychloride cyclisation of 2-arylaminobenzoic acids. The chlorine atom in 1-chloro-4-nitroacridone also undergoes displacement and a range of amines have been prepared in this manner (J. Romanowski and Z. Eckstein, Rocz. Chim., 1977, 51, 2455). The nitro group is essential for activation of the halogen, since only the 9-halogen is displaced when 6,9-dichloroacridone is hydrolysed.

9-Acridone is aromatised to 9-methoxyacridine, m.p. 197°C, by dimethyl sulphate under phase transfer conditions (I. Willner and M. Halpern, Synthesis, 1979, 177). Substituted acridones give mainly N-alkyacridones on treatment with simple alkyl halides under phase transfer conditions, but branched alkyl halides favour formation of the alkoxyacridine (A. Mahamoud *et al.*, J. heterocyclic Chem., 1982, 19, 503).

The reaction of acridones with organolithium reagents provides a route to acridinium salts which has been utilised to prepare some 10-ethynyl salts (A.R. Katritzky and W.H. Ramer, J. org. Chem., 1985, 50, 852).

Tetraphosphorus decasulphide in hexamethylphosphoric triamide converts 9-acridones into the corresponding thiones, complementing the established routes from acridine and sulphur and from a 9-haloacridine and various sulphur reagents. The method is an improvement on the previous techniques for O–S exchange in acridones (R.R. Smolders *et al.*, Synthesis, 1982, 493). Various thioacridones have been *S*-alkylated and *S*-acylated under mild phase transfer conditions, utilising the tautomeric nature of the thione (M. Vlassa, M. Kezdi and I. Goia, Synthesis, 1980, 850).

[13]C-Nmr data for 9-acridone are given below (R. Faure *et al.*, Spectroscopy Letters, 1981, <u>14</u>, 223) and the ^{13}C-nmr spectra of a range of substituted acridones have been discussed in terms of steric and substituent effects and the electronic structure of the acridone ring system. Data for thioacridone (R. Faure *et al.*, Spectroscopy Letters, 1983, <u>16</u>, 431) and for various 10-methylacridones and several acridone alkaloids (D. Bergenthal *et al.*, Phytochem., 1979, <u>18</u>, 161; Z. Naturforsch., 1979, <u>34</u>, 516) are also available.

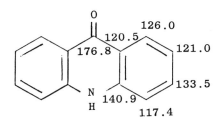

[13]C-chemical shifts (δ) for 9-acridone

The ion produced from acridone in CF_3SO_3D exchanges at H-2 and H-7 at room temperature (R.J. Smith, T.M. Miller and R.M. Pagni, J. org. Chem., 1982, __47__, 4181).

The difference between the experimental and calculated dipole moments for a series of *N*-alkyl derivatives of 9-acridone is attributed to a slight folding of the molecules. For 10-methyl-9-acridone, μ is 5.20 D (A.-M. Galy *et al.*, Farmaco Ed. Sci., 1981, __36__, 38).

2. *Phenanthridine and its derivatives*

In addition to the survey of phenanthridine in the second edition of Rodd, a comprehensive account of the synthesis, properties and reactions of phenanthridines covers the period 1950-1970 (B.R.T. Keene and P. Tissington in 'Advances in Heterocyclic Chemistry', ed. A.R. Katritzky and A.J. Boulton, Academic Press, New York, 1971, vol. 13, p. 315). Reviews of alkaloids based on phenanthridine regularly update knowledge in this area (S.D. Phillips and R.N. Castle, J. heterocyclic Chem., 1981, __18__, 223).

(i)   Phenanthridines

Further examples of the cyclisation of biphenyls to phenanthridines include carbanion based methods. In the case of the 2-aminobiphenyl (38), metallation is directed to the 2'-position by the 3'-methoxy group, and cyclisation to the phenanthridine follows formylation (N.S. Narasimhan, P.S. Chandrachood and N.R. Shete, Tetrahedron, 1981, <u>37</u>, 825).

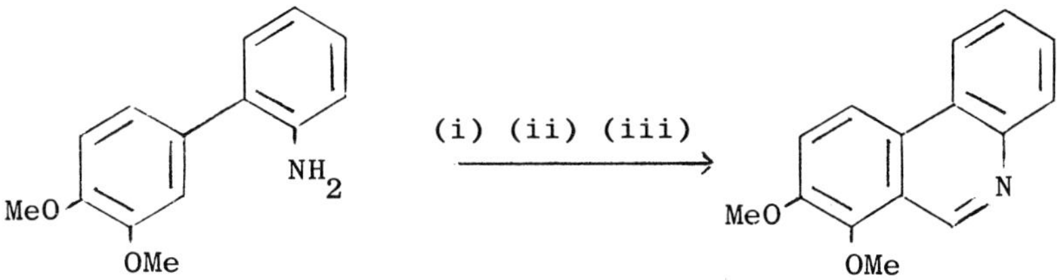

(38)

(i)   n-BuLi;   (ii)   DMF;   (iii) H_2O

The amide (39) affords a low yield of the 6-aryl-5-methyl phenanthridinium salt on treatment with *t*-butyl lithium (D. Hellwinkel, R. Lenz and F. Lammerzahl, Tetrahedron, 1983, <u>39</u>, 2073). Formanilides yield phenanthridines on treatment with phosphorus pentachloride (N.S. Narasimhan, *loc. cit.*)

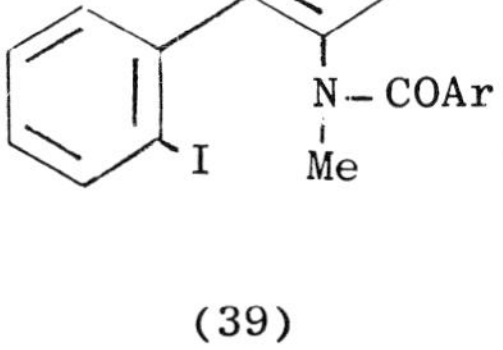

(39)

A different approach to the cyclisation of biphenyl derivatives involves the generation of imidoyl radicals from Schiff bases (40) using diisopropyl peroxydicarbonate. The intramolecular homolytic substitution is apparently not regiospecific since the methoxy derivative (40; R=OMe) affords a mixture of isomers (R. Leardini, A. Tundo and G. Zanardi, Synthesis, 1985, 107).

(40)

Schmidt rearrangement of 9-arylfluoren-9-ol using sodium azide in polyphosphoric acid gives the 6-arylphenanthridine, whereas in sulphuric acid migration of the more electron rich aryl ring leads to the fluorenylidene aniline. It is suggested that PPA complexes to the methoxy groups thereby reducing the electron richness of the aryl ring. The intermediate azide yields the phenanthridine on either pyrolysis or photolysis (S. I. Clarke and R. H. Prager, Austral. J. Chem., 1982, 35, 1645).

Photochemical methods of synthesis of the phenanthridine system involve the cyclisation of benzanilides and *N*-benzylideneanilines. The unfavourable geometry of these species and the competing n ⟶ π* excitation of the latter, generally result in low yields of the heterocycle. These adverse steric and electronic effects have been largely overcome in the photocyclisation of the boron complexes of *N*-arylbenzohydroxamic acids, a process which is both fast and high yielding (S. Prabhakar, A. M. Lobo and M. R. Tavares, Chem. Comm., 1978, 884).

Photolysis of aromatic Schiff's bases in strongly acidic media results in cyclisation to phenanthridines (A. Padwa, Chem. Rev., 1977, 77, 37). Irradiation of anils of polyfluoroaromatic ketones in trifluoroacetic acid also yields phenanthridines by oxidative cyclisation; examples are quoted which involve cleavage of a C–F bond (N.I. Danilenko *et al.*, Izvestia, 1980, 1606).

Further examples of the reactivity of a 6-chlorine atom in phenanthridine have been noted (M.S. Manhas and S.G. Amin, J. heterocyclic Chem., 1976, 13. 903; D.F. Pipe and C.W. Rees, Chem. Comm., 1982, 520).

The [1]H-chemical shifts for phenanthridine are as shown below. Tritium is incorporated especially at the 3- and 8-positions, whilst steric hindrance largely prevents exchange at the 1- and the 10-position (J.A. Elvidge *et al.*, J. chem. Soc. Perkin II, 1979, 386).

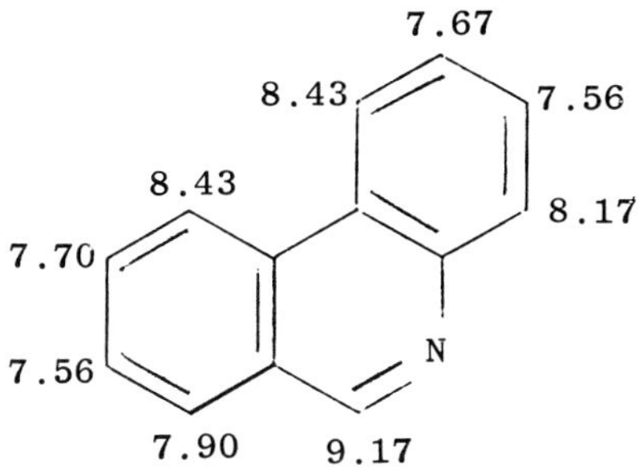

Proton chemical shifts (δ) for phenanthridine

(ii)   Phenanthridinium salts

Ethidium bromide (41), 3,4-diamino-5-ethyl-6-phenyl phenanthridinium bromide, is of value as a trypanocidal agent and, together with propidium iodide (42), is used as a probe for characterising nucleic acid function and structure. These are the only intercalating dyes which are capable of separating closed circular DNAs from linear DNAs (E. Gurr *et al.*, in 'The Chemistry of

Synthetic Dyes' ed. K. Venkataraman, Academic Press, New York, 1974. vol. VII, p. 277). Binding of ethidium and related species to DNAs has been surveyed (H.W. Zimmermann, Angew. Chem. intern. Edn., 1986, 25, 115).

(41)   R = $-C_2H_5$ (43)   R = $-CH_3$

(42)   R = $-CH_2CH_2CH_2N$—Et, with Me and Et branches

Acylation of ethidium occurs exclusively at the 8-amino group, although diazotisation results in attack at each amine function; azido and acetamido derivatives of ethidium are thereby available (W.J. Firth III et al., J. heterocyclic Chem., 1983, 20, 759).

The [1]H- and ^{13}C-nmr spectral data for ethidium bromide are given below. The assignments of the ^{13}C resonances for this salt and for dimidium bromide (43) and its des-phenyl derivative are derived from spectra simplified by using the inversion-recovery technique, since in the normal off-resonance decoupled spectra many signals overlap (B.G. Griggs *et al.*, Org. mag. Res., 1980, <u>14</u>, 371).

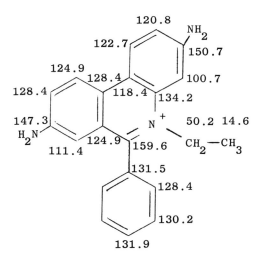

Proton chemical shifts (δ) for ethidium bromide

^{13}C-chemical shifts (δ) for ethidium bromide

(iii)   Reduced phenanthridines

The photo-stimulated reaction between 2-iodobenzylamine and the enolate derived from cyclohexanone gives 1,2,3,4-tetrahydrophenanthridine, presumably *via* the hexahydro derivative (R. Bengelmans, J. Chastenet and R. Roussi, Tetrahedron, 1984, __40__, 311). Cyclohexanone also features in the synthesis of 5-substituted 7,8,9,10-tetrahydrophenanthridinium salts involving its reaction with a secondary aromatic amine and formaldehyde in the presence of a mineral acid (B.M. Gutsulyak and Z.L. Novitskii, Zh. org. Khim., 1978, __48__, 1872), whilst the use of ammonium acetate and aqueous formaldehyde solution converts cyclohexanone into 1,2,3,4,7,8,9,10-octahydrophenanthridine, b.p. 110°C at 0.5 mm, in a most acceptable yield (R.P. Thummel and D.K. Kohli, J. org. Chem., 1977, __42__, 2742).

A neat construction of the phenanthridine system is based on the reaction of an *ortho*-dilithiated aniline (45) with an α-exomethylene ketone, such as pulegone (44). The reaction affords the 5,6,7,8,9,10-hexahydrophenanthridine directly under non-acidic conditions (P. Pedaja, C. Westerlund and A. Hallberg, J. heterocyclic Chem., 1986, __23__, 1353).

(44)   (45)

Acid catalysed dehydration of the alcohol (46) affords 1,2,3,4-tetrahydrophenanthridine *via* the *o*-cyclohexenylanilide (D.P. Curran and S.-C. Kuo, J. org. Chem., 1984, _49_, 2063).

(46)

Reduction of phenanthridones with either diborane in tetrahydrofuran (R.K.-Y. Zee-Cheng, S.-J. Yan and C.C. Cheng, J. med. Chem., 1978, _21_, 199) or borane-dimethyl sulphide (S.I. Clarke and R.H. Prager, Austral. J. Chem., 1982, _35_, 1645) yields 5,6-dihydrophenanthridines, whilst sodium bis(2-methoxyethoxy)aluminium hydride brings about partial reduction to 5,6-dihydrophenanthridin-6-ol. Both phenanthridine and *N*-methylphenanthridinium iodide undergo facile reduction to the respective 5,6-dihydro derivative on reaction with diborane (P.C. Keller, R.L. Marks and J.V. Rund, Polyhedron, 1983, _2_, 595).

5,5-Dialkyl-5,6-dihydrophenanthridinium salts (47) undergo a Stevens rearrangement to 5,6-dialkyl-5,6-dihydrophenanthridines (48); there is no evidence for benzyl migration which would lead to a dihydrobenzazepine. The migratory aptitude $PhCH_2 > Et > Me$ is consistent with the radical–pair mechanism accepted for Stevens rearrangements (R. Bedford *et al.,* Tetrahedron Letters, 1983, 1553).

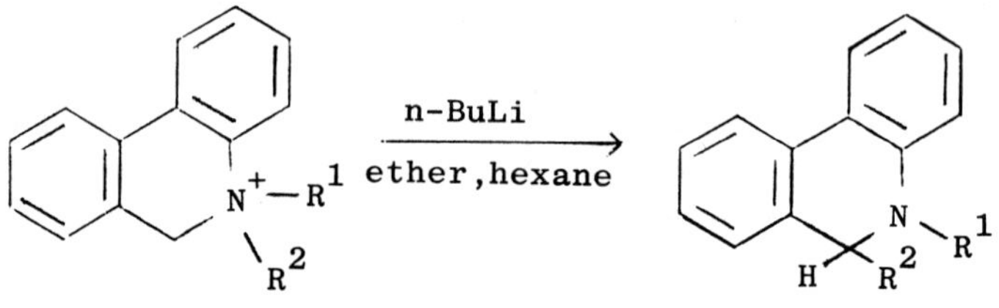

(47)　　　　　　　　　　　　　(48)

(iv)　Phenanthridones

Oxyacetic acids derived from *o*-phenylbenzohydroxamic acids are oxidised by persulphate to phenanthridone through the intermediacy of an amidyl radical (A.R. Forrester *et al.,* Tetrahedron Letters, 1977, 3601).

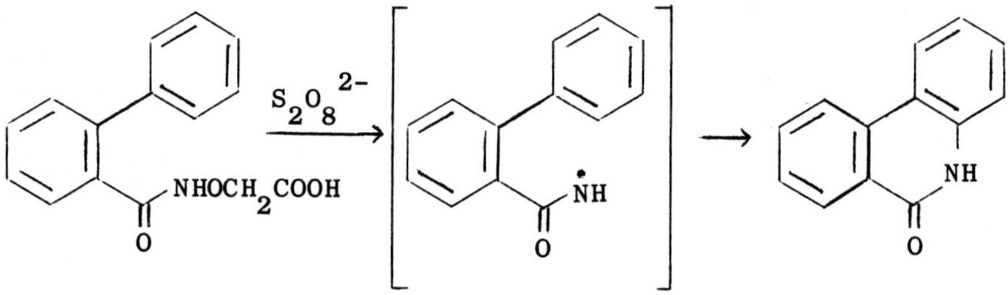

Extension of this method to the generation of N methoxybenzamidyls from N-methoxybenzamides allows the synthesis of N-methoxyphenanthridones, albeit in variable yields and admixed with several by-products (A.M. Forrester, E.M. Johansson and R.H. Thompson, J. chem. Soc. Perkin I, 1979, 1112).

Intramolecular aromatic substitution leads to the formation of N-methoxyphenanthridone when N-chloro-N-methoxybiphenyl-2-carboxamide is treated with silver tetrafluoroborate in the dark; a nitrenium ion is implicated (S. A. Glover *et al.*, J. chem. Soc. Perkin I, 1984, 2255).

Carbamoyl radicals initiate the cyclisation of 2-formamidobiphenyls to phenanthridones in high yield (R. Leardini, A. Tundo and G. Zamardi, J. chem. Soc. Perkin I, 1981, 3164).

$$\xrightarrow[\text{PhCl},110^{\circ}\text{C}]{(\text{t-BuO})_2}$$

Phenanthridones also result from the cyclisation of the ethoxycarbonyl derivative of 2-aminobiphenyls with phosphorus pentachloride, whilst 2-amino-3',4'-dimethoxybiphenyl-2'-carboxylic acid cyclises spontaneously to the phenanthridone (N. S. Navarsimhan, P. S. Chandrachood and N.R. Shete, Tetrahedron, 1981, 37, 825).

Cyclodehydrohalogenation of 2-halobenzamides to phenanthridones occurs in dimethylacetamide in the presence of palladium catalysts. When the halogen is present in the benzoyl ring, phenanthridone formation does not take place probably as a result of complex formation. Prior *N*-methylation allows the cyclisation to proceed with the formation of *N*-methylphenanthridone (D.E. Ames and A. Opalko, Tetrahedron, 1984, 40, 1919).

$$\xrightarrow[\substack{\text{CH}_3\text{CONMe}_2 \\ 160-170^{\circ}\text{C}}]{\text{Pd(OAc)}_2,\text{Na}_2\text{CO}_3}$$

The photochemical cyclisation of *o*-bromoanilides provides practical yields of phenanthridones (R.K.-Y. Zee-Cheng, S.-J. Yan and C.C. Cheng, J. Med. Chem., 1978, 21, 199 and B.R. Pai *et al.*, Indian J. Chem., 1979, 17B, 503). However, only the 2-chlorobenzanildes cyclise to phenanthridones in cyclohexane; the bromo and iodo analogues undergo dehalogenative reduction, yielding salts of 2-aminobenzophenone. The cyclisation fails in polar solvents and is retarded by triplet quenchers in hydrocarbon solvents (J.A. Grimshaw and A.P. de Silva, J. chem. Soc. Perkin II, 1982, 857).

5-Methylphenanthridone is formed from the reaction of either phenanthridine-5-oxide or 5-methylphenanthridinium fluorosulphonate with potassium superoxide (A. Picot *et al.*, Tetrahedron Letters, 1977, 3811). Photochemical rearrangement of the phenanthridine-5-oxide (49) yields a phenanthridone and the same product results from the photoisomerisation of the nitrone (50) (S.I. Clarke and R.H. Prager, Austral. J. Chem., 1982, 35, 1645).

A similar photorearrangement occurs with 6-cyanophenanthridine-5-oxide, although the product, 5-cyanophenanthridone, is accompanied by the oxazepine (51) (C. Kaneko *et al.*, Tetrahedron Letters, 1978, 2799). It is suggested that the photoreaction involves an oxaziridine intermediate, the fate of which varies with the solvent (K. Tokumura *et al.*, J. Amer. chem. Soc., 1980, 102, 5643). It is of interest to note that only photolytic deoxygenation occurs in the presence of triphenylphosphine.

The melting points of some phenanthridones are listed in Table 1.

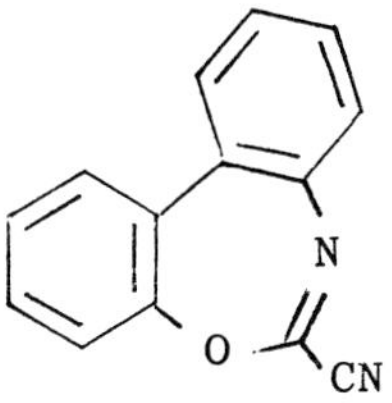

(51)

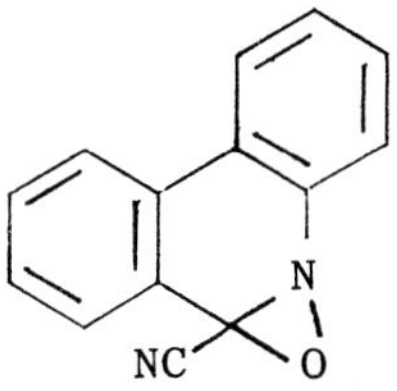

TABLE 1

PHENANTHRIDONES

Substituent	m.p. (^{o}C)	Ref.
2-Trifluoromethyl	291-293	1
2-Methyl	250-252	1
2-Methoxy	228-230	1
2-Carbethoxy	271-272	1
3-Chloro	295-297	2
3-Hydroxy	290	3
3-Methoxy	248-250	2
7,9-Dimethyl	277-278	2
8-Nitro	320-322	2
8-Hydroxy	305	3
8,9-Dimethoxy	259	3
8,9,10-Trimethoxy	195	3

References

1. J. Grimshaw and A.P. de Silva, J. chem. Soc. Perkin II, 1982, 857.

2. R. Leardini, A. Tundo and G. Zanardi, J. chem. Soc. Perkin I, 1981, 3164.

3. B.R. Pai, H. Suguna, B. Geetha and K. Sarada, Indian J. Chem., 1979, 17B, 503.

The photocylisation of *N*-benzoylenamines yields *trans-* hexahydrophenanthridones (I. Ninomiya *et al.*, J. chem. Soc. Perkin I, 1979, 1723). If the enamine bears an *o-* methoxy group, a [1,5]-shift of the substituent occurs and loss of methanol follows to give the 1,2,3,4-tetrahydrophenanthridone. Irradation of *N*- α,β - unsaturated acylanilides gives a mixture of the *cis* and *trans* octahydrophenanthridones (I. Ninomiya *et al.*, J. chem. Soc. Perkin I, 1980, 197). However, when an electron withdrawing substituent is present in the *ortho* position, the reaction proceeds stereoselectively to give the *trans* –product through [1,5]-migration of the *o*-substituent to the 6'-position.

In the presence of a chiral dibasic acid, photocylisation of the unsaturated acylanilide yields an optically active phenanthridone; the optical yield can be as high as 42% (T. Naito, Y. Tada and I. Ninomiya, Heterocycles, 1984, 22, 237).

A vinyl isocyanate, which may be considered as a 1,4-dipolar system, will react with an enamine to give a pyridone. Octahydrophenanthridone results when cyclohexene isocyanate and a cyclohexene enamine are used (J.H. Rigby and H. Balasubramanian, J. org. Chem., 1984, 49, 4569).

1-Cyano-1,2-dihydrocyclobuta[c]quinoline-3(4H)-one (52) undergoes cycloadditions with electron deficient alkenes to afford a mixture of the *cis* and the *trans* phenanthridone. If an electron rich alkene is incorporated into the cyclobutane moiety, an intramolecular cycloaddition leads to the phenanthridone (C. Kaneko, T. Naito and M. Ito, Tetrahedron Letters, 1980, 1645.

(52)

3. Benzoquinolines

(i) Benzo(f)quinolines

Cyclisation of 1,3-dimethyl-1,2,5,6-tetrahydro-2-phenylethylpyridine (53) under acidic conditions yields the *cis* -octahydrobenzo[f]quinoline with 96% stereoselectivity. On the other hand, a mixture of the *cis* and the *trans*-isomer results from the ring closure of 1-(2-cyanoethyl)-1-methyl-2-tetralone (54). Isolation of the *trans*-isomer involves *N*-methylation and fractional crystallisation of the picrates. The ^{13}C-nmr signal for the angular methyl group is near to δ 32 for the *cis*-isomer and δ 23 for the *trans*-compound (E. Reimann and U. Thyroff, Arch. Pharm., 1983, 316, 1024).

(53)   (54)

(i) HBr; (ii) H$_2$, PtO$_2$, AcOH; (iii) HCHO, NaBH$_3$CN

A novel approach to benzo[f]quinoline involves the liberation of the *N*-oxide function from 4,7-phenanthroline-7-oxide during reaction with the methylsulphinyl carbanion. Presumably C-8 is subject to nucleophilic attack and subsequent ring opening and ring closure on to the methylsulphinyl carbon leads to expulsion of the N-oxide group (Y. Hamada *et al.*, Chem. pharm. Bull. Japan, 1979, <u>27</u>, 1535).

The non-oxidative photocyclisation of N-2-naphthylacrylamides provides a general route to 1,2-dihydrobenzo[f]quinolin-3(4H)-ones (I. Ninomiya *et al.*, J. chem. Soc. Perkin I, 1983, 2967). Nitration of the ester of benzo[f]quinoline-6-carboxylic acid occurs at the 8- and 10-positions.

A partially reduced benzo[f]quinolin-3-one is the initial product of the reaction between the enamine of 2-tetralone and acrylamide (I. Ninomaya *et al.*, Chem. Comm., 1971, 451). Further elaboration and reduction of the 1a-4a double bond by a variety of reagents gives a mixture of the *cis* - and the *trans* -octahydrobenzo[f]-quinoline (55). The isomers are separable by fractional crystallisation (J.G. Cannon *et al.*, J. med. Chem., 1984, 27, 190). However, triethylsilane in trifluoroacetic acid provides the *trans*-fused lactam exclusively (J.G. Cannon *et al.*, Synthesis, 1986, 494). Catalytic hydrogenation of the quinolinone followed by lithium aluminium hydride reduction of the amide leads to the *cis*-ring fused compound (J.G. Cannon *et al.*, J. med. Chem., 1979, 22, 341).

(55)

In a useful approach to partially reduced benzo[f]quinoline-1-ones, 1-tetralone is converted into 1-acetyl-3,4-dihydronaphthalene which undergoes a Mannich reaction in acetic acid to yield the benzoquinolinone directly (J.J. Salley, Jr. and R.A. Glennon, J. heterocyclic Chem., 1982, 19, 545).

1-(2-Naphthyl)azetidin-2-one (56) undergoes acyl migration in trifluoroacetic acid to give 1,2,3,4-tetrahydrobenzo[f]quinolin-1-one. m.p. 140–141°C, in high yield (S. Kano, T. Ebata and S. Shibuya, J. chem. Soc. Perkin I, 1986, 2105).

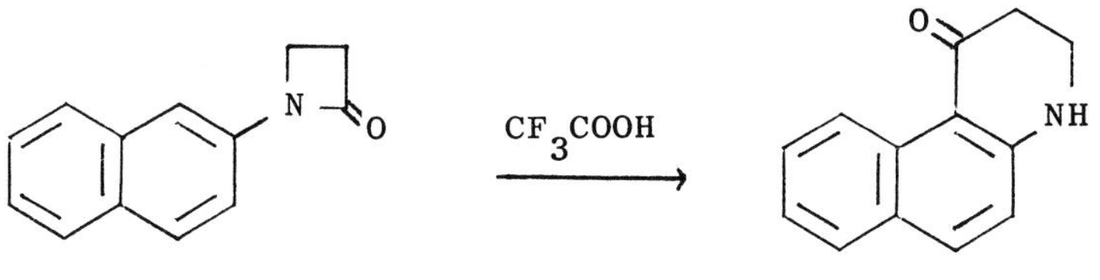

(56)

Catalytic hydrogenation of benzo[f]quinoline over platinum in trifluoroacetic acid and subsequent treatment with acetic anhydride affords a mixture of 4-acetyl-1,2,3,4,7,8,9,10-octahydrobenzo[f]quinoline (57) m.p. 68–69°C, 7,8,9,10-tetrahydrobenzo[f]quinoline (58) m.p. 55–56°C, and 5,6,6a,7,8,9,10,10a-octahydrobenzo[f]quinoline (59) m.p. 146–147°C. These three compounds can be separated (M. Cardellini *et al.*, J. org. Chem., 1982, 47, 688). Selective reduction of the heterocyclic ring occurs under mild conditions using hydrogen in the presence of chlorotris(triphenylphosphine) rhodium (I) (Ph$_3$P)$_3$RhCl (R.H. Fish, J.L. Tan and A.D. Thormodsen, J. org. Chem., 1984, 49, 4500).

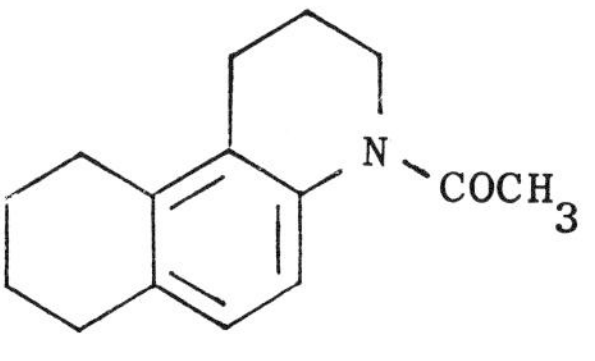

(57)

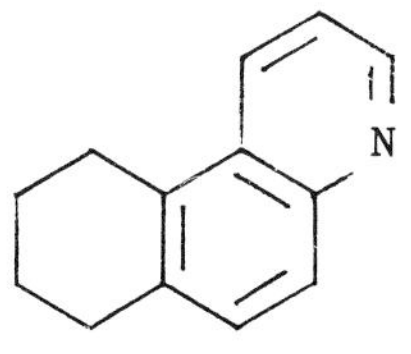

(58)

(59)

The 5,6-bond in benzo[f]quinoline is the reactive centre
of the molecule.   Under phase transfer conditions, sodium
hypochlorite   solution   yields   benzo[f]quinoline-5,6-
epoxide, m.p. 167–168°C (S. Krishnan, D.G. Kuhn and
G. A. Hamilton, J. Amer. chem. Soc., 1977, 99, 8121).
N –Bromoacetamide in acetic   acid   gives the *trans*
–bromoacetate by addition to the 5,6-double bond; the
low yield is attributed to protonation of the heteroatom
(P.J. van Bladeren and D.M. Jerina, Tetrahedron Letters,
1983, 4903).

The reaction of benzo[f]quinoline with methylsulphinyl
carbanion prepared from dimethylsulphoxide and sodium
hydride leads to the 5-methyl and the 6-methyl
derivatives in a 1:4 ratio.   Although benzo[f]quinoline-
4-oxide gives a high yield of phenanthrene under these
conditions, the carbanion generated using potassium *t*-
butoxide as the base leads to alkylation at C–3 and to
simultaneous deoxygenation (Y. Hamada and I. Takeuchi, J.
org. Chem., 1977, 42, 4209).

Although both a 1– and a 3–chlorine atom in
benzo[f]quinoline are subject to nucleophilic
displacement (R.P. Tyagi and B.C. Joshi, Bull. chem. Soc.
Japan, 1972, 45, 2507; 1974, 47, 1786), it is suggested

that the 3-halogen is the more reactive towards the thiophenoxide ion (R.B. Bahuguna *et al.*, J. heterocyclic Chem., 1982, 19, 957).

(ii) Benzo(g)quinolines

Flash vacuum pyrolysis at 800°C of a 2-benzylpyridine-1-oxide gives rise to a benzo[g]quinoline as the major product provided that a methyl group is present at either the 3- or the 2'-position. In the absence of such a substitutent, a pyrido[1,2-a]indole results (A. Ohsawa. T. Kawaguchi and H. Igeta. J. org. Chem., 1982, 47, 3497).

The *trans*-fused octahydrobenzo[g]quinoline (60) is formed when the related 2-benzylpiperidine-3-carboxylic acid is heated in polyphosphoric acid. The 5-carbonyl group in the product is reduced to a methylene group by lithium aluminium hydride-aluminium chloride (J.G. Cannon *et al.*, J. med. Chem., 1984, 27, 190). The stereochemical assignment is based on the strong absorption at 2780 cm^{-1} in the infrared spectrum of (60) and the large (65 Hz) chemical shift difference exhibited by the benzyl

methylene protons of the *N*-benzyl derivative of (60),
both characteristic features of *trans*-fused rings (Z.
Horii, T. Kurihara and I. Ninomiya, Chem. pharm. Bull.
Japan, 1969, 17, 1733; D.A. Walsh and E.E. Smissman, J.
org. Chem., 1974, 39, 3705). The *N*-alkylation of
octahydrobenzo[g]quinolines has been described (J.G.
Cannon *et al.*, J. med. Chem., 1980, 23, 1).

(60)

A *trans*-octahydrobenzo[g]quinoline is also formed by the
reductive cyclisation of 3-(2-cyanoethyl)-2-tetralone
(J.G. Cannon *et al.*, J. heterocyclic Chem., 1980, 17,
1633.

Both 2,6-dimethylheptan-2,6-diol and 2,6-dimethylhept-5-
en-2-ol react with benzyl cyanide in methanesulphonic
acid to give the air-sensitive hexahydrobenzo[g]quinoline
through intramolecular capture of an intermediate
nitrilium ion. Oxidation leads to the 10-ketone
(M. Shome *et al.*, Tetrahedron Letters, 1980, 2927).

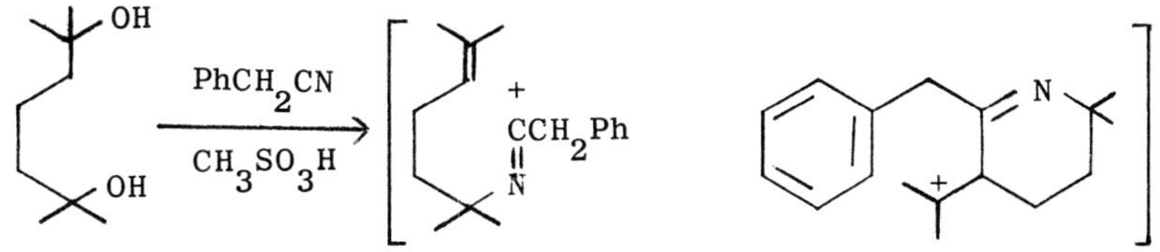

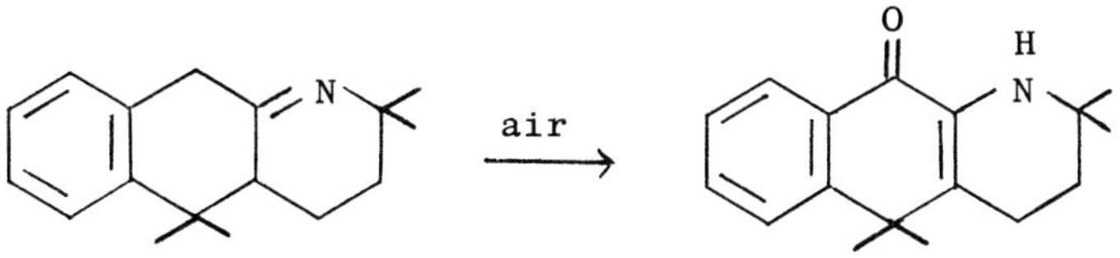

Although 1-aza-1,3-dienes are somewhat reluctant to undergo [4+2] cycloadditions, α,β-unsaturated hydrazones show increased reactivity with electrophilic dienophiles, behaving as 1-amino-1-aza-1,3-dienes. For example, the *NN*-dimethylhydrazone derived from methacrolein gives a high yield of 3-methylbenzo[g]quinolin-5,10-dione (62) via the adduct (61) on reaction with an excess of naphtho-1,4-quinone (B. Serckx-Poncin, A.-M. Hesbain-Frisque and L. Ghosez, Tetrahedron Letters, 1982, 3261).

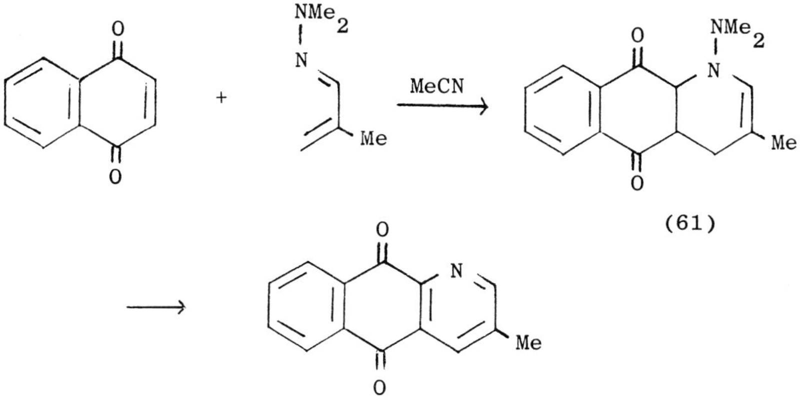

(61)

(62)

Cycloaddition of 1-methoxycyclohexa-1,3-diene to quinoline-5,8-dione results in the regioselective formation of 8-methoxybenzo[g]quinoline-9,10-dione, the structure of which was confirmed by a synthesis involving hetero-atom directed lithiation. The hetero-atom may alternatively be derived from the diene component; 1-dimethylamino-3-methyl-1-azabuta-1,3-diene and 5-methoxynaphthoquinone affording the 5-methoxy-3-methylbenzoquinoline (K.T. Potts, D. Bhattacharjee and E.B. Walsh, Chem. Comm., 1984, 114).

The structure of the orange pigment phomazarin reported in the second edition of C.C.C. (Vol. 4G, p. 65) has been revised on the basis of degradative and spectral studies (A.J. Birch *et al.*, J. chem. Soc. Perkin I, 1979, 807; V. Guay and P. Brassand, Synthesis, 1987, 294). This unique naturally occurring benzo[g]quinoline-5,10-dione (63) contains only one acetate starter molecule and is thought to be biosynthesised from a single nonaketide, probably *via* an anthraquinone (A.J. Birch and T.J. Simpson, J. chem. Soc. Perkin I, 1979, 816).

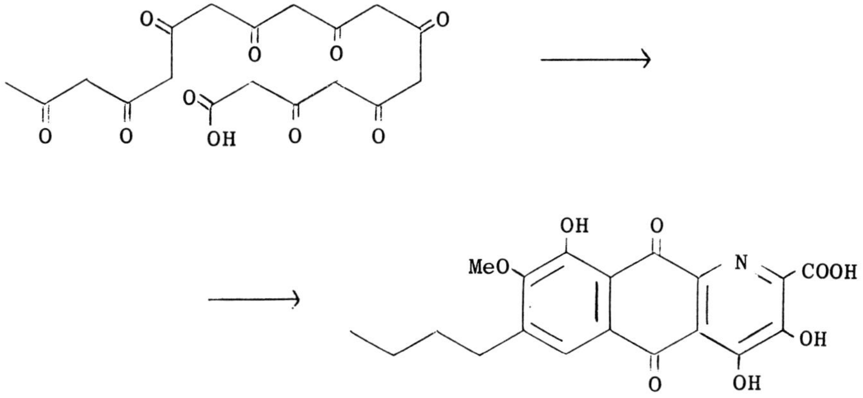

(63)

The 3-methylene group in the benzo[g]quinolinetrione (64) is activated and can be converted into the 3-oxime and hence the 3,4,5,10-tetraone, which is a source of

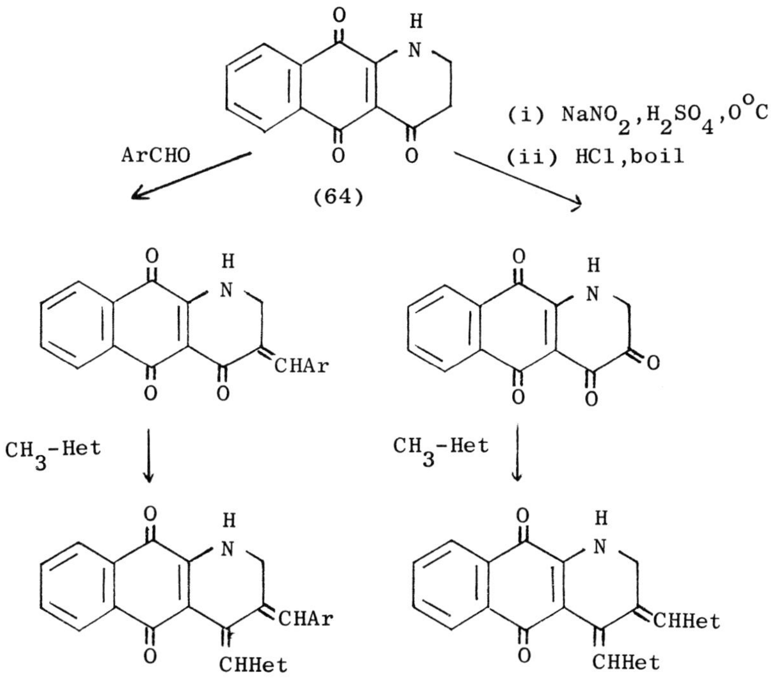

(64)

(i) NaNO$_2$, H$_2$SO$_4$, 0°C

(ii) HCl, boil

ArCHO

CH$_3$-Het

CH$_3$-Het

symmetrical bis-styryl cyanine dyes. Unsymmetrical analogues are available from the 3-arylidene derivative of the trione by reaction with an activated methyl compound. The dyes, which range in colour from red to violet, show a blue fluorescence in polar solvents (M.A. El Maghraby, A.I.M. Koraiem and A.K. Khalafalla. J. chem. Tech. Biotechnol., 1983, 33A, 71).

(iii) Benzo(h)quinolines

An improved Skraup synthesis of benzo[h]quinolines from 1-naphthylamine and an unsaturated carbonyl compound utilises a mixture of fuming sulphuric acid and nitrobenzene (Sulfo-mix), iron(II) sulphate and boric acid (Y. Hamada and I. Takeuchi, J.org.Chem., 1977, 42, 4209).

Although restricted in its applicability, the synthesis of the 1,2-dihydrobenzo[h]quinoline (65) from lithium 1-naphthylamide and phenylacetylene is an interesting example of a tin(IV) chloride catalysed Friedel-Crafts reaction (A. Arduini *et al.*, Synthesis, 1981, 975).

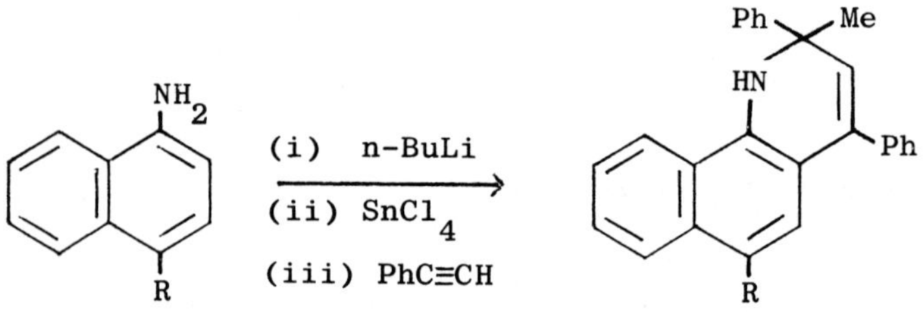

(65)

Several routes to benzo[h]quinolines utilise other
heterocyclic systems as the precursor. The fused
pyrimidopyridine (66), obtained from 4-aminopyrimidine-5-
carboxaldehyde and 1-tetralone, is readily hydrolysed to
2-aminobenzo[h]quinoline-3-carboxaldehyde (67) (T.
Majewicz and P. Caluwe. J. org. Chem., 1979, 44, 531).

(66)   (67)

1,10-Phenanthroline-1-oxide loses the N-oxide function
on brief treatment with the carbanion derived from
dimethylsulphoxide and sodium hydride to give
benzo[h]quinoline in 48% yield. Prolonged reaction
results in methylation of the benzoquinoline at the 5-
and the 6-position (Y. Hamada et al., Chem. pharm. Bull.
Japan, 1979, 27, 1535).

2-Aminonaphtho[1,2-b]pyrans are converted into 5,6-
dihydrobenzo[h]quinolines under basic conditions through
a Dimroth rearrangement. The oxygen heterocycle is
formed by a Michael addition of malononitrile to a 2-
arylidene-1-tetralone and the whole sequence provides an

attractive route to the 3-cyano derivatives of benzo[h]quinolines. The hexahydrobenzo[h]quinolinones (68) can be isolated from treatment of the naphthopyran with hydrogen chloride or can be prepared directly from the arylidenetetralone and cyanacetamide (H.-H. Otto and O. Rinus, Arch. Pharm. 1979, *312*, 548; H.-H. Otto, O. Rinus and H. Schmelz, Monatsh, 1979, *110*, 115).

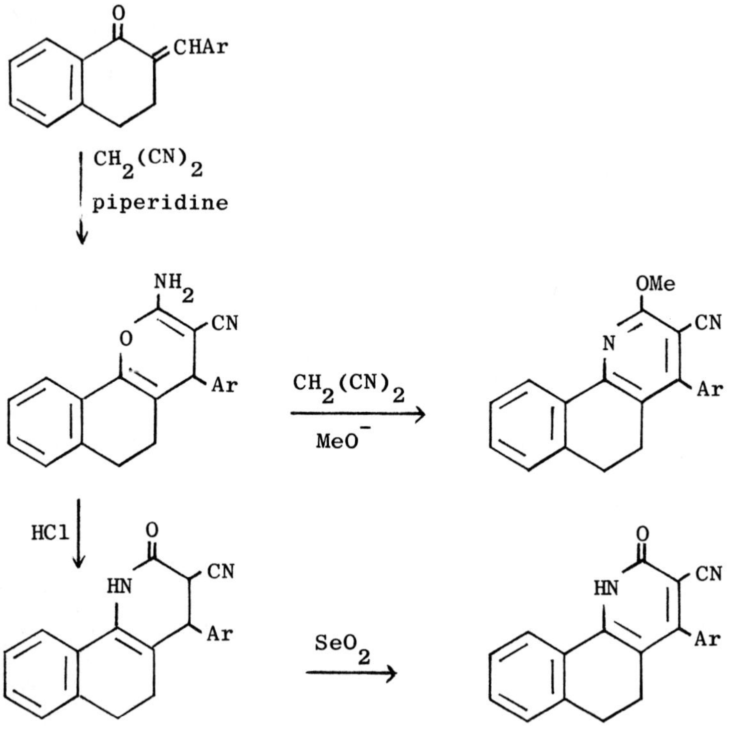

O
CHAr

$CH_2(CN)_2$
piperidine

NH_2
CN
O
Ar

$CH_2(CN)_2$
MeO^-

OMe
CN
N
Ar

HCl

O
HN
CN
Ar

SeO_2

O
HN
CN
Ar

(68)

Arylidenetetralones undergo an ammonium acetate catalysed cycloaddition with *N*-substituted cyanoacetamides to give a mixture of 1-substituted 4-aryl-5,6-dihydrobenzo[h]quinolin-2-one and its 3-cyano derivative (A.H. Moustafa *et al.*, J. prakt. Chem., 1978, *320*, 97).

Reaction of the 2-ethoxynaphthopyran (70) derived from the cycloaddition of vinyl ether to the benzylidene tetralone (69) with hydroxylamine affords 5,6-dihydrobenzo[h]quinolines (M.C. Bellassoued-Fargeau and P. Maitte, J. heterocyclic Chem., 1984, 21, 1549).

(69)   (70)

The extensive studies of the reactions of pyrylium and pyridinium salts by Katritzky and his co-workers have led to an appreciation of the value of these salts in organic synthesis. Chromenylium salts, which are readily available from 1-tetralones, react with aqueous ammonia at room temperature to give 5,6-dihydrobenzo[h]quinolines in very high yield, providing an exceptionally good route to the nitrogen heterocycle. Considerable variation in the substituent pattern is possible and further rings may be annelated (A.R. Katritzky *et al.*, J. chem. Soc. Perkin II, 1984, 857 and earlier papers).

Two syntheses of the structurally unique azaphenanthrene alkaloid eupolauramine (73) have been described, each involving the construction of a suitably functionalised benzo[h]quinoline. The cyclodehydration in polyphosphoric acid of the acetoacetamidonaphthalene (71) prepared from diketene and 4-methoxynaphthylamine affords the benzo[h]quinolin-2-one (72) (P. Karuso and W.C. Taylor, Austral. J. Chem., 1984, 37, 1271). Phosphorus pentachloride chlorinates the quinolinone at the 5-position, whereas phosgene in pyridine efficiently yields the 2-chloro derivative. The dichloro compound is readily dehalogenated to provide the precursor to the alkaloid.

The second approach utilises Meyer's work on oxazolines to synthesise the unsaturated ester (74). The heterodiene undergoes a thermal intramolecular Diels Alder reaction, the adduct oxidatively fragmenting to a mixture of the two partially reduced benzo[h]quinolines (75) and (76). In the presence of DBN, however, the benzoquinoline (77) is the exclusive product, which is subsequently aromatised with N-bromosuccinimide (J.I. Levin and S.M. Weinreb, J. org. Chem., 1984, 49, 4325).

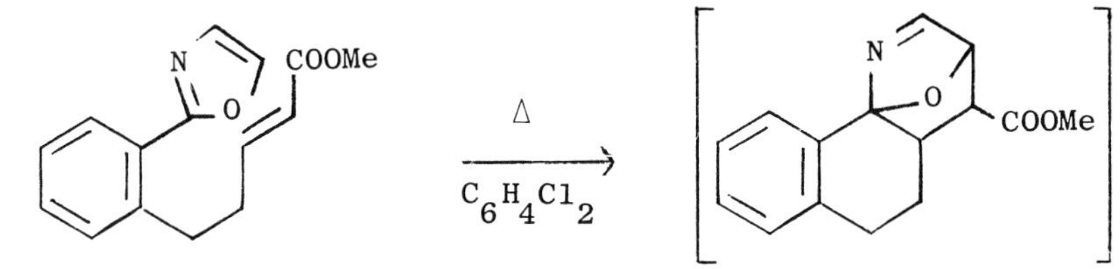

(74)

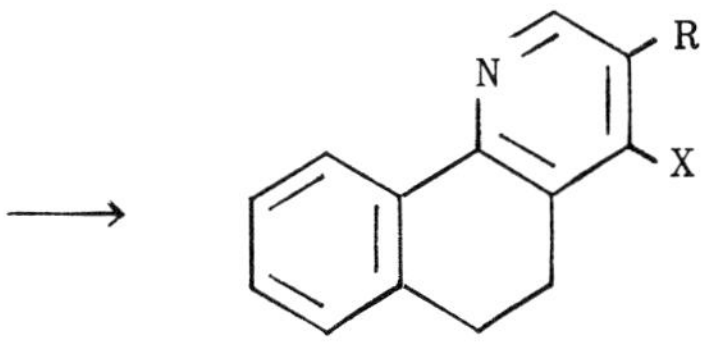

(75: R=OH, X=H)

(76: R=OH, X=COOMe)

(77: R=H, X=COOMe)

Further elaboration to the alkaloid involves epoxidation at the 5,6-double bond. This reaction is in keeping with the formation of 5,6-epoxybenzo[h]quinoline, m.p. 111-112°C, when the parent heterocycle is treated with sodium hypochlorite under phase transfer conditions (S. Krishnan, D.G. Kuhn and G.A. Hamilton, J. Amer. chem. Soc., 1977, 99, 8121).

Some of the methods of synthesis of benzo[f]quinoline are readily adapted to yield the isomeric benzo[h]quinoline, generally involving a 1-naphthyl- in place of a 2-naphthyl-derivative as the precursor. Thus, the *N*-1-naphthylacrylamide (78) gives a benzo[h]quinolin-2-one on photolysis (I. Ninomiya *et al.*, J. chem. Soc. Perkin I, 1983, 2967) and a Fries-like rearrangement of 1-naphthylazetidin-2-one (79) yields benzo[h]quinoline-4-one (S. Kano, T. Ebata and S. Shibuya, *ibid.*, 1980, 2105).

(78)   (79)

Reduction of the heterocyclic ring of the benzo[h]quinoline occurs selectively using chlorotris(triphenylphosphine)rhodium(I) as the hydrogenation catalyst (R.H. Fish. J.L. Tan and A.D. Thormodsen, J. org. Chem., 1984, 49, 4500).

Benzo[h]quinoline is methylated at $70^{\circ}C$ by the methylsulphinyl carbanion at the 4-, 5- and 6-positions. The same reagent converts benzo[h]quinoline-1-oxide into phenanthrene (Y. Hamada and K. Takeuchi, J. org. Chem., 1977, 42, 4209).

The nitration of 4-hydroxy-2-methyl-7,8,9,10-tetrahydro benzo[h]quinoline and various other reactions have been described (M. Abbasi *et al.*, J. heterocyclic Chem., 1978, 15, 649).

(iv) Benzo(d,e)quinolines

The syntheses of this tricyclic system vary both in the nature of the reaction used to construct the third ring and in the actual bond which is formed during cyclisation.

Introduction of a suitable side chain into the 1-position of an isoquinoline and subsequent cyclisation on to the 8-position of the isoquinoline nucleus affords a route to benzo[d,e]quinoline. As an illustration, the isoquinolypropionic acid (80) with fuming sulphuric acid yields the benzo[d,e]quinolin-7-one (81). The ketone has been phenylated indirectly at C-8 and a Clemmensen reduction leads to the benzoquinoline derivative (F.C. Copp *et al.*, J. Chem. Soc. Perkin I, 1983, 909).

(80)   (81)

1-Naphthylamine acts as an alternative precursor of benzo[d,e]quinolines, cyclisation taking place on to the 8-position of the naphthalene ring. A diasteriosomeric mixture of 1-amino-2-aryl-1,2,3,4-tetrahydronaphthalenes results from a Leuckart reaction on 2-aryl-1-tetralones. Separation as the acetals and cyclisation of the individual isomers in acidic conditions leads to 9-aryl-1*H*-2,3,7.8,9,10-hexahydrobenzo[d,e]quinolines of established stereochemistry. The stereochemical assignments are based on chemical shifts and coupling constants, notably large for the C_{10} proton in the *trans*-isomer ($J_{9,10}$ = 11-12 Hz) compared with 6-8 Hz for the *cis*-derivative (S. Yamamura *et al.*, Chem. pharm. Bull. Japan, 1979, **27**, 858).

(i) $BrCH_2CH(OEt)_2, K_2CO_3$

Benzocyclohexenone reacts with 1-aminoprop-2-ene to give the imine (82), formally a derivative of 1-naphthylamine, which after base induced isomerisation to the azadiene (83) undergoes a thermal electrocyclisation to the 3-methylbenzo[d,e]quinoline (84) (C.K. Govindan and G. Taylor, J. org. Chem., 1983, **48**, 5348).

(82)   (83)   (84)

A third variation involves formation of the C–N bond and hence the heterocyclic ring as a result of an intramolecular aza-Wittig reaction. The azidocinnamate (85) yields the substituted benzo[d,e]quinoline (87) on treatment with triethyl phosphite, presumably *via* the iminophosphorane (86) (D.M.B. Hickey *et al.*, Chem. Comm., 1984, 776).

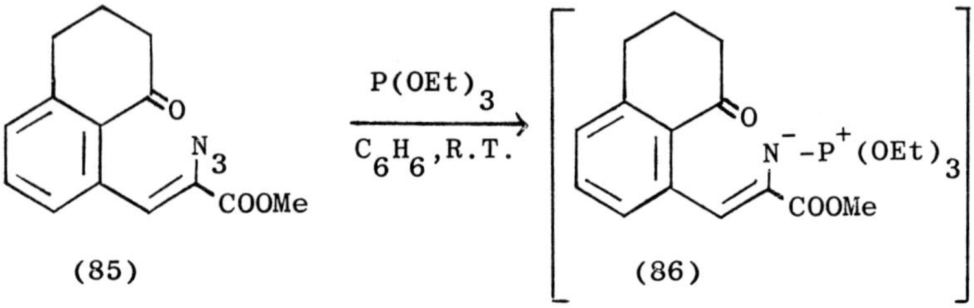

(85)   (86)

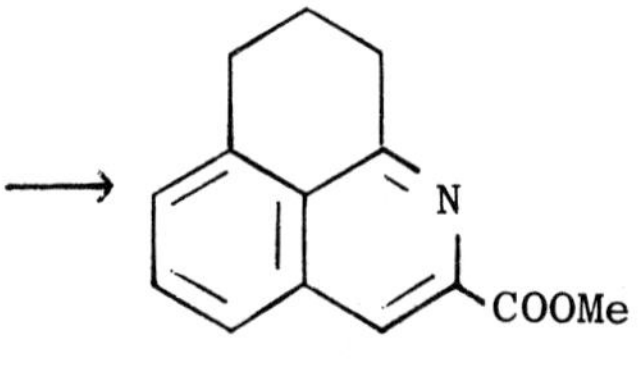

(87)

A totally different approach to the tricycle is based on the desulphurisation of hexahydro–4*H*–benzo[d,e]benzothieno[2,3-g]quinolines (88) which leads to the corresponding 8–phenylbenzo[d,e]quinoline (89) or the partially oxidised derivative (90). The hexahydrobenzoquinolines are obtained as diastereoisomers. The pentacyclic starting materials are

prepared by cyclisation of a 1-(3-benzothienylmethyl)-
isoquinoline (F.C. Copp *et al.*, J. chem. Soc. Perkin I,
1983, 909).

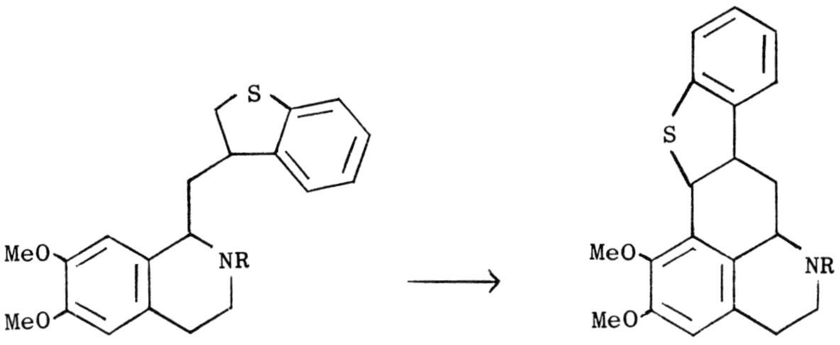

(88)

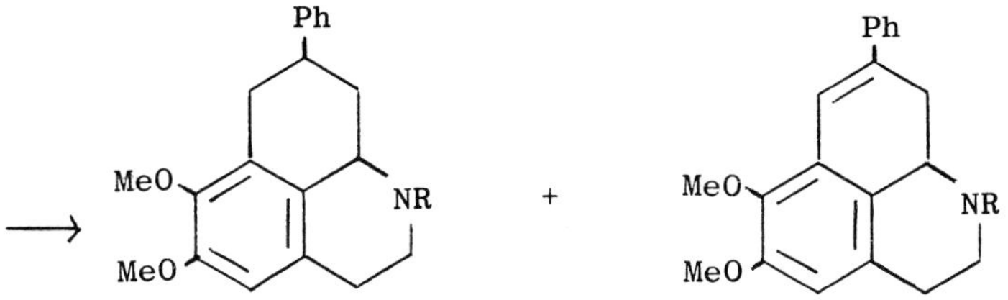

(89)   +   (90)

4. Benzoisoquinolines

The structural relationship between benzoisoquinolines and phenanthroindolizidine alkaloids, examples of which exhibit anti-leukemia activity, has prompted some interest in the former heterocycle. The benzoisoquinoline alkaloids are regularly reviewed (K.W. Bentley, Alkaloids (London) 1983, 13, 99).

(i) Benzo(f)isoquinolines

Electrophilic cyclisation, in which the heterocyclic ring is formed in the manner of the Pomeranz-Fritsch isoquinoline synthesis, offers a potential route to benzoisoquinolines. Cyclisation of the N-2-naphthylmethyl derivative of 2-amino-2-methyl-1-phenylpropan-1-ol (91) in polyphosphoric acid gives the reduced 2,2-dimethylbenzo[f]isoquinoline (92) in 74% yield. The non-equivalence of the gem-dimethyl groups is clearly shown in the ^{1}H nmr spectrum (G. Bobowski, J.M. Gottlieb and B. West. J. heterocyclic Chem., 1980, 17, 1563). N-Methylation occurs under Eschweiler-Clarke conditions.

(91)   PPA
           $\xrightarrow{}$
           110°C

(92)

The intramolecular Friedel-Crafts cyclisation of *N*-2-naphthylalkylalanines proceeds with retention of configuration and subsequent reduction of the resulting isoquinolinone provides a stereospecific synthesis of the diastereoisomeric 1,2,3,4-tetrahydrobenzo[f]isoquinolin-1-ols (93 and 94) (E. Gellert, N. Kumar and D. Tober, Austral. J. Chem., 1983, 36, 157).

(93)   4:1   (94)

The nmr spectra of further examples of compounds obtained by this route (E. Gellert and N. Kumar, Austral. J. Chem., 1984, 37, 819) confirm the significant deshielding of H-10 by the carbonyl group in the dihydrobenzo[f]isoquinolin-1(2*H*)-one (ca. $\delta = 9.1$).

The intramolecular aromatic alkylation of the tetrahydropyridine derivative (95) in hydrobromic acid is stereoselective and leads to *cis*-3,10b-dimethyl-1,2,3,4,4a,5,6,10b-octahydrobenzo[f]isoquinoline (E. Reimann, Ann., 1978, 1963).

(95)

Naphthalene-2-carboxaldehyde forms the naphthyl-2-azabuta-1,3-diene (96) on treatment with 1-aminoprop-2-ene followed by base catalysed isomerisation of the unconjugated azadiene. Pyrolysis brings about an electrocyclisation to the 1,2-dihydrobenzo[f]isoquinoline (C.K. Govidan and G. Taylor, J. org. Chem., 1983, **48**, 5348).

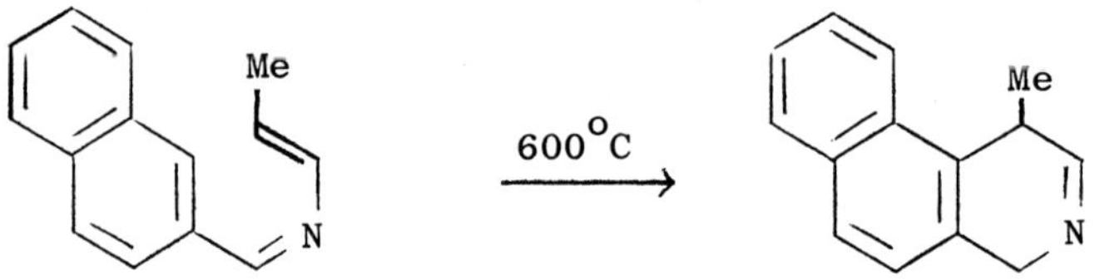

(96)

3,4-Dihydrobenzo[f]isoquinolin-4-one, prepared by the thermolysis of the related acyl azide (F. Eloy and A. Deryckere, J. heterocyclic Chem., 1970, **7**, 1191), is converted into the thione on treatment with phosphorus pentasulphide. Methylation and subsequent reaction with hydrazine hydrate affords the 4-methylthio and thence the 4-hydrazino benzo[f]isoquinolines (I. Lalezari and S. Nabahi, J. heterocyclic Chem., 1980, **17**, 1761).

R=SMe or NHNH$_2$

The methylene group adjacent to the ketone function in 4-phenyl-1,4-dihydrobenzo[f]isoquinolin-2(3H)-one is sufficiently activated to undergo reaction with benzaldehyde in the presence of sodium hydride in dimethylformamide. The 1-benzyl derivative results, together with aromatised starting compound; presumably steric factors hinder the condensation to some extent.

Aromatisation accompanied by methylation at the 1-position occurs using sodium hydride and DMF alone (L. Hazai *et al.*, J. heterocyclic Chem., 1982, 19, 49).

(ii) *Benzo(g)isoquinolines*

A mixture of the 6,8- and 7,9-dimethoxybenzo [g]isoquinoline-5,10-diones (97 and 98) results from the reaction between 1,1-dimethoxyethene and isoquinoline-5,8-diones. The predominance of the former isomer is consistent with control of the nucleophilic addition of the alkene by the more electron deficient of the two carbonyl functions (D.W. Cameron, K.R. Deutscher and G.I. Feutrill, Austral. J. Chem., 1982, 35, 1439). The 9-position of (98) is prone to photosubstitution and utilisation of this reactivity provides a synthesis of the antibiotic bostrycoidin (99). In a related manner, 1-methoxycyclohexa-1,3-diene reacts with isoquinoline-5,8-dione, although this time in a regioselective manner to yield the 5-methoxybenzoisoquinoline-5,10-dione (K.T. Potts, D. Bhattacharjee and E.B. Walsh, Chem. Comm., 1984, 114).

(97: X=OMe, Y=H)

(98: X=H, Y=OMe)

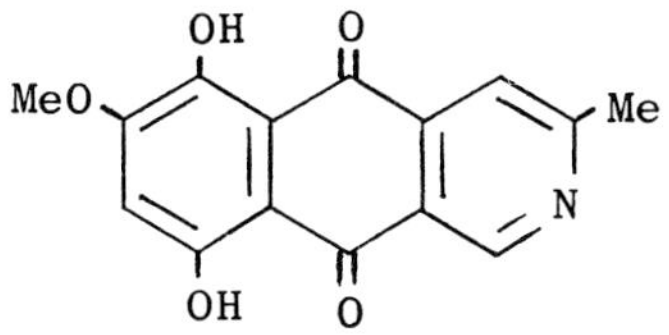

(99)

Cinchomeronic anhydride and 1,4-dimethoxybenzene react in a melt of aluminium chloride and sodium chloride to give the benzo[g]isoquinoline-5,10-dione (100). An alternative approach to this 2-azaanthraquinone system involves acylation of pyridinecarbonitriles with the nucleophilic radicals generated from aromatic aldehydes. The resulting benzoylpyridine carbonitriles cyclise under mild Houben-Hoesch conditions. This route has been successfully applied to the synthesis of bostrycoidin (99) (D. W. Cameron *et al.*, Austral. J. Chem., 1982, 35, 1451.

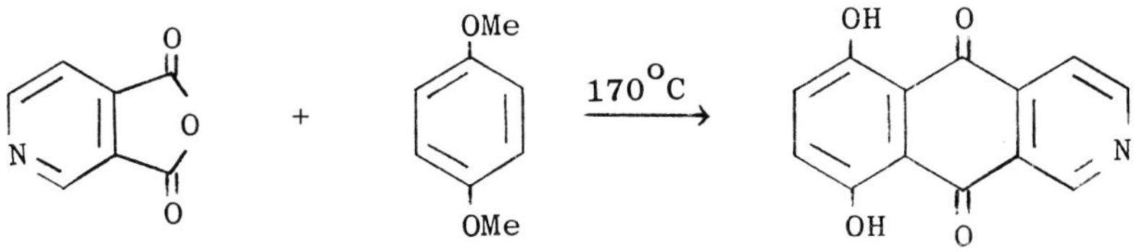

(100)

Cycloaddition of the electrophilic dienophile 1,4-naphthoquinone with a 2-aza-1,3-diene provides access to the benzo[g]isoquinoline system (F. Sainte *et al.*, J. Amer. chem. Soc., 1982, **104**, 1428).

R	%
H	72
Me	44

(iii) Benzo(h)isoquinolines

When the synthetic routes to benzo[f]isoquinolines based on 2-substituted naphthalenes are applied to 1-naphthalene derivatives, similar reactions usually occur and benzo[h]isoquinolines result.

The *N*-1-naphthylmethyl derivative of 2-amino-2-methyl-1-phenylpropan-1-ol (101) cyclises in polyphosphoric acid to 1,2,3,4-tetrahydro-3,3-dimethyl-4-phenylbenzo [h]isoquinoline (102) which is *N*-methylated by formaldehyde and formic acid (G. Bobowski, J.M. Gottleib and B. West, J. heterocyclic Chem., 1980, 17, 1563).

(101)   (102)

Intramolecular acylation of (*S*)-(+)- *N* -(1-naphthylmethyl)- *N* -tosylalanine (103) occurs without racemisation to give (*S*)-(+)-3-methyl-2-tosyl-1,2-dihydrobenzo[h]isoquinolin-4(3 *H*)-one (104). Reaction with lithium aluminium hydride not only reduces the carbonyl function but also removes the protecting group and leads to the diastereoisomeric alcohols (105 and 106) in the ratio of 4:1 (E. Gellert, N. Kumar and D. Tobert, Austral. J. Chem., 1983, 36, 157).

The thermal electrocyclisation of 1-(2-naphthyl)-2-azabuta-1,3-diene, derived from naphthalene-2-carboxaldehyde, leads to the partially reduced benzo[h]isoquinoline (107) (C.K. Govindan and G. Taylor, J. org. Chem., 1983, 48, 5348).

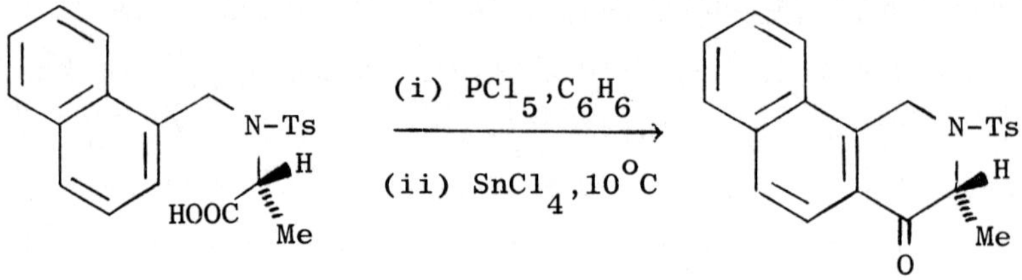

(103)

$$(i) \ PCl_5, C_6H_6$$
$$(ii) \ SnCl_4, 10°C$$

(104)

$$LiAlH_4$$

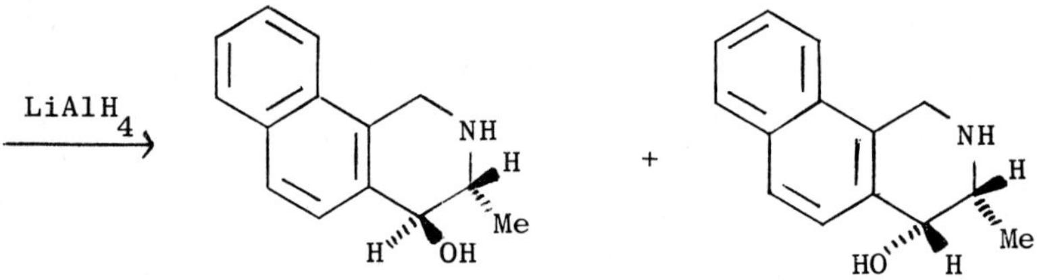

(105)

+

(106)

1-Phenyl-1,4-dihydrobenzo[h]isoquinolin-3(2*H*)-one condenses with benzaldehyde at the 4-position in the presence of sodium hydride in dimethylformamide (L. Hazai *et al.*, J. heterocyclic Chem., 1982, **19**, 49).

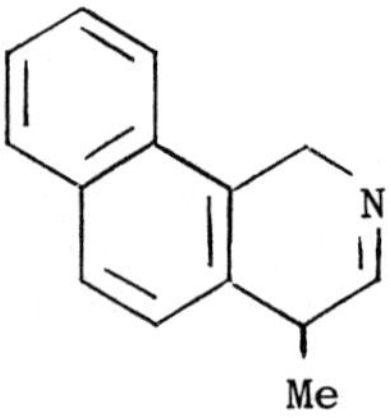

(107)

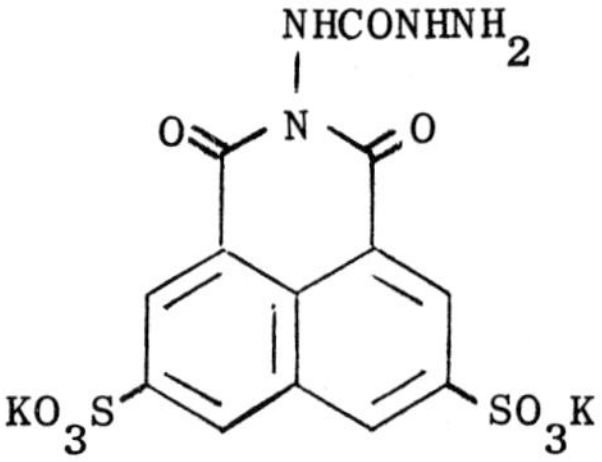

(108)

(iv)   Benzo(d,e)isoquinolines

A general synthesis of 3,6-disulphonated 4-aminonaphthylimides (108), the Lucifer Yellow dyes, is available through the reaction of the dipotassium salt of 4-amino-3,6-disulphonaphthalic anyhydride with amines (W.W. Stewart, J. Amer. chem. Soc., 1981, 103, 7615). The compounds show an intense yellow green fluorescence and are useful as biological tracers, for example enabling the shape of neurons to be revealed and their regeneration pattern to be studied. The N-[3-(vinylsulphonyl)phenyl] derivative reacts rapidly and covalently with proteins under mild conditions.

2,3-Dihydrobenzo[d,e]isoquinolines results from the reaction of 1,8-bisbromomethylnaphthalene with amines (W. Reid and J. Grabosch, Ber., 1958, 91, 2485). The saturated ring of the quaternary salt derived by treatment with an alkyl halide undergoes rapid inversion at room temperature (F. Potmischil and D. Romer, Rev. Rouman. Chem., 1977, 22, 1375).

6-Aminobenzo[d,e]isoquinolin-1,3-diones are yellow to orange and colour polyester a bright fluorescent greenish-yellow with good light fastness (A.J. Peters and M.J. Bide, Dyes and Pigments, 1985, 6, 349).

5. *Benzacridines*

Biological activation of polycyclic-hydrocarbon carcinogens proceeds through dihydrodiols and diol epoxides and such pathways are clearly possible for aza-aromatic compounds. To assist in investigating the existence of such routes, a range of dihydrodiols and diol epoxides of benz[c]acridine and benz[a]acridine (R.E. Lehr and S. Kumar, J. org. Chem., 1981, 46, 367; M. Schaefer-Ridder and U. Engelhardt, *ibid.*, 1981, 46, 2895; S. Kumar and R.E. Lehr, Tetrahedron Letters, 1982, 4523; C.C. Duke, P.T. Murphy and G.M. Holder, J. org. Chem., 1984, 49, 4446) and of dibenz[c,h]acridine (Y. Kitahara, K. Shudo and T. Okamoto, Chem. pharm. Bull. Japan, 1980, 28, 1958) and dibenz[a,j]acridine (C.A. Rosario, G.M. Holder and G.C. Duke, J. org. Chem., 1987, 52, 1064) have been synthesised.

Microsomal metabolism of both benz[a]acridine and benz[c]acridine leads to two dihydrodiols, with oxidation at the K-region predominating. Only benz[c]acridine responds to cytochrome P448 by stimulating K-region oxidation, which may relate to its higher carcinogencicity.

It has also been postulated that the difference in carcinogenic activity of benz[a]acridines and benz[c]acridines may arise because *N*-oxidation accompanies formation of the bay-region epoxide in the former series, whereas attack at the heteroatom is sterically prevented in the latter (U. Engelhardt and M. Schaefer-Ridder, Tetrahedron Letters, 1981, 4687).

Dissimilar behaviour is also shown in the stability of their 5,6-diols. Osmium(VIII) oxide converts benz[c]acridines into the *cis*-5,6-dihydro-5,6-dihydroxybenzacridine (L.J. Boux *et al.*, Tetrahedron Letters, 1980, 2923), but *cis*-5,6-dihydro-5,6-dihydroxy-12-methylbenz[a]acridine is unstable, readily dehydrating to the 6-hydroxy derivative (M. Croisy-Delcey *et al.*, J. med. Chem., 1983, 26, 303).

(i) Benz(a)acridines

Pyrolysis of the Mannich base (109) derived from 2-naphthol leads to generation of a quinone methide which when trapped with an aromatic amine leads to benz[a]acridines (O. Bilgic and D.W. Young, J. chem. Soc. Perkin I, 1980, 1233). Thus, aniline yields the parent compound together with 7,12-dihydrobenz[a]acridine. Formation of the dihydro compound appears to be variable for whereas *N*-methylaniline and *o*-anisidine yield only 7,12-dihydro-7-methylbenz[a]acridine, m.p. 122-123°C and 7,12-dihydro-8-methoxybenz[a]acridine, m.p. 135-136°C, respectively, *o*-phenylenediamine gives 8-amino-benz[a]acridine exclusively.

In a closely related approach, thermal cyclisation of the Mannich bases (110) to benz[a]acridines occurs directly by an intermolecular process (J.L. Asherson, O.Bilgic and D.W. Young, J. chem. Soc. Perkin I, 1981, 3041).

(110)

The reaction of 2-fluoro-5-nitrobenzaldehyde with 2-naphthylamine affords 2-(2-naphthylamino)-5-nitrobenzaldehyde which readily cyclises to the 10-nitrobenz[a]acridine (J. Rosevear and J.F.K. Wilshire, Austral. J. Chem., 1981, 34, 839).

Benz[a]acridine is reduced by lithium in liquid ammonia to the 1,4,7,12-tetrahydro derivative, m.p. 177°C (M. Schaefer-Ridder and U. Engelhardt, J. org. Chem., 1981, 46, 2895). Oxidation with o-chloranil gives 1,4-dihydrobenz[a]acridine, m.p. 135°C, isomerisation of which gives a mixture of the 1,2- and 3,4-dihydrobenzacridines. Oxidation of 7,12-dihydrobenz[a]acridines to the fully aromatic molecules can be accomplished readily using potassium dichromate (O. Bilgic and D.W. Young, J. chem. Soc. Perkin I, 1980, 1233).

TABLE 2

BENZ[a]ACRIDINES

Substituent	m.p. (°C)	Ref
H	131	1
8-NH$_2$	170-171	2
8-OMe	182-183	2
8,10-(OMe)$_2$	129-131	2
10-Me	153-155	2
10-NO$_2$	275-277	3
6-OH-12-Me	149	4

References

1. G.M. Badger, J.H. Seidler and B. Thompson J. chem. Soc., 1951, 3207.

2. O. Bilgic and D.W. Young, J. chem. Soc. Perkin I, 1980, 1233.

3. J. Rosevear and J.F.K. Wilshire, Austral. J. Chem., 1981, 34, 839.

4. M. Croisy-Delcey, A. Croisy, F. Zajdela and J.-M. Lhoste, J. med. Chem., 1983, 26, 303.

The [1]H-nmr spectra of the 7,12-dihydrobenz[a]acridines show the 12-methylene group as a singlet between δ 4 and δ 5, whereas it is strongly deshielded in the fully aromatic compounds and appears near to δ 9.5.

(ii)   Benz(b)acridines

The acid catalysed cyclisation of 2-acetyl-3-arylamino-1,4-naphthoquinones (111) gives benz[b]acridine-6,11-quinones. Oxidative amination of 2-acetyl-1,4-

naphthoquinone in the presence of sodium iodate provides
the starting compounds (K. Joos, M. Pardo and W. Schafer,
J. chem. Research (M), 1978, 4901).

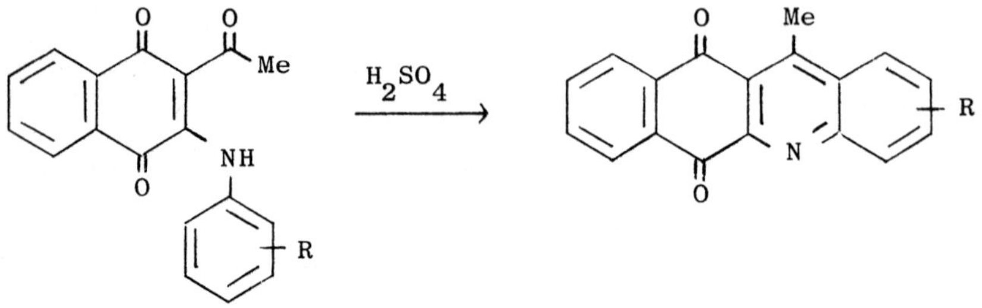

(111)

The addition of the enolate of dimedone to the
aminoquinone (112) generated from the corresponding
azidonaphthoquinone also leads to a benzo[b]acridine-
6,11-quinone (A.J. Hamdan and H.W. Moore, J. org. Chem.,
1985, 50, 3427).

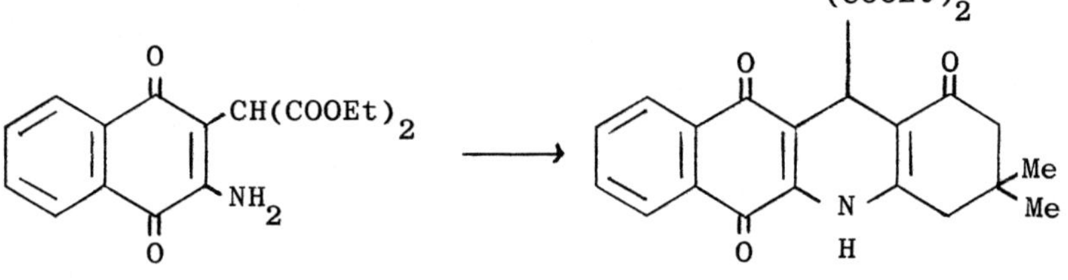

(112)

Benz[b]acrid-12-one m.p. 300–306°C, is the sole product
from the thermolysis of 3-phenylnaphtho[2,3-d]−v−triazin−
4(3H)−one (114) (J.A. Barker et al., J. chem. Soc. Perkin
I, 1979, 2203).

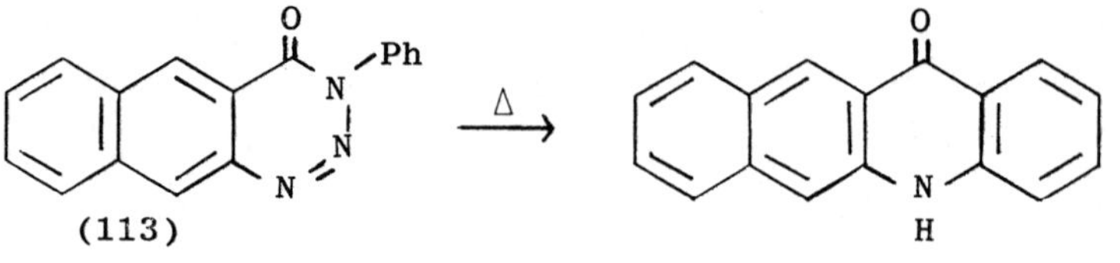

(113)

(iii)   Benz(c)acridines

Several routes to benz[c]acridines involve electrophilic
aromatic substitution to form the heterocyclic ring.
Thus, 9-nitrobenz[c]acridine results from the cyclisation
of the 2-(1-naphthylamino)benzaldehyde derived from 1-
naphthylamine   and   2-fluoro-5-nitrobenzaldehyde   (J.
Rosevear and J.F.K. Wilshire, Austral. J. Chem., 1981,
34, 839).   The reaction between a *N*-aryl-1-naphthylamine
and acetic anhydride has been used to synthesise 9- and
11-hydroxy-7-methylbenz[c]acridines   and   cyclisation   of
partially   reduced   2-(1-naphthylamino)benzoic   acid   by
phosphorus   oxychloride   affords   the   7-chloro-1,2,3,4-
tetrahydro derivative (B.V. Lap *et al.*, J. heterocyclic
Chem., 1983, 20, 281).

R^1 = OMe, R^2 = H R^1 = OH, R^2 = H

R^1 = H, R^2 = OMe R^1 = H, R^2 = OH

An alternative approach to the tetracyclic system forms
the heterocyclic ring by nucleophilic addition of an
amine   to   a   carbonyl   group.   Application   of   the
Friedländer   quinoline   synthesis   to   various   methoxy-1-
tetralones   yields   the   methoxy-5,6-dihydrobenz-
[c]acridines, which are dehydrogenated to the aromatic
compound   by   distillation   from   palladium-charcoal
(M. Croisy-Delcey *et al.*, J. med. Chem., 1983, 26, 303).

When treated with lithium diisopropylamide, 1-tetralone oxime is dimetallated. Upon reaction with 2-aminobenzophenones, the dilithiooxime gives a 7-aryl-5,6-dihydrobenz[c]acridine (D.J. Park, T.D. Fulmer and C.F. Beam, J. heterocyclic Chem., 1981, 18, 649).

Vilsmeier formylation of 6-methoxy-1-tetralone yields the 1-chloronaphthalene-2-aldehyde which on treatment with aniline and subsequent thermolysis affords the 5,6-dihydrobenz[c]acridine. The presence of electron releasing or electron withdrawing groups in the amine component is compatible with the reaction (J.K. Ray, S. Sharma and B.G. Chatterjee, Synth. Comm., 1979, 9, 727).

An Ullmann reaction between 5,6,7,8-tetrahydro-1-naphthylamine and 2-bromobenzoic acid leads to the 2-[1-naphthylamino]benzoic acid (114), which is cyclised to the benz[c]acrid-7-one in polyphosphoric acid (B.V. Lap *et al.*, J. heterocyclic Chem., 1983, 20, 281).

(114)

The thermolysis of 3-(1-naphthyl)-1,2,3-benzotriazin-4(3*H*)-one (115) results in the loss of nitrogen and formation of benz[c]acrid-7-one, m.p. 365-367°C, together with the major product benzo[c]phenanthrid-6-one. The benzacridone is considered to arise by rearrangement of a naphthoazetinone (A.J. Barker *et al.*, J. chem. Soc. Perkin I, 1979, 2203).

The halogen atom of 7-chloro-1,2,3,4-tetrahydrobenz [c]acridine is prone to nucleophilic displacement, as expected of a γ-halogenopyridine system. Ready hydrolysis yields the 7-hydroxy compound and sodiomalononitrile affords the dicyanomethyl derivative; both products exist as the keto-tautomers (B.V. Lap *et al.*, J. heterocyclic Chem., 1983, 20, 281).

Various products result from the reduction of benz[c]acridine depending upon the reagents used. Reduction with sodium in pentanol gives 1,2,3,4,7,12-hexahydrobenz[c]acridine which is oxidised to the 1,2,3,4-tetrahydro compound, m.p. 77-78.5°C in an overall yield of 86% (R.E. Lehr and S. Kumar, J. org. Chem., 1981, 46, 3675). The use of lithium in liquid ammonia affords the 1,4,7,12-tetrahydro derivative, m.p. 125°C (M. Schaefer-Ridder and V. Engelhardt, J. org. Chem., 1981, 46, 2895), whereas hydrogenation of 7-methylbenz-[c]acridine in trifluoroacetic acid over Adam's catalyst yields 8,9,10,11-tetrahydro-7-methylbenz[c]acridine, m.p. 120-121°C (B.V. Lap *et al.*, *loc. cit*).

Allylic bromination followed by dehydrobromination enables both 1,2,3,4- and 8,9,10,11-tetrahydro-7-methylbenz[c]acridines to be converted into the respective dihydro derivative (C.C. Duke, P.T. Murphy and G.M. Holder, J. org. Chem., 1984, 49, 4446).

Benz[c]acridine affords the 5,6-epoxide, m.p. 153-154°C, on treatment with sodium hypochlorite in buffered solution (R.E. Lehr and S. Kumar, J. org. Chem., 1981, 46, 3675); the 7-methyl derivative has m.p. 148-149°C (L.J. Boux *et al.*, Tetrahedron Letters, 1980, 2923).

Metabolic oxidation of 7-methylbenz[c]acridine occurs at the methyl substituent and at the 1,2-, 5,6-, 8,9- and 10,11-positions (L.J. Boux *et al.*, Carcinogenesis, 1983, 4, 1429). Photo-oxidation of 7-methylbenz[c]acridine in methanol is complex, but the identified products involve reaction at the 5,6-position although not *via* the epoxide; this process is an alternative mechanism for the biological activation of the benzacridine (C.D. Burt *et al.*, J. chem. Soc. Perkin I, 1986, 741).

The synthesis of a range of 5,6-dimethylbenz[c]acridines functionalised at the 7-position has been described (N.G. Cromwell *et al.*, J. heterocyclic Chem., 1979, 16, 699). 5,6-Dihydro-5,5-dimethylbenzacridines are readily aromatised by treatment with concentrated sulphuric acid.

A downfield signal at around δ 9 - 9.5 is characteristic for H-1 in the bay-region of benz[c]acridines. In the 5,6-dihydro derivatives this proton absorbs near δ 8.5. Some ^{13}C-nmr data for several 5,6-dihydro-7-methylbenz[c]acridines are available (C.D. Burt *et al.*, J. chem. Soc. Perkin I, 1986, 741).

(iv) Dibenzo(a,c)acridines

Pyrolysis of the tricyclic Mannich base (116) in the presence of aniline leads to dibenzo[a,c]acridine, m.p. 204°C (O. Bilgic and D.W. Young, J. chem. Soc. Perkin I, 1980, 1233). The proton at the 14-position and that at the 8-position are strongly deshielded and appear at δ 9.2 and 9.5, respectively.

(116)

(v) Dibenzo(a,j)acridines and dibenzo(a,i)acridines

6-Aminotetralin and the Mannich base, 1-[(dimethylamino)methyl]-2-naphthol yield a mixture of 1,2,3,4-tetrahydrodibenzo[a,j]acridine (117), m.p. 160-162°C, and 9,10,11,12-tetrahydrodibenzo[a,i]acridine (118), m.p. 145°C, on heating. Dehydrogenation over palladium on carbon affords dibenzo[a,j]acridine, m.p. 216-218°C, and dibenzo[a,i]acridine, m.p. 207-209°C, respectively (C.A. Rosario, G.M. Holder and C.C. Duke, J. org. Chem., 1987, 52, 1064).

(117)   (118)

The reduction of dibenzo[a,j]acridine by sodium in liquid ammonia yields a complex mixture of which the 7,14-dihydro derivative is a key product.

On treatment with *m*-chloroperbenzoic acid, the dibenzoacridine affords the *N*-oxide, m.p. 263-264°C, which undergoes rearrangement in acetic anhydride to 7*H*-dibenzo[a,j]acridone, m.p. 307-309°C.

(vi) Dibenzo(c,h)acridines

The susceptibility of pyrylium salts to attack at C-2 by nucleophiles and the subsequent ring opening and ring closure is of value in the synthesis of a range of heterocyclic compounds. During the course of the transformation of the primary amino group into another functionality, pyrylium salts are converted into pyridinium salts and thence into pyridine derivatives (A.R. Katritzky, Tetrahedron, 1980, 36, 679).

Application of this sequence of reactions to the 5,6,8,9 -tetrahydrodibenzo[c,h]xanthylium salt (119; Z=O) provides useful syntheses of the corresponding dibenz[c,h]acridinium salt (119; Z=$\overset{+}{N}$R) and dibenzacridine (119; Z=N). For example, reaction of the xanthylium fluoroborate with benzylamine gives the *N*-benzylbenzacridinium salt (A.R. Katritzky, J.M. Lloyd and R.C. Patel, J. chem. Soc. Perkin I, 1982, 117), whilst the xanthylium trifluoromethanesulphonate gives the free base on treatment with aqueous ammonia solution (A.R. Katritzky *et al.*, J. chem. Soc. Perkin I, 1983, 487). Pyrolysis of the dibenzacridinium fluorides at their melting points affords the tetrahydrodibenzacridine (A.R. Katritzky *et al.*, Tetrahedron, 1981, 37, 3603).

(119)

6. Benzophenanthridines

Much of the interest in these heterocycles is associated
with their natural occurrence as alkaloids, an area which
has been reviewed (M. Sharma, 'The Isoquinoline
Alkaloids', Academic Press, New York, 1972, p 315; V.
Simanek, Alkaloids (Academic Press), 1985, 26, 1859; S.D.
Phillips and R.N. Castle. J. heterocyclic Chem., 1981,
18, 223). The fully aromatised benzo[c]phenanthridine
alkaloids, exemplified by nitidine (120), show anti-
leukaemic activity (F.R. Stermitz et al., J. med. Chem.,
1975, 18, 708) and have been obtained from
protoberberines by fission of the C-6 - N bond and
recyclisation by forming the C-6 - C-13 bond (M.
Hanaoka et al., Tetrahedron Letters, 1984, 5169).

(120)

Chelidonine and corynoline posses a hexahydrobenzo[c]phen
anthridine nucleus and together with related alkaloids
comprise a separate group.

The preparation and physiological properties of azachrysenes, which includes some benzophenanthridines, naphthoquinolines and naphthoisoquinolines have been reviewed (M.J. Hearn and S.L. Swanson, J. heterocyclic Chem., 1981, 18, 207).

(i) Benzo(c)phenanthridines

The photocyclisation of enamides provides a notably successful approach to this group. It is based on the ready formation of enamides by acylation of the imines derived from 1-tetralone and primary amines (I. Ninomiya et al., J. chem. Soc. Perkin I, 1979, 1723). The photocyclisation is non-oxidative and stereospecific, affording the trans-BC-tetrahydrobenzo[c]phenanthridone (121). Under oxidative conditions, the dehydro derivative results. If a good leaving group is present in the ortho-position of the N-aroyl moiety, the dehydrolactam is again the product as a consequence of regiospecific cyclisation at that position followed by a 1,5-shift of the substituent and its subsequent elimination.

A further significant feature of the chemistry of the tetrahydro compounds is the thermally or photochemically induced facile BC-trans ⟶ BC-cis isomerisation. The importance of this reaction lies in the fact that the known benzophenanthridone alkaloids possess a cis-BC fused configuration.

Lead(IV) acetate oxidises the dihydrobenzo[c]phenanthrid-6-one exclusively at the 12-position; further oxidation gives the 11,12-quinone thereby providing access to the

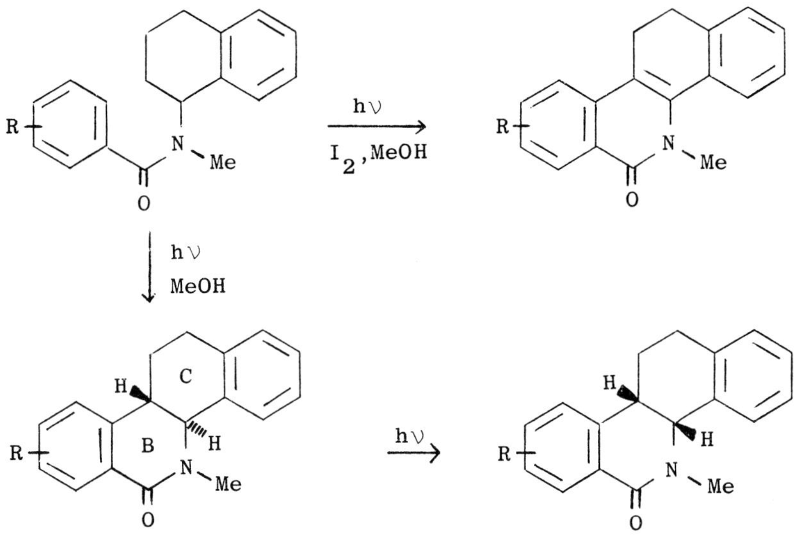

(121)

11,12-diols (122), which have the basic structure of the chelidonine alkaloids (I. Ninomiya, O. Yamamoto and T. Naito, J. chem. Soc. Perkin I, 1983, 2165).

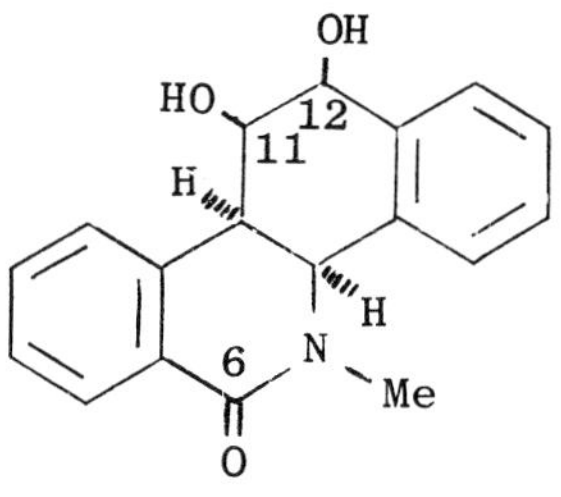

(122)

Photolysis of the mixture of the isomeric Z- and E- enol acetates (123, Ar = $3,4(MeO)_2C_6H_3$) derived from 4-(3,4-dimethoxyphenylacetyl)-2-methylisoquinolin-1-one gives a mixture of 11-acetoxy-5,6-dihydro-2,3-dimethoxy-5-methylbenzo[c]phenanthrid-6-one (124) and the 3,4-dimethoxy isomer (M. Onda, Y. Harigaya and T. Suzuki, Chem. pharm. Bull. Japan, 1977, 25, 2935).

(123)   (124)

4-Styrylisoquinolines also undergo photolytic cyclisation to benzo[c]phenanthridines (S.F. Dyke and M. Sainsbury, Tetrahedron, 1967, 23, 3161, whilst formation of the boron complex of *N*-1-naphthyl-*o*-bromobenzohydroxamic acid facilitates the photolysis to the benzophenanthridine (S. Prabhakar, A.M. Lobo and M.R. Tavares, Chem. Comm., 1978, 884).

A number of chemical as opposed to photochemical approaches to the benzo[c]phenanthridine nucleus have been successful, some of which are based on conventional cyclisation techniques. An improved yield of the parent compound has been reported (J.H. Boyer and J.R. Patel, Synthesis, 1978, 205) based on the cyclisation of 1-formamido-2-phenylnaphthalene, and the *N*-methyl derivative is also accessible *via* a Bischler-Napieralski reaction (H. Ishii *et al.*, J. chem. Soc. Perkin I, 1984, 2283). Reduced benzo[c]phenanthridines are obtained *via* a Mannich reaction on 1-amino-2-aryltetrahydronaphthalenes (T. Kametani *et al.*, Chem. pharm. Bull. Japan, 1971, 19, 1150).

An improvement on the cyclisation of bromo imines *via* benzynes (S.V. Kessar, R. Gopal and M. Singh, Tetrahedron, 1973, <u>29</u>, 167) obviates the need for preparation of the bromo compound by converting the imine directly to the tetracyclic derivative using vanadium oxyfluoride in trifluoroacetic anhydride. Simultaneous trifluoroacetoxylation occurs at C-6 (J.M. Quante, F.R. Stermitz and L.L. Miller, J. org. Chem., 1979, <u>44</u>, 293.

The thermal isomerisation in tetradecane of 1-isocyano-2-phenylnaphthalenes to benzo[c]phenanthridine is accompanied by the 6-tetradecyl derivative formed by solvent participation. Irradiation of the isocyanide at 300 nm also produces the benzophenanthridine (J.H. Boyer and J.R. Patel, J. chem. Soc. Perkin I, 1979, 1070).

The addition of a carbene to the 3,4-double bond of N-methylisoquinolin-2-one yields a cyclopropane adduct, which undergoes thermal transformation to the β,γ-unsaturated ester. Alkylation and oxidative photocyclisation gives a mixture of benzo[c]phenanthridones (U.K. Pandit, Symposium on Heterocycles, 1977, 22).

Isoquinoline-1,3,4-triones are useful intermediates for the synthesis of benzo[c]phenanthridones in about a 30% yield in a three step sequence which utilises phosphonates to construct the side chain which eventually forms rings C and D of the product (C. Pollers-Wieers *et al.*, Tetrahedron, 1981, 37, 4321).

The highly substituted 5,6-dihydrobenzo[c]phenanthridinium salt (126) is obtained from the 2-benzopyrylium salt (125) through a ring-opening ring-closing sequence which yields the isoquinoline derivative (A. Carty, I.W. Elliot and G.M. Lenior, Canad. J. Chem., 1984, 62, 2435).

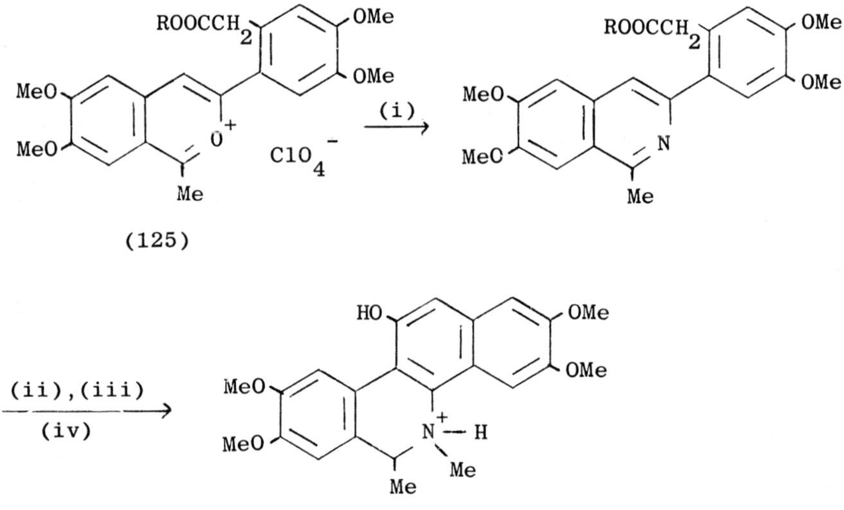

(125)

(126)

(i) aq.NH$_3$,EtOH; (ii) MeI; (iii) NaBH$_4$; (iv) HCl,MeOH

Benzo[c]phenanthridines show three groups of aromatic ^{13}C-nmr signals: C-1, C-4, C-7 and C-10 resonate between

δ 99 and 110; C-4a, C-6a, C-10a, C-10b, C-11 and C-12 in
the region δ 115 to 130; and C-2, C-3, C-4b, C-6, C-8 and
C-9 occur between δ 149 and 155. The [1]H-nmr spectrum
shows two downfield singlets assigned to H-6, just below
δ 9, and H-4, slightly above δ 9 (J.M. Quante, F.R.
Stermitz and L.L. Miller, J. org. Chem., 1979, 44, 293).

(ii)   Benzo(a)phenanthridines and Benzo(b)phenanthridines

Although electrochemical reduction of 2-halogeno- N -
methyl- N -(1-naphthyl)benzamides does not yield any
benzo[c]phenanthridone, the corresponding 2-naphthyl
derivative gives a mixture containing N -
methylbenzo[a]phenanthridone and the isomeric
benzo[b]phenanthridone (J. Grimshaw and R.J. Haslett, J.
chem. Soc. Perkin I, 1980, 657).

Cyclisation of either *cis-* or *trans* -2-benzylaminotetra-
hydro-1-naphthol affords (-)-(6a,S,12aS)-5,6,6a,7,8,12a-
hexahydrobenzo[a]phenanthridine. The presence of the
cis ring-junction follows from the low value, 4 Hz, of
the coupling constant between H-12b and H-6a, and has
been confirmed by X-ray crystallography. The absolute
configuration derives from the formation from (-)-(1R,
2S)-2-aminotetrahydro-1-naphthol (S. Hagashita, M. Shiro
and K. Kuriyama, J. chem. Soc. Perkin I, 1984, 1655).

(iii)   Benzo(i)phenanthridines

Several of the synthetic routes to benzo[c]phenanthridines can be successfully applied to the benzo[i]phenanthridine system. Thus photocyclisation of the *N*-cyclohexenyl-*N*-methyl-1-naphthamide (127) gives the *trans*-benzo[i]phenanthridone, which undergoes photoisomerisation to the *cis*-isomer (I. Ninomiya *et al.*, Chem. Comm., 1981, 692).

(127)

Two syntheses of benzo[i]phenanthridones make use of the reaction of the enamine from 2-tetralone with an isocyanate.   4-Methoxyphenylisocyanate affords the dihydrobenzo[i]phenanthridone (128) after cyclisation in phenylhydrazine and acetic acid (S.D. Sharma and V. Rani, Indian J. Chem., 1976, 14B, 132).   A good yield of the hexahydrobenzo[i]phenanthridone (129) is obtained when 1-isocyanatocyclohexene is heated in toluene with the enamine (J.H. Rigby and N. Balasubramanian, J. org. Chem., 1984, 49, 4569).

(128)   (129)

Benzo[i]phenanthridine results from the photolysis of 2-
(2-isocyanophenyl)naphthalene but pyrolysis affords very
little of the tetracyclic compound (J.H. Boyer and J.R.
Patel, J. chem. Soc. Perkin I, 1979, 1070).

(iv) Benzo(k)phenanthridines

The photocyclisation of 4-phenyl-3-vinylquinolin-2(H)-
ones provides a facile route to benzo[k]phenanthridones.
Under non-oxidative conditions, the tetrahydro derivative
results, but photolysis in the presence of iodine gives
the fully aromatic heterocycle.   Introduction of a 2-
chlorine substituent into the starting material results
in   the   formation   of   6-chloro-7,8-dihydrobenzophen-
anthridones (K. Veeramani *et al.*, Synthesis, 1978, 855).

Photolysis of styrylquinolones (130) affords benzo-
[k]phenanthridones whilst the benzylidene lactones (131),
from which the styrylquinolones may be prepared, give 6-
oxobenzo[k]phenanthridine-7-carboxylates on irradiation
in methanol (V. Arisvaran *et al.*, Synthesis, 1981, 821).

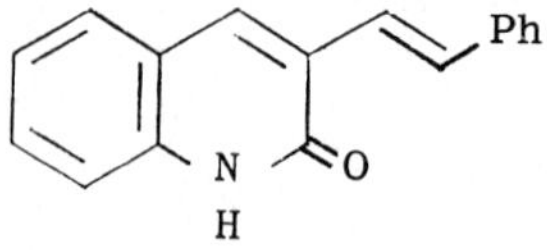

(130)

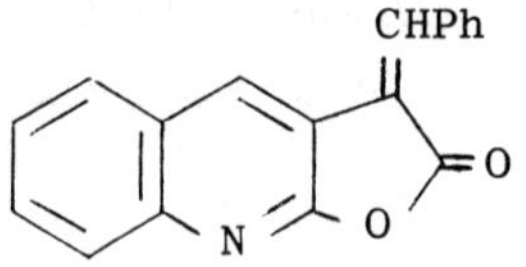

(131)

The benzo[k]phenanthridine system is accessible by the rearrangement of dihydrofuro[2,3-b]quinolines brought about by anhydrous aluminium chloride. The furoquinolines are prepared from 2-aminobenzophenones and hence this route to the tetracycle is attractive (K. Paramasivan, K. Ramasamy and P. Shanmugam, Synthesis, 1977, 768; K. Paramasivam and P. Shanmugam, Indian J. Chem., 1984, 23B, 311).

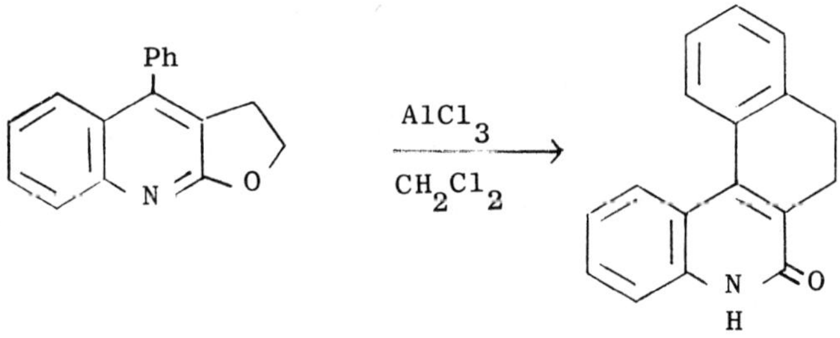

Both benzo[k]phenanthridone and the 7,8-dihydro derivative afford the corresponding 6-chlorobenzo-[k]phenanthridine on treatment with phosphorus oxychloride.

7. *Naphthoquinolines and Naphthoisoquinolines*

Interest in the naphthoquinolines and naphtho-isoquinolines has been stimulated by their physiological properties, their occurrence in liquefied coal products and, of course, by their structural relationship to steroids. This last feature not only promotes interest

in their biological activity, but also identifies them as
intermediates for aza steroid synthesis.

(i) Naphtho(2,3-g)quinolines

The early syntheses of this ring system, based on the
Skraup reaction on 2-aminoanthraquinones, have been
augmented by a pericyclic approach. The [4+2]-
cycloaddition of dimethylenecyclohexane to quinoline-5,8-
quinones gives high yields of the adducts, which are
readily oxidised to the tetrahydronaphtho[2,3-g]-
quinolines (N. Oda *et al.*, Heterocycles, 1981, 15, 857).

(ii) Naphtho(2,1-f)quinolines

A totally different approach from the Skraup and Conrad-
Limpach syntheses of naphtho[2,1-f]quinoline from 2-
aminophenanthrene involves expansion of the five-membered
ring of a steroid to a six-membered heterocyclic ring.

Thus, 17-oximinosteroids undergo a Beckmann rearrangement to the 17a-D-homolactam (S.H. Larsen and R.H. Williams, J. Antibiot., 1975, **28**, 102). For example, the 17-oxime of adrenosterone gives a 91% yield of the lactam on treatment with 4-acetamidobenzenesulphonyl chloride (W. Nagata, M. Narisada and T. Sugasawa, J. chem. Soc. (C), 1967, 648). 2,2'-Dipyridyl disulphide effects a similar conversion (D.H.R. Barton *et al.*, Chem. Comm., 1984, 337).

In the case of 4-oestrene-3,17-dione, protection of the 3-carbonyl group as its enol ether and subsequent oximation, rearrangement and deprotection gives the lactam with a functionality at C-3 which allows further elaboration to a neuromuscular blocker (R. J. Marshall *et al.*, Eur. J. med. Chem.-Chim. Ther., 1984, **19**, 43).

Rearrangement of 17-ketosteroids occurs under Schmidt conditions, but yields a mixture of 17a- and 17-D-homolactams (B. Matkovics, B. Tarodi and L. Balaspori, Acta Chim. (Budapest), 1974, **80**, 79).

Photochemical transpositions of steroids into the naphtho[2,1-f]quinoline system include a low-yielding photo-Beckmann (H. Suginome and T. Uchida, Bull. chem Soc. Japan, 1974, 47, 687) and the conversion of 17-nitrite esters to the 17-a hydroxamic acids, reduction of which affords the lactam (S.H. Imam and B.A. Marples, Tetrahedron Letters, 1977, 2613).

Oxidative photolysis of the acetylhydrazones of 17-oxosteroids gives the lactam together with its 13-α isomer (H. Suginome and T. Uchida, J. chem. Soc. Perkin I, 1980, 1356).

[1]H- and ^{13}C-nmr data for a number of 17a-aza- and 17-aza-
D-homosteroids are available (T.A. Crabb *et al.*, J. chem.
Soc, Perkin I, 1981, 1041 and 1982, 571; V.S.
Bogdanov *et al.*, Izv. Akad. Nauk. SSSR, 1984, 1045; D.
Marcano *et al.*, Org. mag. Res., 1984, 22, 736).

(iii) Naphtho(1,2-h)quinolines

Beckmann rearrangement of the 15-oxime derived from
ergosterol affords the lactam which yields the imine and
dienimine on successive reduction and dehydrogenation
(D.H.R. Barton *et al.*, Tetrahedron, 1983, 39, 2201).

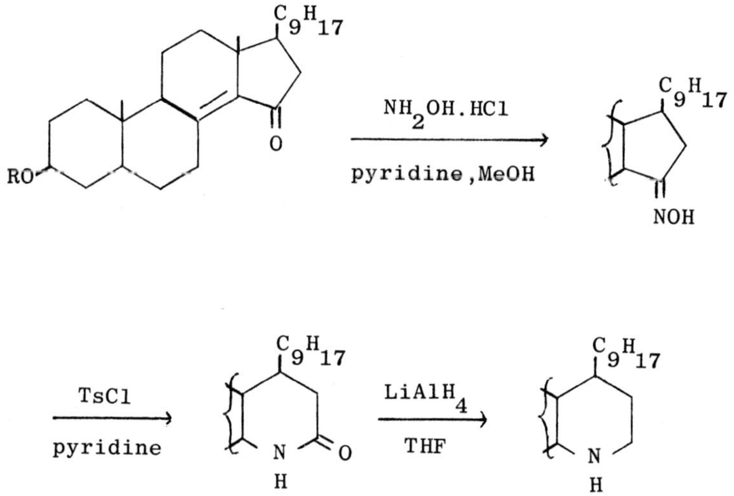

(iv) Naphtho(2,1-f)isoquinoline

Naphtho[2,1-f]isoquinoline, m.p. 224-226°C, is derived
from phenanthrene-1-aldehyde by a Bischler-Napieralski
reaction through the sequence shown below (W.M. Whaley
and M. Meadow, J. org. Chem., 1954, 19, 661).

This ring system is also available from a 16-oximinosteroid *via* a Beckmann transformation, although it appears that only the α-oxime affords the [2,1-f] isomer. In a variant of this approach, a 16-hydroximino-17-ketosteroid yields a dione (E.R.H. Jones, G.D. Meakins and K.Z. Tuba, J. chem. Soc. (C), 1969, 1597).

Photolysis of 17-azidosteroids results in their rearrangement to the naphthoisoquinoline (A. Pancrazi and K.H. Qui, Tetrahedron, 1975, 31, 2041 and 2049).

(v) Naphtho(1,2-h)isoquinolines

The classical syntheses of quinolines and isoquinolines such as the Skraup and Bischler–Napieralski reactions have been extended to the naphthologues. These routes have been supplemented by syntheses based on rearrangements of derivatives of steroidal ketones The heteroatom is incorporated either at the position originally occupied by the carbonyl function or at a contiguous position.

Formylation of 2-(2-phenanthryl)ethylamine and Bischler–Napieralski cyclisation affords 1,2-dihydronaphtho-[1,2-h]isoquinoline, isolated as its hydrochloride (W.M. Whaley and M. Meadow, J. org. Chem., 1954, 19, 661).

Beckmann rearrangement of 16-β-oximinosteroids affords the 16-aza-17-oxo-D-homosteroid (K. Tsuda and R. Hayatsu, J. Amer. chem. Soc., 1956, 78, 4107).

(vi) Naphtho(2,3-h)isoquinolines

The parent molecule, m.p. 162–163°C, is a gold-coloured solid which is derived from the Diels–Alder adduct of 1,4-naphthoquinone and N-benzoyl-1,2,3,4-tetrahydro-4-vinylpyridine. The diene is not very reactive and Lewis acid catalysis is necessary (M.J. Tanga and E.J. Reist, J. Org. Chem., 1982, 47, 1365).

(vii) Naphtho(1,8-f,g)quinolines

Quaternization of the pyridylnaphthalene derivatives (132) followed by reduction affords a mixture of diastereoisomers of which only one undergoes acid catalysed cyclisation to the partially reduced *cis*-fused naphtho[1,8-f,g]quinoline (E. Reimann and G. Bauer, Arch.Pharm., 1984, 317, 517). The ^{13}C-nmr shifts for the angular methyl group lie between δ 28 and 35 in the six examples quoted and these confirm the stereospecific nature of the cyclisation.

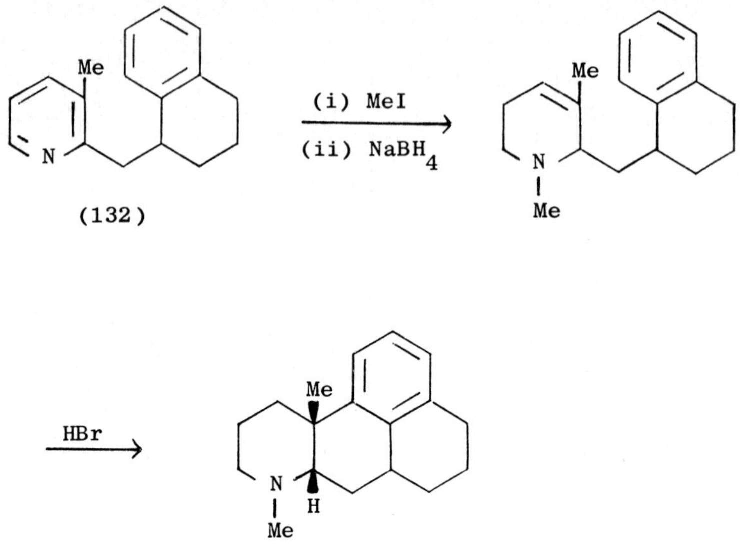

(132)

(i) MeI

(ii) NaBH₄

HBr

Chapter 29

SIX-MEMBERED HETEROCYCLES CONTAINING PHOSPHORUS, ARSENIC,
ANTIMONY, AND BISMUTH AS A SINGLE HETEROATOM

R. LIVINGSTONE

Introduction

Since the publication of the second edition there has
been a large increase in the number of six-membered heterocycles containing either phosphorus or arsenic, which have been
reported. The synthesis of bismabenzene made complete the
group 5 heterobenzenes, which previously contained pyridine,
phosphorin (phosphabenzene), arsabenzene, and stibabenzene.

1. *Phosphorus compounds*

(a) *Phosphorinane (phosphacyclohexane, hexahydrophosphabenzene,
and its derivatives*

(i) *Phosphorinanes*

The [1]H-nmr spectrum of phosphorinane (1) indicates that
the proton on phosphorus is axial and gives no evidence of the
presence of an equatorial isomer. Similar conclusions are
drawn from the spectrum of phosphorinane 1-sulphide (2), but no
conclusions can be reached concerning the proton on the
phosphorus of phosphorinane methiodide (3) because of second-
-order spectral complications (J.B. Lambert and W.L. Oliver,
Tetrahedron, 1971, 27, 4245).

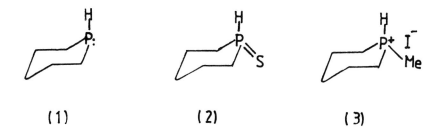

(1)   (2)   (3)

Investigation of the conformational equilibrium of 1-methyl-phosphorinane (4) by low temperature [1]H- and ^{31}P-nmr shows a temperature dependence, favouring the equatorial methyl conformer at low temperature and the axial methyl one at room temperature (S.I. Featherman and L.D. Quin, J. Amer. chem. Soc., 1973, 95, 1699).

(4)   (5)   (6)

A mixture of *cis*- (5) and *trans*- (6) 1-phenyl-4-*tert*--butylphosphorinane 1-oxide, m.p. 160-161° and 88.5-95°, respectively, is obtained on boiling 1,1-diphenyl-4-*tert*--butylphosphorinanium bromide with M sodium hydroxide for a long time. The isomers have been separated and converted by phenylsilane reduction into *cis*- and *trans*-1-phenyl-4-*tert*--butylphosphorirane, each with a b.p. 110-120°/0.1 mm. 1,1--Diphenyl-4-*trans*-butylphosphorinanium bromide on treatment with aqueous sodium hydroxide yields a mixture of (5; 60%) and (6; 40%) (K.L. Maris *et al.*, J. org. Chem., 1977, 42, 1306). Studies of phosphorus inversion in and conformational analysis of *cis*- and *trans*-1-phenyl-4-*tert*-butylphosphorinane have been made using [13]C- and ^{31}P-nmr spectroscopy (G.D. Macdonell *et al.*, J. Amer. chem. Soc., 1978, 100, 4535). ^{13}C-nmr chemical shifts for 1-methyl- and 1-phenyl-phosphorinane, 1--methylphosphorinane 1-oxide and 1-sulphide, and 1-methyl- and 1,1-dimethyl-phosphorinanium iodide (Lambert *et al.*, J. Amer. chem. Soc., 1976, 98, 3778); and ^{13}C chemical shifts and [13]C-^{31}P coupling constants for 1-phenylphosphorinane and its 1-oxide, and 1,1-diphenyl- and 1-benzyl-1-phenyl--phosphorinanium bromide (G.A. Gray, S.E. Cremer, and Marsi, *ibid.*, p.2109) have been reported. A [13]C- and ^{31}P-nmr study has been made of the stereochemical consequences of C--methylation of 1-methylphosphorinane 1-oxide and 1-sulphide

(Quin and S.O. Lee, J. org. Chem., 1978, $\underline{43}$, 1424) and the
^{13}C spin-lattice relaxation times have been measured for 1-
-methylphosphorinane (Lambert and D.A. Netzel, J. Amer. chem.
Soc., 1976, $\underline{98}$, 3783).

Radical addition of trimethylsilylphosphine to 1,4-
-pentadiene affords the synthon 1-trimethylsilylphosphorinane,
hydrolysis of which gives phosphorinane in quantitative yield
(D.M. Schubert and A.D. Norman, Inorg. Chem., 1984, $\underline{23}$, 4130).
1-Methylphosphorinane forms a 1:1 and a 1:2 adduct with
bromine and with iodine, but only a 1:1 adduct with chlorine
(Lambert and H.-n. Sun, J. org. Chem., 1977, $\underline{42}$,1315).

1-Methyl-1-methylenephosphorinane (7) reacts with oxirane
and oxetane to form spirobicyclic phosphoranes (8; n = 3 and 4
respectively) (H. Schmidbaur and P. Holl, Ber., 1979, $\underline{112}$,
501).

n = 3 or 4

(7)   (8)

Spirobicyclic ylide (9) is prepared in 59% yield by the
reaction of 1-methylphosphorinane with 1,4-dibromobutane
followed by cyclizing the intermediate phosphonium salt. The
presence of the isomeric ylides (10) and (11) in the reaction
mixture is shown by nmr data. The spirobicyclic ylide (9) on
treatment with hydrogen chloride at 3° affords the
spirobicyclic salt (12; n=5) (Schmidbaur and A. Moertl, Z.
Naturforsch., 1980, $\underline{35B}$, 990).

114

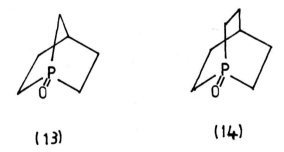

(9)

(10)   (11)   (12)

Salt (12; n=4) has been synthesized (Schmidbaur and Moertl, Z. Chem., 1983, 23, 249).

The electron impact induced fragmentation and rearrangements of 1-ethylphosphorinane 1-oxide, 1-phosphabicyclo[2.2.1]-heptane 1-oxide (13), and 1-phosphabicyclo[2.2.2]octane 1-oxide (14) have been investigated (G.L. Kenyon, D.H. Eargle, Jr., and C.W. Koch, J. org. Chem., 1976, 41, 2417), and a study has been made of the syntheses, reactions, and stereochemistry of 1,4--dimethyl-2-phosphabicyclo[2.2.1]heptane and 4-*tert*--butylphosphorinane derivatives (A. Gamliel, Diss. Abs. Int. B, 1984, 45, 558).

(13)   (14)

Acetylenic derivatives of 2,5-dimethyl-1-phenylphos-
phorinane 1-oxide and 1-sulphide have been synthesized and
their stereochemistry studied (B.M. Butin *et al.*, Izv. Akad.
Nauk Kaz. SSR, Khim., 1977, 27, 49).

1-Phosphaadamantane (15) and a number of related
compounds (H.J. Meeuwissen, T.A. Van der Knaap, and F.
Bickelhaupt, Tetrahedron, 1983, 39, 4225; Phosphorus Sulphur,
1983, 18, 109), and 2-phenyl-2-phosphaadamantane-4,8-dione 2-
-oxide (16) have been synthesized (V.P. Kuhhar, V.N. Zemlyanoi,
and A.M. Aleksandrov, Zh. obshch. Khim., 1984, 54, 220).

(15)   (16)

Single crystal X-ray analysis of 1-benzyl-1-phenylphos-
phorinanium bromide (17) and 1,1-diphenyl-4-methylphosphorin-
anium bromide (18) shows that in both, the ring is in the chair
form and in (17) the benzyl group is equatorial (J.C. Gallucci
and R.R. Holmes, J. Amer. chem. Soc., 1980, 102, 4379).

(17)   (18)

(ii)   Phosphorinanones

A number of mono-, di-, tri-, and tetra-methyl-1-
-phenylphosphorinane-4-ones have been prepared with the methyl
groups in the 2-, 3-, and 5-positions;   the tetramethyl
derivative being 1-phenyl-2,2,3,5-tetramethylphosphorinan-4-
-one (Yu. G. Bosyakov *et al.*, Tr. Inst. Khim. Nauk, Akad. Nauk
Kaz. SSR, 1977, 46, 125).   2,5-Dimethyl-1-phenylphosphorinan-
-4-one 1-oxide and 1-sulphide are converted into their 4-
-hydroxy and 4-ethynyl derivatives by treatment with reducing
agents, such as, lithium tetrahydridoaluminate and sodium
tetrahydridoborate, and by ethynylation in liquid ammonia,
respectively (Bosyakov *et al.*, Zh. obshch. Khim., 1978, 48,
1299;   1980, 50, 1712).   Studies have been made of the reaction
between ethynylmagnesium and the *cis-* and the *trans-*isomer   of
2,5-dimethyl-1-phenylphosphorinan-4-one 1-oxide, 1-sulphide,
and 1-selenide, with reference to the conformation of the
resulting derivatives (A.P. Logunov *et al.*, Izv. Akad. Nauk
Kaz. SSR, Ser. Khim., 1981, 58).

The reaction of 2-methyl-, 2,5-dimethyl-, and 2,2,5-
-trimethyl-phosphorinan-4-one with hydrogen peroxide or
potassium permanganate, or with sulphur or selenium in a
suitable solvent on boiling affords the corresponding 1-oxide,
-sulphide, or -selenide in good yield (I.N. Azerbaev *et al.*,
ibid., 1976, 26, 47).   For the preparation, reactions, and
stereochemistry of these compounds see Bosyakov *et al.*, (Tr.
Inst. Khim. Nauk, Akad. Nauk Kaz. SSR, 1980, 52, 171).

A considerable amount of spectral data, as follows, has
been reported on a number of phosphorinanone derivatives and in
some instances it has been related to their conformation.

(1)   The ^{13}C-nmr spectra of 1-methyl- and 1-ethyl-phosphor-
anin-4-one and derivatives and also that of phosphorinane-4,4-
-diols in water at 30° have been recorded, along with the
^{31}P-nmr spectra and equilibrium compositions for the
hydration of some phosphorinan-4-ones.   Also recorded the ir-
and uv-spectra for 1-ethylphosphorinan-4-one and its 1-oxide
and 1-sulphide (J.J. Breen, S.O. Lee, and L.D. Quin, J. org.
Chem., 1975, 40, 2245). ^{31}P-nmr spectral data of 1-phenyl-
phosphorian-4-one (K. Ramarajan, M.D. Herd, and K.D. Berlin,
Phosphorus Sulphur, 1981, 11, 199).

(2) ^{13}C-nmr spectral data have been reported for 1-phenyl-

phosphorinan-4-one and some derivatives and a single crystal
X-ray diffraction analysis made of 1-phenylphosphorinan-4-one
1-oxide and 1-sulphide. These compounds exist as a flattened
chair form in the solid state (S.D. Venkataramu *et al.*,
Phosphorus Sulphur, 1979, 7, 133) with the phenyl substituent
axial in the parent ketone (A.T. McPhail, J.J. Breen, and
L.D. Quin, J. Amer. chem. Soc., 1971, 93, 2574). Configuratio-
nal and conformational studies have been made of 1-phenyl-
phosphorinan-4-ones (1; R=H, Me) and their selenides (2; R=H,
Me), by application of their ^{13}C- and ^{31}P-nmr spectral data
(K.M. Pietrusiewicz, Org. mag. Res., 1983, 21, 345).

(1)   (2)

^{13}C-nmr spectral data have been used to show that 1-methyl-
and 1-phenyl-phosphorinane and their 4-ones, related 1-oxides
and 1-sulphides and 1,1-dimethylphosphorinanium salts, possess
similar chair conformations to those of the analogous S, O, and
N six-membered heterocycles (J.A. Hirsch and K. Banasiak,
ibid., p.457). From its crystal structure 2,5-dimethyl-1-
-phenylphosphorinan-4-one 1-sulphide has a chair conformation
with equatorial methyl groups and an axial phenyl group
(A.L. Yanovskii *et al.*, Zh. struckt. Khim., 1984, 25, 79).

(3) ^{1}H- and ^{13}C-nmr spectral data for the *trans* isomers of
1-phenyl-2,5-dimethylphosphorinan-4-one 1-oxide, 1-sulphide,
and 1-selenide (L.P. Krasnomolova *et al.*, Zh. fiz. Khim., 1980,
54, 1447) and the dipole moments, Kerr constants, and ^{31}P-nmr
chemical shifts for a number of these compounds and related
compounds have been reported (I.I. Patsanovskii *et al.*, Zh.
obshch. Khim., 1980, 50, 527; Tr. Inst. Khim. Nauk, Akad. Nauk
Kaz. SSR, 1980, 53, 175). ^{1}H- and ^{13}C-nmr spectral data
for the *cis*-isomers of the above 1-sulphide and 1-selenide

indicate that these compounds exist in a twist conformation
(Logunov, Krasnomolova and Bosyakov, Izv. Akad. Nauk Kaz. SSR,
Ser. Khim., 1981, 55).

(4) A series of 1,2,6-triphenylphosphorinan-4-ones has been
prepared and an analysis of the ^{1}H-, ^{13}C-, and ^{31}P-nmr
spectral data of all these compounds indicates that a flattened
chair is the major conformation in each case (J.B. Rampel *et
al.*, J. org. Chem., 1981, 46, 1156). Similar analysis has been
carried out on a series of substituted phosphorinanones and all
the compounds appeared to be chair forms in solution as
indicated by the chemical shifts and coupling constants *(idem,
ibid.*, p.1166).

Nmr spectral data indicate that 3-carbomethoxy and 3-cyano-1-
-phenylphosphorinan-4-one (3) exist as a mixture of keto and
enol forms in a number of solvents and when neat liquids. Also
the carbomethoxy, cyano, and phenyl substituents prefer the
equatorial orientation (B.A. Arbuzov *et al.*, Doklady Akad.
Nauk SSSR, 1977, 233, 858; Izv. Akad. Nauk SSSR, Ser. Khim.,
1978, 1533).

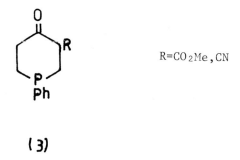

R=CO$_2$Me,CN

(3)

The ms of *cis*- and *trans*-1-phenyl-2,5-dimethylphosphorinan-4-
-one 1-oxide and 1-sulphide and 2,2-dimethyl-1-phenylphos-
phorinan-4-one 1-oxide and 1-sulphide have been determined.
They show that the intensity of the molecular ion peak is
greater for the sulphides than for the oxides. Two types of
molecular ion peak are formed, one with charge localized on the
carbonyl group and the other with the charge round the
phosphorus atom. The effects of the carbonyl group on the
fragmentation is appreciately lowered by the presence of the
sulphur atom (A.E. Lyuts *et al.*, Izv. Akad. Nauk Kaz. SSR, Ser.
Khim., 1979, 20).

Boiling a methanolic solution of 1-phenylphosphorinan-4-
-one and methyl orthoformate saturated with dry hydrogen
chloride affords 4,4-dimethoxy-1-phenylphosphorinane (4),
which in the solid state is shown by X-ray analysis to have a
chair conformation with an axial phenyl substituent. Its [1]H-
nmr spectrum in solution suggests that one conformation,
presumably that for the solid, is predominant. [13]C- and 31$_P$
chemical shifts have also been reported (A.T. McPhail *et al.*,
Chem. Comm., 1971, 1020).

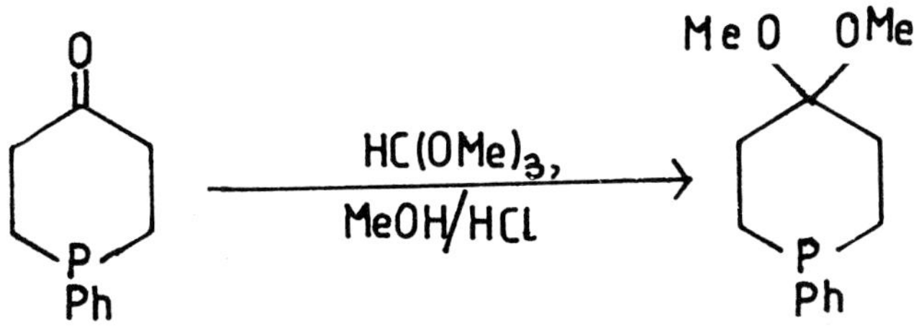

$$ (4) $$

(iii) Phosphorinanols

Reduction of 2,2-dimethyl-1-phenylphosphorinan-4-one 1-
-sulphide with lithium tetrahydridoaluminate, sodium tetrahydridoborate, aluminium isopropoxide, or sodium in ethanol
gives a mixture of two stereoisomers (1) and (2) of 2,2-
-dimethyl-1-phenylphosphorinan-4-ol 1-sulphide (Z.A. Abramova,
Yu. G. Bosyakov, and K.B. Erzhanov, Izv. Akad. Nauk Kaz. SSR,
Ser. Khim., 1978, 28, 33).

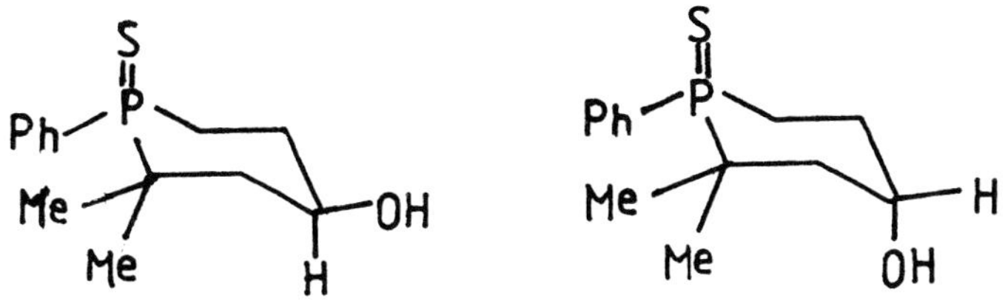

(1)   (2)

2,2,6,6-Tetramethylphosphorinan-4-ol (4), b.p. 60-64°/0.3 mm., m.p. 47-50°, (4) is prepared from 1-phenyl--2,2,6,6-tetramethylphosphorinan-4-one (3) by first protecting the carbonyl group as the ethylene ketal, then reductively removing the 1-phenyl substituent with lithium in tetrahydrofuran, and subsequently removing the protecting group followed by hydride reduction of the keto group (S.D. Pastor, P.A. Odorisio, and J.D. Spivack, J. org. Chem., 1984, 49, 2906). Its [1]H-nmr spectrum suggests that the ring has a biased conformation in solution where the proton on the phosphorus assumes an axial ring position.

(3)   (4)

(a) $(CH_2CH)_2/MeC_6H_4SO_3H-4$ (b) Li/THF (c) (i) 3M HCl/THF , (ii) $LiAlH_4$/THF

(b) Dihydro- and tetrahydro- phosphorins

1-Substituted 4-alkyl- and 4-phenyl-1,4-dihydro-4--methoxy-λ^3-phosphorin (2) are obtained by cyclization of the dilithium compound (1) with the appropriate substituted phosphorus dichloride. The acid-catalysed rearrangement of (2) yields the 1,4-substituted 4-methoxy-λ^5-phosphorin (3), which on acid-catalysed hydrolysis affords the 1-oxide (4) and on oxidative methylation the 4-substituted 1,1-dimethoxy-λ^5--phosphorin (5) (G. Märkl et al., Ann., 1981, 870).

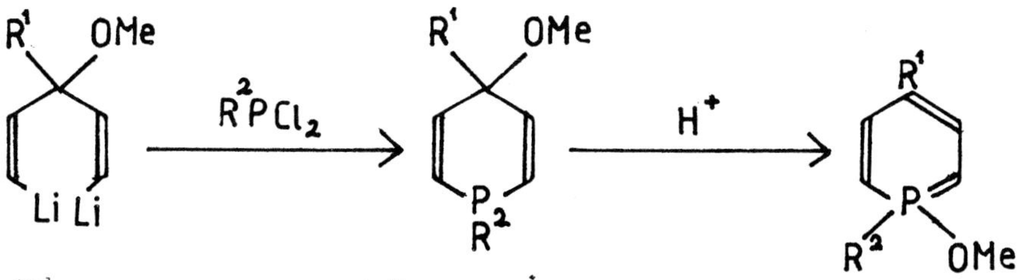

$(R^1 = alkyl, Ph)$

(1)

$(R^2 = alkyl, Ph, NEt_2,$
$\quad BuO, PhS)$

(2)

(3)

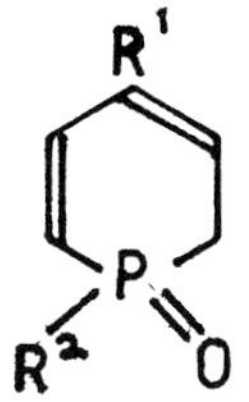

(4)

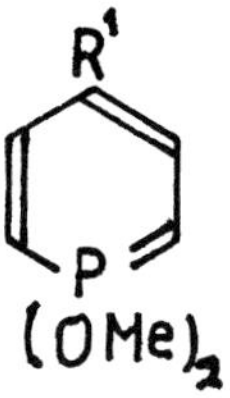

(5)

The rearrangement of some *cis*- and *trans*- 1,4-disubstituted 4-
-alkoxy-1,4-dihydrophosphorins has been discussed (R. Liebl
and A. Huettner, Angew. Chem., 1978, 90, 566). Some η^5-1,2-
-dihydrophosphorin complexes with iron and manganese, for
example, 1,6-dihydro-4,5-dimethyl-1,3-diphenylphosphorin 1-
-sulphide, have been studied (E. Deschamps *et al.*,
Organometallics, 1984, 3, 1144).

Phosphorin reacts with methyllithium to yield anion (6),
which on quenching affords 1,2-dihydro-1-methylphosphorin (7)
(A.J. Ashe and T.W. Smith, Tetrahedron Letters, 1977, 407).

$$\text{MeLi} \longrightarrow \qquad \xrightleftharpoons[\text{base}]{H_2O} \qquad$$

(6)　　　　　　　　　　　　　(7)

2,2-Dichloro-1,2,3,6-tetrahydro-1,4,5-trimethyl-λ^5-
-phosphorin 1-oxide, m.p. 84°, and some related compounds
have been prepared (Y. Kashman and A. Rudi, Tetrahedron
Letters, 1979, 1077).

(c)   Phosphorins, phosphabenzene, phosphinine

The name phosphorin is used to describe compounds which
follow below, because it is still in general use, but it must
be noted that in the revision of the extended Hantzsch-Widman
system of nomenclature for heteromonocycles, phosphorin becomes
phosphininc (Pure appl. Chem., 1983, 55, 409).   An investigation of the nature of the bonding in λ^5-phosphorins
(W. Schäfer *et al.*, J. Amer. chem. Soc., 1976, 98, 4410) and
studies supporting the phosphonium ylide structure (1) for
λ^5-phosphorins (A.J. Ashe and T.W. Smith, *ibid.*, p.7861;
K. Dimroth, S. Berger, and H. Kaletsch, Phosphorus Sulphur,
1981, 10, 305) have been made.

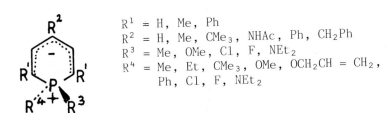

R^1 = H, Me, Ph
R^2 = H, Me, CMe_3, NHAc, Ph, CH_2Ph
R^3 = Me, OMe, Cl, F, NEt_2
R^4 = Me, Et, CMe_3, OMe, $OCH_2CH = CH_2$,
　　　　Ph, Cl, F, NEt_2

(1)

The ESCA-spectroscopic examination of λ^3- and λ^5-
-phosphorins supports the theory that the former are to be
described as aromatic compounds and the latter as cyclic
phosphonium ylides. Contrary to simple electro-negativity
consideration, the P atom in λ^3-phosphorins is nearly neutral
(J. Knecht, Z. Naturforsch., 1984, 39B, 795).

4-Cyclohexylphosphorin (3) is obtained on reacting 4-
-cyclohexyl-1,1-dibutyl-1,4-dihydro-4-methoxystannabenzene (2)
with phosphorus tribromide in the presence of triphenylphosphine (G. Märkl and F. Kneidl, Angew. Chem. internat. Edn.,
1973, 12, 931).

(2)   (3)

The reaction between 1,2,5-triphenylphosphole (4) and
tolane at 230°C affords 2,3,6-triphenylphosphorin (5), m.p.
150°, directly in a 80% yield, along with diphenylmethane,
which is known to be one of the main products resulting from
the formation of diphenylcarbene at high temperature (F. Mathey
et al., J. Amer. chem. Soc., 1981, 103, 4595). This one-step
synthesis of a phosphorin from a phosphole offers numerous
possibilities and supplements earlier procedures (Märkl,
Phosphorus Sulphur, 1977, 3, 77; Ashe, Acc. chem. Res., 1978,
11, 153; Mathey, Tetrahedron Letters, 1979, 1753).

(4)   (5)

Unsymmetrical alkynes afford only one phosphorin with the less bulky substituent at the α-position (C. Charrier, H. Bonnard, and Mathey, J. org. Chem., 1982, *47*, 2376). The initial conversions of phospholes into phosphorins involve a number of stages, the first being the treatment of the phosphole with benzoyl chloride in ether in the presence of triethylamine, for example, the preparation of 4,5-dimethyl-2- -phenylphosphorin (Mathey, *loc. cit.*; Tetrahedron Letters, 1978, 133; J.M. Alcaraz, A. Breque, and Mathey, *ibid.*, 1982, 1565). 4,5-Dimethyl-2-pyridylphosphorin is prepared in a similar manner. For the conversion of 1-phenylphosphole to 2- -phenylphosphorin and 3,4-dimethyl-1-phenylphosphole to 4,5- -dimethyl-2-(furoyl or thenoyl)phosphorin see Alcaraz, E. Deschamps and Mathey (Phosphorus Sulphur, 1984, *19*, 45).

[4+2]-Cycloaddition of 2H-pyrones (6) or cyclopenta-dienones (7) with PhC≡P yield phosphorins (8) (Märkl, G.Y. Jin, and E. Silbereisen, Angew. Chem., 1982, *94*, 383).

(6) (7) (8)

R^1 = H, Me, Et, Ph; R^2 R^3 = H, Ph; R^4 = H, Me, Ph

A number of 3-aryl-λ^3-phosphorins, 3-aryl-λ^5- -phosphorins, and 3-aryl-λ^4-phosphorinium salts have been synthesized from oxaphosphorinium bromides (9). Spectral data indicate that in polar solvents 1-chloro-3-phenyl-1-*tert*-butyl- -λ^5-phosphorin (10) is in equilibrium with 3-phenyl-1-*tert*- -butyl-λ^4-phosphorinium chloride (11). Thermolysis of derivative (10; Ph=aryl) yields the 3-aryl-λ^3-phosphorin (12), which on oxidative alkoxylation with $Hg(OAc)_2$/methanol affords the 3-aryl-1,1-dimethoxy-λ^5-phosphorin (13) (Märkl, K. Hock, and D. Matthes, Ber., 1983, *116*, 445).

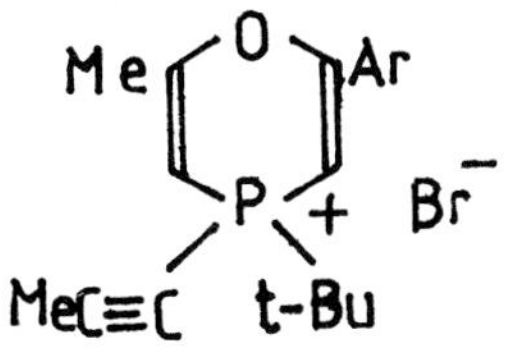

(9) 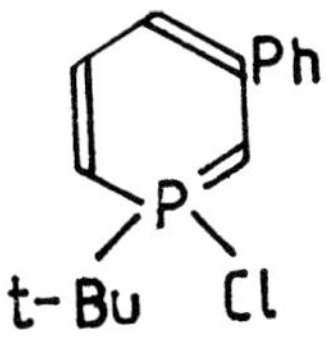 (10) 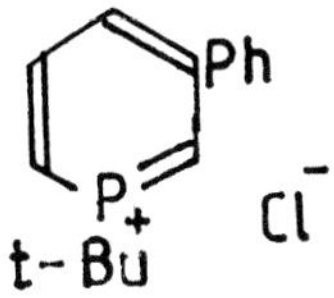 (11)

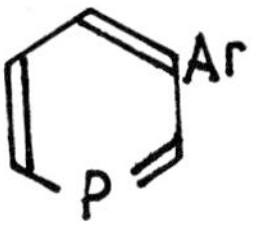

(12) 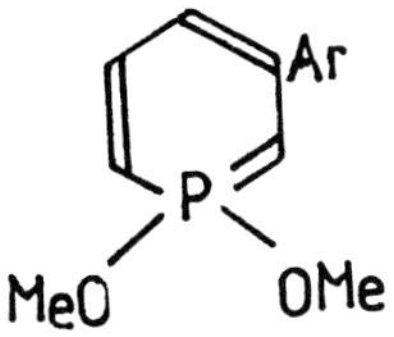 (13)

3,4- And 3,5-disubstituted λ^3- and λ^5-phosphorins (e.g.17 and 18) are synthesized from 5-aryl-1,2,3,6-tetrahydro-4- -methyl-1-*tert*-butylphosphorin-3-one 1-oxides (15) obtained from the oxaphosphorinium bromides (14). Reduction of ketone (15) using silicochloroform gives 3-aryl-1-chloro-4-methyl-1- -*tert*-butyl-λ^5-phosphorin (16), which on thermolysis gives 3- -aryl-4-methyl-λ^3-phosphorin (17). Depending on the nature of the aryl group, during the reaction with silicochloroform some 1,2-migration of the Me group occurs to give 3-aryl-1- -chloro-5-methyl-1-*tert*-butyl-λ^5-phosphorin (18) and hence, 3-aryl-5-methyl-λ^3-phosphorin (19) (Märkl and Hock, Ber., 1983, 116, 1756).

A number of 3,5-disubstituted and 3,4,5-trisubstituted λ^3-
-phosphorins have been prepared from 1,2,3,6-tetrahydro-1-
-*tert*-butylphosphorin-3-ones (Märkl, Hock, and L. Merz, Ber.,
1984, 117, 763).

The reaction of 2,4,6-triphenyl-λ^3-phosphorin (20) with
2-thiophenyl-, 2-benzofuryl-2, 2-benzo-1,3-thiazolyl-, and
ferrocenyl-lithium affords the corresponding 1-substituted
1,2-dihydro-λ^3-phosphorin (21), which on treatment with
mercury (II) acetate in methanol gives the related λ^5-
-phosphorin, for example, 1-methoxy-1-(2-thiophenyl)-2,4,6-
-triphenyl-λ^5-phosphorin (22) (Märkl, C. Martin, and
W. Weber, Tetrahedron Letters, 1981, 1207).

(20)   (21)   (22)

The arylation of 2,4,6-trisubstituted λ^3-phosphorins
with benzenediazonium tetrafluoroborates in methanol gives the
trisubstituted 1-aryl-1-methoxy-λ^5-phosphorin and a second
product with aryl substitution in the 4-aryl group of the
original phosphorin. Related investigations have been carried
out using other benzenediazonium salts (O. Schaffer and
Dimroth, Ber., 1975, 108, 3271, 3281). 2,4,6-Triphenyl-or 4-
-benzyl-2,6-diphenyl-λ^3-phosphorin reacts with diazoalkanes
in the presence of protic nucleophiles to form 1-substituted 1-
-alkyl-2,4,6-triphenyl- or 1-substituted 1-alkyl-4-benzyl-
-2,6-diphenyl-λ^5-phosophorin (P. Kieselack, C. Helland, and
Dimroth, Ber., 1975, 108, 3656). 1-Methoxy-1-phenylphosphorins
readily couple with benzenediazonium tetrafluoroborates in
methanol-benzene to give blue to blue-violet disazo dyes (23)
(Märkl and R. Liebl, Synth., 1978, 846).

$\left(R^1 = c-C_6H_{12}, Ph, t-Bu; R^2 = H, Me, Cl, OMe \right)$

(23)

The conversion of 5-phenyl-1-*tert*-butyl-1,2,3,6-
-tetrahydrophosphorin-3-one (24) into an enol silyl ether
(25), followed by thermolysis and desilylation yields 3-
-hydroxy-5-phenyl-phosphorin (26) (a phospha-phenol), m.p. 48-
50° (Märkl *et al.*, Tetrahedron Letters, 1977, 3445).

The 1-ethoxycarbonylmethyl group of 1-(ethoxycarbonyl-
methyl)-1-methoxy-2,4,6-triphenylphosphorin is readily
hydrolysed to $-CH_2CO_2H$, esterified to $-CH_2CO_2Me$ and
reduced to $-CH_2CH_2OH$ without destroying the ring system
(Dimroth and Kieselack, Ber., 1975, 108, 3671).

The addition of bromine or chlorine to 2,4,6-trialkyl-
and 2,4,6-triaryl-λ^3-phosphorin affords the corresponding
1,1-dihalogeno-2,4,6-tri(alkyl or aryl)-λ^5-phosophorin. The
dihalogeno derivatives may be used to obtain λ^5-phosphorins
with alkyl or aryl groups bound by a heteroatom to the
phosphorus. Physical and chemical properties indicate that the
dihalogeno derivatives are better formulated as aromatic 6π
delocalized heterocycles with d-orbital participation of the
phosphorus than as cyclic 6π delocalised phosophorus ylides
(H. Kanter, W. Mach, and Dimroth, Ber., 1977, 110, 395). 5-
-Phenyl-1-*tert*-butyl-1,2,3,6-tetrahydro-λ^5-phosphorin-3-one

1-oxide (27) reacts with phosphorus pentachloride to give 3-
-chloro-5-phenyl-λ^3-phosphorin (28) after distillation.
Treatment with 5-6 equivalents of phosphorus pentachloride
affords a mixture of tetra-, penta- and hexa-chloro
derivatives. 3-Bromo-5-phenyl-λ^3-phosphorin has also been
prepared (Märkl and Hock, Tetrahedron Letters, 1983, 2645).

(28)

3-Chloro- or 3-bromo-5-phenyl-λ^3-phosphorins undergo
nucleophilic substitution with lithium piperidide in piperidine
to yield the 3-piperidino derivative. Similar results occur
with lithium di-isopropylamide (Märkl and Hock, *ibid.*, p.5055).

A two phase oxidation of 1,1-dimethoxy- and 1,1-
-diphenoxy-2,6-diphenyl-4-(prop-1-enyl)-λ^5-phosphorin (29)
with potassium permanganate affords 1,1-dimethoxy- and 1,1-
-diphenoxy-2,6-diphenyl-λ^5-phorphorin-4-carboxaldehyde (30)
together with a small amount of the related carboxylic acid.
The aldehyde group may be reduced to a CH_3 group, oxidised to
a CO_2H group, or with a suitable Grignard reagent converted
to a $CH=CHC_6H_4R$-4 (R=H, NO_2, Me) group. Amines besides
reacting with the aldehyde group also attack one of the 1-
-methoxy groups to give derivative (31) (Dimroth, J.H. Pohl,
and K.H. Wichmann, Ber., 1979, 112, 1272).

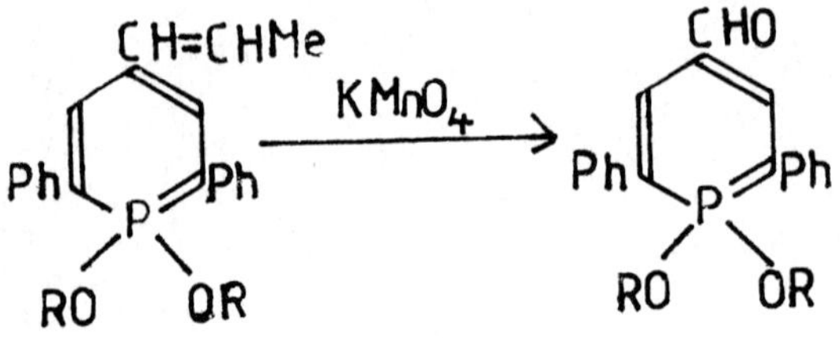

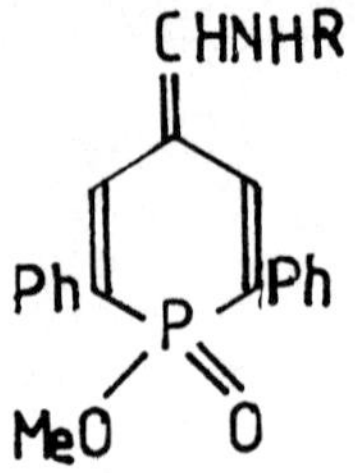

$(R = Me, Ph)$

(29)

(30)

$(R = Ph, C_6H_4OMe-4, CHMe_2)$

(31)

1,1-Dimethoxy-2,6-diphenyl-λ^5-phosphorin-4-carboxaldehyde oxime, m.p. 141-142° (decomp.), 4-cyano-1,1-dimethoxy-2,6--diphenyl-λ^5-phosphorin (Pohl and Dimroth, Angew. Chem. internat. Edn., 1975, 14, 111). A number of λ^5-phosphorin-2--carboxylate esters, and λ^5- and λ^3-phosphorinyl ketones have been prepared (Märkl and Hock, Tetrahedron Letters, 1983, 5051).

The hydrolysis of 4-acetamido-1,1-dimethoxy-2,6-diphenyl--λ^5-phosphorin gives 2,6-diphenyl-1-methoxy-1,2,3,4-tetrahydro-λ^5-phosphorin-4-one 1-oxide and 1,4-dihydro-2,6--diphenyl-1-methoxy-λ^5-phosphorin-4-one 1-oxide, which can be converted into 2,6-diphenyl-1,1,4-trimethoxy-λ^5--phosphorin and characterized as the stable crystalline tricarbonylchromium complex (32) (Dimroth and M. Lückoff, Ber., 1980, 113, 3313).

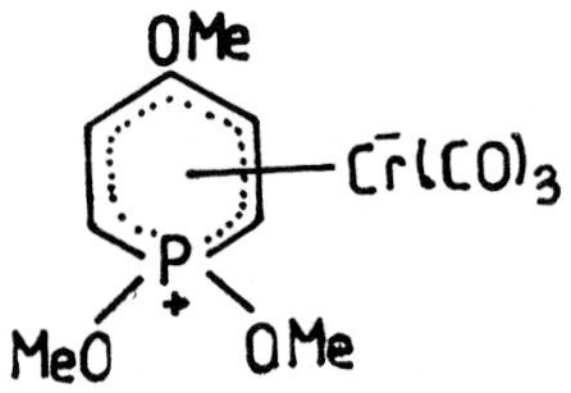

(32)

The crystal structure of 4-acetamido-1,1-dimethoxy-2,6-
-diphenyl-λ^5-phosphorin has been determined
(T. Debaerdemaeker, Cryst. Struct. Comm., 1979, __8__, 309).

A number of tricarbonylchromium-λ^5-phosphorin complexes
have been prepared and shown to possess a phosphonium ylide
structure (Dimroth, Berger, and Kaletsch, Phosphorus Sulphur,
1981, __10__, 295). For the preparation of tricarbonylchromium-,
tricarbonylmolybdenum-, and tricarbonyltungsten-λ^5-phosphorin
see Dimroth, Lückoff, and Kaletsch (*ibid.*, p.285); for the
conversion of tricarbonylchromium-$\eta^6\lambda^3$-phosphorin
complexes into tricarbonylchromium-$\eta^6\lambda^5$-phosphorin
complexes, Dimroth and Kaletsch (J. organometallic. Chem.,
1983, __247__, 271); and for the reactions of some λ^5-phosphorins
and their tricarbonyl complexes, Dimroth and Kaletsch (Angew.
Chem., 1981, __93__, 898). The reactivity of phosphorins as dienes
and dienophiles is increased by complexation of the phosphorus
to pentacarbonyltungsten. Thus with 2,3-dimethylbutadiene the
complex (33) reacts as a dienophile through its 1,2-positions
to give derivative (34).

(33)   (34)

It reacts as a diene through its 1,4-positions with N-
-phenylmaleimide, dimethyl acetylenedicarboxylate, and
cyclopentadiene (Alcaraz and Mathey, Tetrahedron Letters, 1984,
207).

4,5-Dimethyl-2-phenyl-λ^3-phosphorin (35) on heating
with sulphur probably forms a transient P-sulphide (36), which
reacts with 2,3-dimethylbutadiene as a dienophile and with
dimethyl acetylenedicarboxylate as a diene (*idem*, Chem.
Comm., 1984, 508).

1,2,4,6-Tetraphenylphosphorinium tetrachloroaluminate (38) the first phosphorinium (phosphininium) salt analogous to the pyridinium salts is obtained by treating 1-fluoro-1,2,4,6--tetraphenyl-λ^5-phosphorin (37) with aluminium trichloride in methylene dichloride at -78°. Salt (38) on treatment with MeOH, EtOH, PhLi, or Cl$^-$ affords the corresponding derivative (37; F=MeO, EtO, Ph, or Cl) (T.N. Dave, Kaletsch, and Dimroth, Angew. Chem., 1984, 96, 984).

(37)   (38)

Thermal rearrangements of some 1-allyloxy- and 1-
-propargyloxy-λ^5-phosphorins have been investigated, for
example, the rearrangement of 1-allyloxy-1-methyl-2,4,6-
-triphenyl-λ^5-phosphorin (39) to 4-allyl-1,4-dihydro-1-
-methyl-2,4,6-triphenyl-λ^5-phosphorin 1-oxide (40) proceeds
via an anti-Woodward-Hoffman [3s5s] allyl shift.
Rearrangements to give 2-allyl-1,2-dihydro- related λ^5-
-phosphorin 1-oxides have been studied (Dimroth, O. Schaffer,
and G. Weiershaeuser, Ber., 1981, 114, 1752).

(39)   (40)

Oxidation of some 2,4,6-trisubstituted λ^3-phosphorins
by either chemical or electrochemical means gives short-lived
cation-radical intermediates, which *via* the addition of water
or methanol from the solvent afford radicals of the λ^3-

134

-phosphorins. These on further oxidation yield very stable
radicals of λ^5-phosphorins, for example radical (41). The esr
spectra have been discussed. (Dimroth and W. Heide, Ber.,
1981, <u>114</u>, 3019).

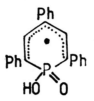

(41)

Also reported are radicals formed by the oxidation of λ^5-
-phosphorin derivatives (*idem*, *ibid.*, p.3004).

Studies or reports have been made of:- the temporary
anion states of phosphorin, arsabenzene, and stibabenzene
(P.D. Burrow *et al.*, J. Amer. chem. Soc., 1982, <u>104</u>, 425);
the anomalously low basicity of phosphorin and arsabenzene
(Ashe *et al.*, *ibid.*, 1979, <u>101</u>, 1764); the ir and Raman
spectra of phosphorin and arsabenzene (Ashe, G.L. Jones, and
F.A. Miller, J. mol. Struct., 1982, <u>78</u>, 169); the photochemical
rearrangements of 1-acyloxy-1-alkoxy-λ^5-phosphorin
derivatives (M. Constenla and Dimroth, Ber., 1976, <u>109</u>, 3099);
the angular and energy dependence of band intensities in the
photoelectron spectra of phosphorin and arsabenzene (Ashe *et
al.*, Helv., 1976, <u>59</u>, 1944); calculations of spin-orbital
interactions in a phosphorin molecule (B.F. Minaev,
D.M. Kizhner, and Kh. T. Akhmetov, Izv. Vyssh. Uchebn. Zaved.,
Fiz., 1976, <u>19</u>, 160); comparative sp and spd-INDO-FPT
calculations of phosphorus-carbon and phosphorus-phosphorus
nuclear spin coupling constants of phosphorins and diphosphines
(V. Galasso, J. mag. Res., 1979, <u>36</u>, 181); [1]H-, [13]C-, and
[31]P-nmr spectral data (Ashe, R.R. Sharp, and J.W. Tolan,
J. Amer. chem. Soc., 1976, <u>98</u>, 5451; T.C. Wong and Ashe,
J. mol. Struct., 1978, <u>48</u>, 216). A number of reviews have been
published [J.E. Kassner and H. Zimmer, Method. Chim., 1978, <u>7</u>
(Part B), 537; Dimroth, ACS Symp. Ser., 1981, <u>171</u> (Phosphorus
Chem.), 463; Acc. chem. Res., 1982, <u>15</u>, 58; Märkl, Chem.
Unserer Zeit, 1982, <u>16</u>, 139; and in part D.W. Allen,
Organophosphorus Chem., 1982, <u>13</u>, 1).

(d) Phosphinolines and dibenzophosphorins

(i) Phosphinolines, phosphanaphthalenes, benzophosphorins

2-Phenylphosphinoline (2-phenyl-1-phosphanaphthalene, 2-phenylbenzo[b]phosphorin) (5), m.p. 101-102°, stable in air, is obtained from 1,1-dibenzyl-2-phenyltetrahydrophosphinolinium tetrafluoroborate (1), after hydrolysis to 1-benzyl-2-phenyl-tetrahydrophosphinoline 1-oxide (2), followed by bromination to the 4-bromo-derivative (3) and subsequently dehydrobromination and reduction to yield 1-benzyl-2-phenyl-1,2-dihydrophosphinoline (4), which on thermolysis affords 2-phenylphosphinoline (5) (G. Märkl and K. -H. Heier, Angew. Chem., internat. Edn., 1972, 11, 1017). 1,1-Dibenzyl-2-phenylphosphinoline, m.p. 123°, is moderately stable in solution, very stable in the solid state (idem, ibid., p.1016) and on thermolysis rearranges to 1,4-dibenzyl-1,4-dihydrophosphinoline, m.p. 126-128°.

4-Benzyl-, 4-methoxycarbonyl-, and 4-phenylcarbonyl-2-
-phenylphosphinoline have been obtained from 2-phenylphos-
phinoline (5) (Märkl and K. Hock, Tetrahedron Letters, 1983,
5051).

3-Butyl-1,2-diphenylphosphindole (6) on reaction with
benzoyl chloride undergoes ring expansion to yield 4-butyl-1,2-
-dihydro-2-hydroxy-1,2,3-triphenylphospinoline 1-oxide (7;
R^1=Ph, R^2=OH). Similarly the phosphindole (6) under
different conditions with ethyl propiolate yields the
phosphinoline derivative (7; R^1=CH$_2$CO$_2$Et, R^2=H)
(A.N. Hughes *et al.*, J. heterocyclic Chem., 1976, 13, 937).

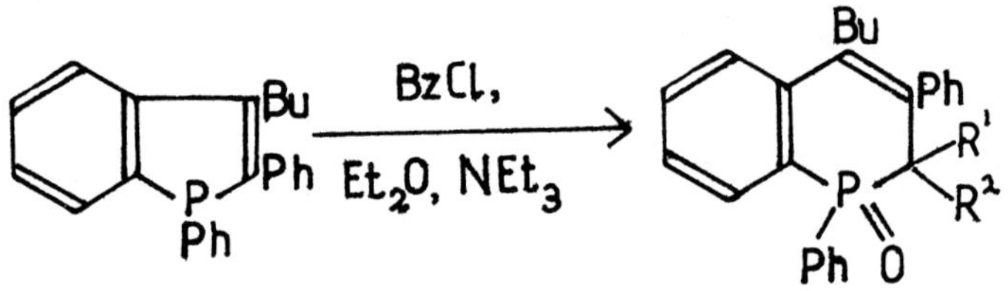

(6)   (7)

3-Methylisophosphinoline (3-methyl-2-phosphanaphthalene,
3-methyl-benzo[c]phosphorin) (8), m.p. 64.5-69°, is prepared
by the route illustrated below (H.G. de Graaf *et al.*,
Tetrahedron Letters, 1973, 2397). The Diels-Alder reaction
between 3-methylisophosphinoline (8) and hexafluorobut-2-yne
yields adduct (9), which on heating gives tarry material and
2,3-bis(trifluoromethyl)naphthalene (10) (30%) (T.C Klebach,
L.A.M. Turkenburg, and F. Bickelhaupt, *ibid.*, 1978, 1099).

(10)   (9)   (8)

(a) (i) NaBH₄, (ii) 10% H₂SO₄ (b) (i) SOCl₂, (ii) LiAlH₄
(c) (i) COCl₂, (ii) DBU (d) F₃CC≡CCF₃ (e) Δ250°

1-Phenylisophosphinoline (14) is obtained from 1-
-benzylphosphindole (11), which on treatment with benzoyl
chloride followed by hydrolysis affords the dihydroisophos-
phinoline 2-oxide (12). The latter is converted into the 2-
-sulphide (13) which on heating with nickel powder gives the
isophosphinoline (14) (F. Nief *et al.*, Nouv. J. Chem., 1981,
5, 187).

(11) → (12) → (13) → (15), (14)

1,2-Dihydro-3-methylisophosphinoline or 1,2-dihydro-2-
-hydroxy-3-methylisophosphinoline 2-oxide on treatment with
phosphoryl chloride and then with triethylamine gives 2-chloro-
-2-hydroxy-3-methyisophosphinoline (15), which unlike other P-
-hydroxy-λ^5-phosphorins is thermally stable (Klebach,
C. Jongsma, and Bickelhaupt, Rec. trav. Chim., 1979, 98, 14).

Heating $(PhCH_2)_2P(O)CH(Me)CO_2H$ with 85% phosphoric
acid yields 5a,6-dihydro-11-methylisophosphinolino[3,2-b]indane
(16) (K.A. Petrov et al., Zh. obshch. Khim., 1983, 53, 56).

(16)

A number of 2-alkoxy-1,2-dihydroisophosphinoline 2-oxides and derivatives have been prepared (J.A. Houbion US Pat, 4,397,790/1983).

The reaction of $Ph(CH_2)_3PCl_2$ with zinc chloride at 170°, followed by hydrolysis with hot hydrochloric acid and oxidation with bromine affords 1-hydroxy-1,2,3,4-tetrahydrophosphinoline 1-oxide (17) in very good yield, incomplete oxidation yields 1,2,3,4-tetrahydrophosphinoline 1-oxide. Also reported are the preparation of 1-chloro-, 1-ethyl-, and 1-phenyl-1,2,3,4-tetrahydrophosphinoline 1-oxide, 1-chloro- and 1-ethyl-1,2,3,4-tetrahydrophosphinoline, and 1-ethyl-1,2,3,4-
-tetrahydrophosphinoline 1-sulphide (L.E. Rowley and J.M. Swan Austral. J. Chem., 1974, 27, 801). 1-Hydroxy-1,2,3,4-tetra-hydrophosphinoline 1-oxide has been converted into 4-propyl-
-1,2,3,4,5,6-hexahydro-1,5-methano-4,1-benzazaphosphocine 1-
-oxide (18) (D.J. Collins, Rowley, and Swan, *ibid.*, p.815).

(17)

(18)

Diphenyl(but-2-enyl)phosphine 1-oxide and diphenyl(but-3-
-enyl)phosphine 1-oxide cylize in the presence of 115% poly-phosphoric acid at 118° to give 4-methyl-1-phenyl-1,2,3,4-
-tetrahydrophosphinoline 1-oxide, also obtained on treating 1,1-diphenyl-4-methyl-1,2,3,4-tetrahydrophosphinolinium hexafluorophosphate with boiling methanol-water containing sodium hydroxide. Diphenyl(3-methylbut-2-enyl)phosphine 1-oxide gives 4,4-dimethyl-1-phenyl-1,2,3,4-tetrahydrophosphinoline 1-oxide (M. El-Deek *et al.*, J. org. Chem., 1976, 41, 1403). A route for the preparation of 6,7-dimethoxy-1,2-dimethyl-
-1,2,3,4-tetrahydroisophosphinoline has been designed with the help of the computer program PASCOP (C. Laureno and G. Kaufmann, Tetrahedron Letters, 1980, 2243).

The structures of 1-phenyl-1,2,3,4-tetrahydro-1,4,4-
-trimethylphosphinolinium hexafluorophosphate (K.K. Wu and
D. Van der Helm, Cryst. Struct. Comm., 1977, 6, 143), 1,4-
-dimethyl-1-phenyl- and 1-ethyl-4-methyl-1-phenyl-1,2,3,4-
-tetrahydrophosphinolinium hexafluorophosphates (R. Fink. Van
der Helm, and K.D. Berlin, Phosphorus Sulphur, 1980, 8, 325),
and (1R, 1´S)-1,1´-(1,2-ethanediyl)bis(4,4-dimethyl-1-phenyl-
-1,2,3,4-tetrahydrophosphinolinium) diperchlorate have been
determined by X-ray diffraction [N. Gurusamy *et al.*, ACS Symp.
Ser., 1981, 171 (Phosphorus Chem.), 561]. A number of
substituted 1,1´-(α,ω-alkaneyl)bis(1,2,3,4-tetrahydrophos-
phinolinium) salts have been synthesized. Certain of these
salts display antimicrobial, antihelminitic, and anti-
cholinergic activities (*idem,* J. Amer. chem. Soc., 1982,
104, 3107). For the synthesis, separation, and resolution of
stereoisomers of 1,1´-(1,2-ethanediyl)bis(4,4-dimethyl-1-
-phenyl-1,2,3,4-tetrahydrophosphinolinium) diperchlorate,
including the use of ^{31}P-nmr analysis to monitor the
resolution see Gurusamy and Berlin (*ibid.,* p.3114).

1,2,3,4,6,7,8,8a-Octahydro-2-phenylisophosphinolin-6-one
2-sulphide (19) has been synthesized (J.B. Rampal, Berlin, N.
Satyamurthy, Phosphorus Sulphur, 1982, 13, 179).

(19)   (20)   (21)

Cyclization of certain cyclohexenyl ketones with
phenylphosphine affords decahydrophosphinolin-4-ones, for
example, ketone (20) yields 1,2-diphenyldecahydrophosphinolin-
-4-one (21). Also prepared by this method are 1-phenyl- and 2-
-methyl-1-phenyl-decahydrophosphinolin-4-ones. All have been
converted into their 1-oxides, -sulphides, and -selenides
(Yu. G. Bosyakov *et al.*, Zh. obshch. Khim., 1983, 53, 1050).

Cyclization of certain 1,5-diketones with phenylphosphine
affords decahydrophosphinoline derivatives, for instance,

diketone (22) gives 1,2-diphenyl-8a-hydroxydecahydrophosphino-
line 1-oxide (23) (V.I. Vysotskii *et al.*, *ibid.*, p.2206).

(22)   (23)

(24)   (25)

Two isomeric forms of 3-benzoyl-1,4-diphenyldecahydrophos-
phinoline (25) are obtained on cyclizing the 2-methylene-1,5-
-diketone (24) with phenylphosphine (Yu. V. Prikhod'ko *et
al.*, Zh. obshch. Khim., 1984, **54**, 1427).

(ii)   Dibenzo[b,e]phosphorins(9-phosphaanthracenes)

Dibenzo[b,e]phosphorin (3) is obtained from the di-
-Grignard agent from bis(2-bromophenyl)methane (1), which on
treatment with dichlorodiethylaminophosphine affords 5-chloro-
-5,10-dihydrodibenzo[b,e]phosphorin (2). The latter is
dehydrochlorinated, in degassed toluene using 1,5-diaza-
bicyclo[4.3.0]non-5-ene, to yield a solution containing
dibenzo[b,e]phosphorin (3), which remains stable for several
days. It has not been isolated and its uv spectrum disappears
on the admission of atmospheric oxygen, sodium hydroxide, or
anhydrous hydrogen chloride, and slowly on addition of dilute
acid. The chloro-compound (2) may also be obtained from the
Grignard reagent (4) from (2-bromophenyl)phenylmethane by
reaction with chlorobis(diethylamino)phosphine to yield product
(5), followed by cyclisation to give compound (2). Oxidation

of chloro-compound (2) with hydrogen peroxide in sodium
hydroxide solution gives 5,10-dihydro-5-hydroxydibenzo-
[b,e]phosphorin 5-oxide. (P. de Koe and F. Bickelhaupt, Angew.
Chem. internat. Edn., 1967, $\underline{6}$, 567).

(a) (i) Et_2NPCl_2,THF, -80^0,
     (ii) HCl,C_6H_{12} (b) DBN
(c) (i) $(Et_2N)_2PCl$,
     (ii) HCl,C_6H_{12} (d) $\Delta AlCl_3$,CS_2

10-Phenyldibenzo[b,e]phosphorin, m.p. 173-176°, is a stable
compound, but it reacts with oxygen faster than do the
monocyclic phosphorins. 5-Chloro-5,10-dihydro-10-
-phenyldibenzo[b,e]phosphorin, b.p. 131°/-10^{-3} torr, m.p.
94-101° (*idem*, *ibid.*, 1968, $\underline{7}$, 889).

   5,10-Dihydro-5-phenyl-10-*tert*-butyldibenzo[b,e]phosphorin
on subjection to pyrolysis at 500° or to electron impact
gives 10-phenyldibenzo{be}phosphorin, *via* the 1,4-migration
of a phenyl group (C. Jongsma, R. Lourens and Bickelhaupt,
Tetrahedron, 1976, $\underline{32}$, 121). The quaternization of 5,10-
-dihydro-5-phenyldibenzo[b,e]phosphorin with benzyl bromide
followed by treatment with sodamide affords 5-benzyl-5-
-phenyldibenzo[b,e]phosphorin, m.p. 161-164°, a λ^5-
-phosphorin (Jongsma, F.J.M. Freijee, and Bickelhaupt,
Tetrahedron Letters, 1976, 481). 5,10-Dihydro-10-hydroxy-10-

-10-methyl-5-phenyldibenzo[b,e]phosphorin and related
derivatives (K. -C. Chen *et al.*, J. org. Chem., 1977, 42,
1170); 5,10-dihydro-5-methyl(phenyl)dibenzo[b,e]phosphorin-10-
-ones (K.A. Petrov, V.A. Chauzov, and N. Yu. Mal'kevich, Zh.
obshch. Khim., 1977, 47, 2516) and 5-oxides and some related
derivatives [Y. Segall, R. Alkabets, and I. Granoth, J. chem.
Res., (S), 1977, 310]; and 3,7-bis(dimethylamino)-5,10-
-dihydro-5,10-diphenyldibenzo[b,e]phosphorin (P. Yu. Ivanov,
et al., Zh. obshch. Khim., 1981, 51, 1533) have been
prepared.

The stereospecific reduction of the 5,10-dihydro-5-methyl-
dibenzo[b,e]phosphorin-10-one by $NaH_2Al(OC_2H_4OMe)_2$ and
$LiAlH(OBu^t)_3$ gives the pseudoaxial and pseudoequatorial
alcohols (6) and (7) (Granoth, H. Segall, and H. Leader,
J. chem. Soc. Perkin I, 1978, 465).

(6)   (7)

The crystal and molecular structure of 5,10-dihydro-10-hydroxy-
-10-methyl-5-phenyldibenzo[b,e]phosphorin 5-oxide (S.E. Ealick
et al., Acta Crystallogr., 1979, B35, 1107) and the structure
of *cis*- 5,10-dihydro-5,10-dimethyl-5-phenyldibenzo[b,e]phos-
phorinium iodide (K.K. Wu *et al.*, Cryst. Struct. Comm., 1977,
6, 405) have been determined, and the acid isomerism of 5,10-
-dihydro-5,10-diphenyl-10-hydroxydibenzo[b,e]phosphorin to
5,10-dihydro-5,10-diphenyldibenzo[b,e]phosphorin 5-oxide
(Petrov, Chauzov, and N. Yu. Lebedeva, Zh. obshch. Khim., 1981,
51, 2142) and the internal rotation in a 3,7-bis(dimethylamino)-5,10-dihydro-5,10-diphenyl-5-thionodibenzo[b,e]phos-
phorin-10-yl cation (V.V. Negrebetskii *et al.*, *ibid.*, 1982,
52, 1930) have been studied.

(iii) Dibenzo[b,d]phosphorins(9-phosphaphenanthrene)

Dibenzo[b,d]phosophorin (2) is obtained from 2-
-phenylbenzylphosphonic acid (1) by the route indicated below
(P. de Koe, R. van Veen, and F. Bickelhaupt, Angew. Chem.
internat. Edn., 1968, _7_, 465).

(a) Ph_2SiH_2 (b) $COCl_2,CH_2Cl_2,N_2$ (c) DBU,MePh

It cannot be isolated in a pure form and its stability is
comparable with that of dibenzo[b,e]phosophorin.

5-Substituted dibenzophospholes (3) on treatment with
benzoyl chloride in the presence of trimethylamine, followed by
hydrolysis undergo ring expansion to yield 5,6-dihydro-
dibenzo[b,d]phosphorin 5-oxides (4) in high yield (D.W. Allen
and A.C. Oades, J. chem. Soc., Perkin I, 1976, 2050).

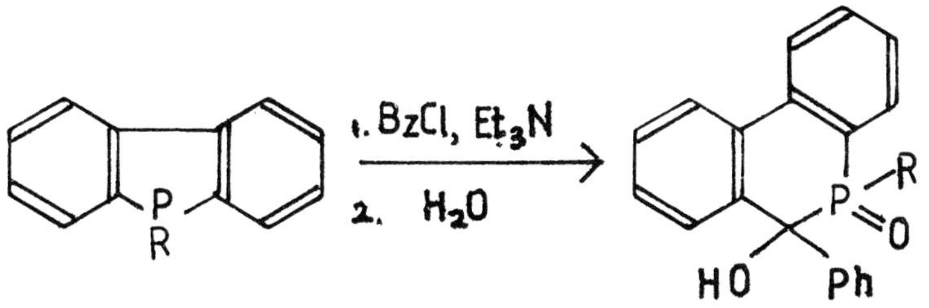

(3)   (4)

Substituent R in compound 4	m.p. (°C)	yield (%)
Ph	276	65
Me	>177 (decomp.)	98
Et	267	92
Pr^i	275	60

5-Benzyldibenzophosphole has been converted *via* ring expansion into 6-phenyldibenzo[b,d]phosphorin (F. Nief *et al.*, Tetrahedron Letters, 1980, 1441). For the preparation of 5,5--dimethyldibenzo[b,d]-λ^5-phosophorin see T. Costa and H. Schmidbaur (Ber., 1982, 115, 1367).

(iv)   Phosphaphenalene derivatives

The reaction between 1,8-bis(bromomethyl)naphthalene and Ph_2PSiMe_3 affords a cyclic phosphonium salt (1), which can be rearranged with $Me_3P=CH_2$ to give the cyclic ylide (2). Its metallation with Bu^tLi yields 2,2-diphenyl-2-λ^5--phosphaphenalenyllithium (3) (H. Schmidbaur and A Mörtl, J. organometallic. Chem., 1983, 250, 171).

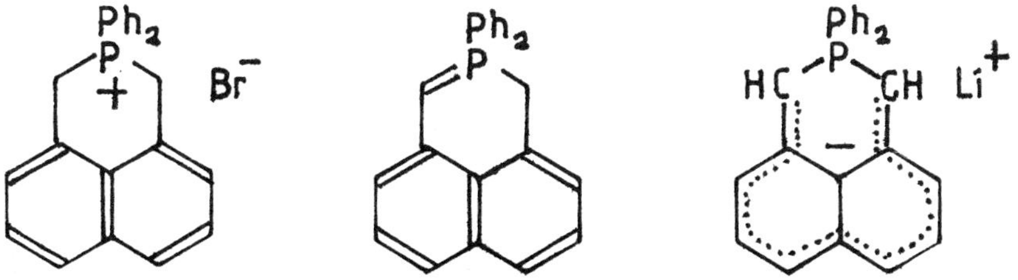

(1)   (2)   (3)

2. *Arsenic Compounds*

(a) *Arsabenzene (arsenin) and its derivatives*

(i) *Arsabenzenes*

Cycloaddition occurs between arsabenzenes and reactive alkynes to give arsabarrelenes, for instance, 2,3,6-triphenyl-arsabenzene (1) reacts with diethyl acetylenedicarboxylate to yield 2,3-diethoxycarbonyl-5,6,7-triphenyl-1-arsabarrelene (2; R=CO$_2$Et) (G. Märkl, J. Advena, and H. Hauptmann, Tetrahedron Letters, 1972, 3961).

(R=CF$_3$,CN,CO$_2$Et)

(1)

(2)

It has been shown that 1-arsabarrelene derivatives are not exclusively formed, but that also some of the 2-arsa derivatives are obtained, giving rise to benzene derivatives on flash pyrolysis by loss of HC≡As (A.J. Ashe and H.S. Friedman, *ibid.*, 1977, 1283).

4-Substituted arsabenzenes (6) are obtained directly by treating the appropriate 4-substituted 4-methoxy-1,1-dibutyl--1,4-dihydrostannabenzene (3) with arsenic trichloride in boiling tetrahydrofuran. Their formation probably goes *via* the 1,4-dihydroarsabenzene (4) and the 1-chloro-1-methoxyarsa-benzene (5) intermediates. The presently known 4-substituted arsabenzenes, with the exception of 4-phenylarsabenzene (crystalline solid), are colourless, distillable, air-sensitive oils (Märkl and F. Kneidl, Angew. Chem. internat. Edn., 1973, 12, 931).

4-Phenyl- and 4-cyclohexyl-arsabenzene, m.p. 51°, and b.p. 107-108°/ 0.02 torr, respectively. Nmr studies relating to the molecular structure of 4-methylarsabenzene have been reported (T.C. Wong, M.G. Ferguson, and Ashe, J. mol. Struct., 1979, 52, 231).

(R=Ph,C₆H₁₁,t-Bu)

(3)   (4)   (5)   (6)

4-Alkyl- and 4-phenyl-arsabenzenes have been obtained from 1,5-
-dilithio-3-alkyl(or phenyl)-3-methoxypent-1,4-diene (Märkl
and R. Liebl, Ann., 1980, 2095).

Treatment of 4-hydroxyarsabenzene (7) with benzyl bromide
gives the dienone (8), which on heating with diphenylketene
affords the methylene derivative (9). Heating to a higher
temperature causes rearrangement to 4-benzyldiphenylmethyl-
arsabenzene (10) (Märkl and J.B. Rampal, Tetrahedron Letters,
1977, 2569).

(7)   (8)   (9)   (10)

(a) PhCH₂Br, K₂CO₃, Me₂CO   (b) Ph₂C=CO, 135-140°   (c) Δ150°

   A number of 2-alkylarsabenzenes (11) have been obtained
by heating the appropriate 2-alkyl-1,1-dibutyl-1,4-dihydro-
stannabenzene with arsenic tribromide in tetraglyme. 2,6-
-Dimethylarsabenzene has also been prepared (Ashe and
W.-T. Chan, J. org. Chem., 1979, 44, 1409).

(R=Me,Et,t-Bu)

(11)

A number of 2-aryl-, 2,6-diaryl-, 2,4-diaryl-, 2-aryl-4-alkyl-,
and 2,4,6-triaryl-arsabenzenes have been prepared (Märkl,
A. Bergbauer, and Rampal, Tetrahedron Letters, 1983, 4079).
2,4-Diaryl- and 2-aryl-4-alkyl-arsabenzenes have also been
obtained by Märkl and Liebl (Angew. Chem., 1977, 89, 670) and
by Märkl, Liebl, and H. Baier (Ann., 1981, 1610).

   4-Substituted 1-aryl-4-methoxyarsacyclohexadienes (12)
rearrange, on reaction with catalytic amounts of strong acids,
for example, 4-toluenesulphonic acid in boiling benzene, or
with boron trifluoride etherate in benzene at room temperature,
to give a 2-arylarsabenzene (13) (Märkl and Liebl, Angew. Chem.
internat. Edn., 1977, 16, 637).

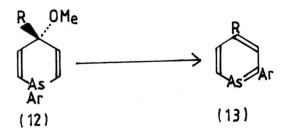

(12)   (13)

2,4-Diphenyl-, m.p. 75-77°; 2-(4-methylphenyl)-4-phenyl-,
m.p. 47-48°; 4-methyl-2-phenyl-, oil; 4-ethyl-2-phenyl-,
oil; 4-cyclohexyl-2-phenyl-, oil; 2-phenyl-4-*tert*-butyl-,
oil; 2-(4-methylphenyl)-4-*tert*-butyl-arsabenzene, m.p. 51-
53°.

The addition of methyllithium to an ether—tetrahydrofuran solution of arsabenzene affords a dark green
solution of the anion (14) of lithium 1-methylarsacyclohexa-
2,4-dienide. Quenching with water yields 1-methylarsacyclo-
hexa-2,4-diene (15), which on treatment with base regenerates
the anion (14). The chemical shifts of the protons of anion
(14) are similar to those of its carbocyclic analogue,
indicating that little or no stabilizing interaction is derived
from the presence of the arsenic atom (Ashe and T.W. Smith,
Tetrahedron Letters, 1977, 407).

$$
\text{(arsabenzene)} \xrightarrow[\text{Et}_2\text{O} \; \text{THF}]{\text{MeLi}} \underset{\substack{\text{As} \\ \text{Me}}}{}{}^{-} \; \underset{\text{base}}{\overset{\text{H}_2\text{O}}{\rightleftharpoons}} \; \underset{\substack{\text{As} \\ \text{Me}}}{}
$$

(14)   (15)

1-Methylarsacyclohexa-2,4-diene (16) may be quaternised
with methyl iodide to yield salt (17), m.p. 195-197°, which
reversibly yields 1,1-dimethyl-λ^5-arsabenzene (18) on
treatment with dimsyl anion in dimethylsulphoxide (Ashe and
Smith, J. Amer. chem. Soc., 1976, 98, 7861).

$$
\underset{\substack{\text{As} \\ \text{Me}}}{} \xrightarrow{\text{MeI}} \underset{\substack{\text{Me}^+\text{Me} \\ \text{I}^-}}{\overset{\text{As}}{}} \; \underset{+\text{H}^+}{\overset{-\text{H}^+}{\rightleftharpoons}} \; \underset{\text{Me}_2}{\overset{\text{As}}{}}
$$

(16)   (17)   (18)

1-Methyl- and 1-phenyl-arsenane (19) form 1:1 and 1:2 adducts with bromine and iodine, but only 1:1 adducts with chlorine (J.B. Lambert and H.-n. Sun, J. org. Chem., 1977, 42, 1315).

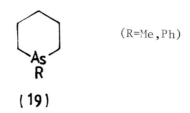

(R=Me,Ph)

(19)

A study has been made of the conformational properties of 1--methylarsenane (*idem*, J. organometallic. Chem. 1976, 117, 17) and a number of methylarsenan-4-ones have been synthesized (Yu. G. Bosyakov *et al*., Tr. Inst. Khim. Nauk, Akad. Nauk Kaz. SSR, 1977, 46, 125). ^{1}H- and ^{13}C-nmr spectral data of arsabenzene (Ashe, R.R. Sharp, and J.W. Tolan, J. Amer. chem. Soc., 1976, 98, 5451) and the ir and Raman spectra of arsa- and phospha- benzene have been recorded (Ashe, G.L. Jones, and F.A. Miller, J. mol. Struct., 1982, 78, 169). Arsabenzene gives a molybdenum-carbonyl complex (Ashe and J.C. Colburn, J. Amer. chem. Soc., 1977, 99, 8099) and some 4-substituted 2--arylarsabenzenes afford 6-tricarbonylchromium (molybdenum, tungsten) (0) complexes (Märkl *et al*., J. organometallic. Chem., 1981, 217, 333). The molecular structure of arsa- and phospha- benzene has been determined by analysis combining electron diffraction and microwave data (T.C. Wong and L.S. Bartell, J. mol. Struct., 1978, 44, 169) and by nmr studies (Wong and Ashe, *ibid.*, 48, 219). Electron transmission spectroscopy has been employed to study temporary anion formation in arsabenzene (see p. 24). For the synthesis of substituted arsabenzenes see S.T. Abu-Orabi (Diss. Abs. Int. B, 1982, 43, 1846); for the synthesis and electronic structure of arsa- and phospha-benzene see W.-T. G. Chan (*ibid.*, 1978, 38, 5375); and for a review of arsa- and phospha-benzene see Märkl (Chem. Unserer Zeit., 1982, 16, 139).

(ii) Derivatives of arsabenzene possessing a functional group

4-Hydroxyarsabenzene (4), m.p. 102-105° (decomp.), is

obtained by reacting 4-acetoxy-1,1-dibutyl-1,4-dihydrostanna-
benzene (1) with arsenic trichloride to give a mixture of
cis/trans isomers of 4-acetoxy-1-chloro-1,4-dihydroarsabenzene
(2) one isomer of which, spontaneously eliminates hydrogen
chloride to give the acetoxy derivative (3). The other isomer
yields 4-acetoxyarsabenzene (3) on treatment with triethylamine in benzene. Hydrolysis of the acetoxy derivative (3)
affords 4-hydroxyarsabenzene (4) and spectral data indicates
that it exists as the arsaphenol and not as 1-arsacyclohexa-
-2,5-dien-4-one (5) (Märkl, H. Baier, and S. Heinrich, Angew.
Chem. internat. Edn., 1975, 14, 710).

Alkylation of 4-hydroxyarsabenzene with methyl or ethyl iodide gives the 1-alkyl-4-one (p.37), but with propyl iodide a trace of 4-propoxyarsabenzene is also obtained and with ethyl and propyl bromide mixtures of the corresponding 1-alkyl--4-ones and alkoxyarsabenzenes are formed (Märkl and Rampal, Tetrahedron Letters, 1976, 4143). The acetylation of 1--arylarsacyclohexa-2,5-dien-4-one (6) results in a dienone--phenol type of rearrangement to yield 4-acetoxy-2-aryl-arsabenzene (7), which on hydrolysis affords the 2-aryl-4--hydroxyarsabenzene (8) (Märkl and Rampal, *ibid.,* 1977, 3449).

$$(Ar=C_6H_5, \quad 4-MeC_6H_4, \quad 4-ClC_6H_4)$$

(6)   **(7)**   **(8)**

The reaction between 2-aryl-4-hydroxyarsabenzene (8; Ar= C_6H_5, 4-MeC$_6$H$_4$) and phenyl radicals gives the ketone (9), which on acetylation yields the 4-acetoxy-2-aryl-6--phenylarsabenzene (10). Hydrolysis of (10) affords a product, which in ethanol and chloroform appears to contain both tautomers (11) and (12) (Märkl and Rampal, *ibid.,* 1978, 1471).

$(Ar=C_6H_5,\ 4\text{-MeC}_6H_4)$

1,3,5-Triallylarsacyclohexa-2,5-dien-4-one has been prepared
by treating 4-hydroxyarsabenzene with allyl bromide. 1-Allyl-
arsacyclohexa-2,5-dien-4-one is first formed and undergoes a
hetero Cope rearrangement to yield 3-allyl-4-hydroxyarsa-
benzene, which is then converted into the 1,3,5- triallyl-
arsacyclohexa-2,5-dien-4-one (Märkl and Rampal Angew. Chem.,
1976, 88, 728). Thermolysis of 4-alkoxy-1-chloro-4-(diethoxy-
methyl)-1,4-dihydroarsabenzenes gives 4-alkoxyarsabenzenes
(Märkl and Rampal, Angew. Chem. internat. Edn., 1974, 13, 668).
4-Methoxy- and 4-ethoxy- arsabenzene (*idem ibid.*, p.667).

The oxime of 1-phenylarsacyclohexa-2,5-dien-4-one (13) on
boiling with acetic anhydride rearranges to give the diacetyl
derivative of 4-amino-2-phenylarsabenzene (an arsaaniline
derivative) (14) (Märkl and Rampal, Tetrahedron Letters, 1978,
1175).

Reduction of the 1-chloro-1,4-dihydroarsabenzene (15) with trialkyltin hydride gives the 1,4-dihydroarsabenzene (16), which under the conditions for its formation undergoes 1,4-elimination of methanol or ethanol to yield the diethyl acetal of arsabenzene-4-carboxaldehyde (4-diethoxymethylarsabenzene) (17). Treatment of acetal (17) with an acidic ion exchanger in moist acetone affords arsabenzene-4-carboxaldehyde (4-arsabenzaldehyde) (18). It shows the normal electrophilic reactions of the aldehyde group, for example, it undergoes an aldol condensation with acetone to give the alcohol (19) and with acetone on addition of 20% sodium hydroxide solution affords the arsabenzylideneacetone (20) (Märkl and F Kneidl, Angew. Chem. internat. Edn., 1974, 13. 668).

Arsabenzene-4-carboxaldehyde (18) undergoes a Knoevenagel condensation with malonic acid in pyridine to give the related 4-arsacinnamic acid (arsabenzene-4-acrylic acid). Aldehyde (18) also condenses with cyclic ketones and reacts with hydroxylamine to yield an oxime, which dehydrates in boiling acetic acid to the corresponding arsabenzonitrile. On reaction with phenyl- or 1-naphthyl-magnesium bromide, followed by hydrolysis of the resulting Grignard complex,

aldehyde (18) gives the respective 1-phenyl- or 1-(1-naphthyl)-
-1,2-dihydroarsabenzene-4-carboxaldehyde (Märkl, Rampal and
V. Schöberl, Tetrahedron Letters, 1979, 3141). Arsabenzene-4-
-carboxaldehyde undergoes the Wittig reaction to give 4-
-vinylarsabenzenes (*idem, ibid.*, 1977, 2701). [1]H-nmr
spectral data indicates that acetylation of arsabenzene affords
a mixture of 4-acetyl- (80%) and 2-acetyl-arsabenzene (20%)
(A.J. Ashe, W.-T. Chan, and T.W. Smith, *ibid.*, 1978, 2537).
Similarly nitration gives a mixture of the 2- and the 4-
-substituted products and proton-deuterium exchange takes place
in trifluoroacetic acid-*d* in the 2-and the 4-position (Ashe
et al., J. org. Chem., 1981, **46**, 881).

   4-Ethoxycarbonylarsabenzene (ethyl 4-arsabenzoate) is
prepared in a number of steps from ethyl dichloroethoxyacetate
(Märkl, H. Kellerer, and Kneidl, Tetrahedron Letters, 1975,
2411), and is hydrolysed with sodium hydroxide under nitrogen
to yield the corresponding sodium salt, which can be converted
into 4-carboxyarsabenzene (4-arsabenzoic acid), stable under
nitrogen. Its ir, uv and nmr spectral data and pK value have
been reported (Märkl and Kellerer, *ibid.*,1976, 665). pKa
Values of 2-, 3- and 4-carboxyarsabenzenes have been recorded
(Ashe and Chan, J. org. Chem., 1980, **45**, 2016).

(b) Arsanaphthalenes (benzoarsenins)

   Arsabenzene reacts with benzenediazonium-2-carboxylate to
give 1,4-etheno-1,4-dihydro-1-arsanaphthalene (1), m.p. 45°,
which on treatment with an acetylene abstraction agent 3,6-
-di(2-pyridyl)-s-tetrazine, at low temperature, results in the
evolution of N_2 and the formation of 1-arsanaphthalene (2).
It is very air-sensitive yellow oil, but it can be trapped with
reactive dienophiles, for example, hexafluorobut-2-yne to
yield the 1:1 adduct (3), m.p. 58.5° (A.J. Ashe,
D.J. Bellville, and H.S. Friedman, Chem. Comm., 1979, 880).

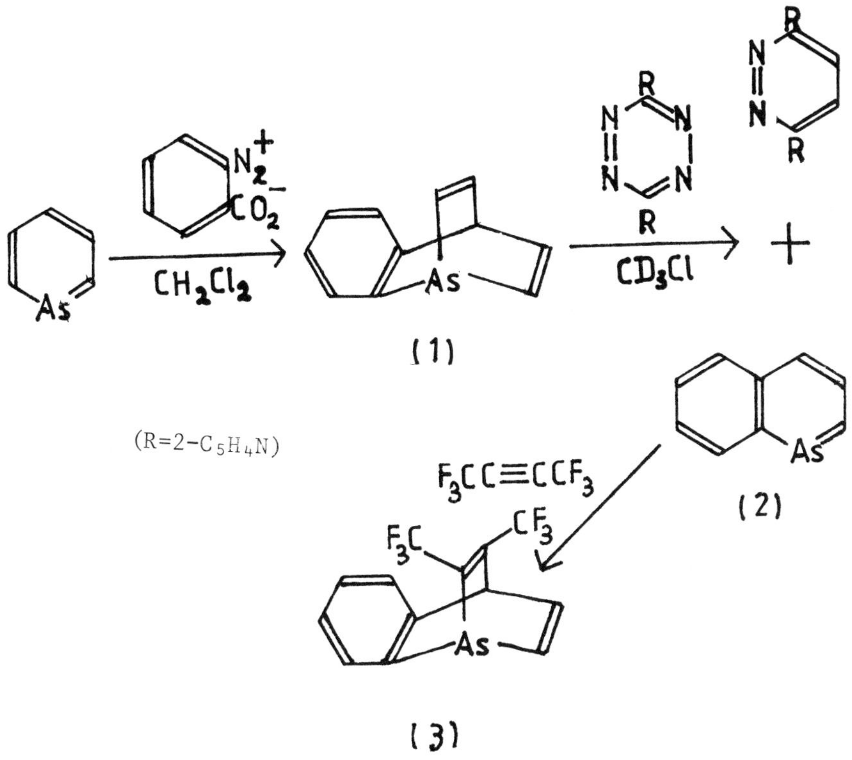

(1)

(R=2-C_5H_4N)

(2)

(3)

(c) Dibenzoarsenins

(i) Dibenzo[be]arsenins (arsaanthracenes)

5,10-Dihalogeno-5,10-dihydrodibenzo[be]arsenins (2) and
5,10-dihydro-5-halogeno-10-methoxydibenzo[be]arsenins (3) are
obtained from 5,10-dihydro-5-hydroxydibenzo[be]arsenin-5-one
(1). Treatment of the dihydrodibenzo[be]arsenins (2) and (3)
with 1,5-diazabicyclo[5.4.0]undec-5-ene in tetrahydrofuran in
a high-vacuum sealed vessel results in the elimination of
hydrogen halide and formation of the corresponding dibenzo-
[be]arsenin derivatives (4), which due to their instability
resist isolation. 10-Chloro- and 10-methoxy-dibenzo[be]-
arsenin (4, R=Cl and OMe) besides being identified by their
spectral data have been characterised by their Diels-Alder

adducts (5) formed with maleic anhydride (R.J.M. Weustink, C. Jongsma, and F. Bickelhaupt, Tetrahedron Letters, 1975, 199).

10-Aryldibenzo[be]arsenins (7) have been prepared by the flash vacuum pyrolysis at 500° of 5-aryl-10-benzyl-5,10-
-dihydrodibenzo{be}arsenins (6) (*idem*, Rec. trav. chim., 1977, 96, 265) and a radical mechanism involving intramolecular 1,4-
-migration of the aryl group has been suggested for their formation (Weustink, R. Lourens, and Bickelhaupt, Ann., 1978, 214).

$(R^1=H,Me,OMe$
$R^2=H,Me,Ph,CH_2Ph)$

(6)   (7)

10-Mesityldibenzo[be]arsenin (10-mesityl-9-arsaanthracene) is
the most stable of the known dibenzo[be]arsenins (Weustink,
P.J.A. Geurink, and Bickelhaupt, Heterocycles , 1978, 11,
299). The structure of *trans*-10-benzyl-5-phenyl-5,10-dihydro-
dibenzo[be]arsenin has been determined and it has been shown
that the dibenzo[be]arsenin ring system is folded with an angle
of 133° between the benzene rings (C.H. Stam, Acta
Crystallog., 1980, B36, 455).

Table 1

5,10-Dihydrodibenzo[be]arsenin and dibenzo[be]arsenin
derivatives

	M.p.(^{O}C)	Ref.
5,10-Dichloro-5,10-dihydrodibenzo[be]arsenin	167-170	1
5,10-Dibromo-5,10-dihydrodibenzo[be]arsenin	179-181	1
5-Chloro-5,10-dihydro-10-methoxydibenzo[be]-arsenin	120-122	1
5-Bromo-5,10-dihydro-10-methoxydibenzo[be]-arsenin	116-118	1
Maleic anhydride adduct of:-		
10-Chlorodibenzo[be]arsenin	255-256	1
10-Methoxydibenzo[be]arsenin	252-254	1

[1]R.J.M. Weustink, C. Jongsma, and F. Bickelhaupt, Tetrahedron
Leters, 1975, 199.

(ii)   Dibenzo[bd]arsenin

Treatment of two moles of 2-phenylbenzyl bromide (1) with
one mole of the di-Grignard reagent, from phenylarsine gives
phenylbis(2-phenylbenzyl)arsine (2), which after conversion
into the dichloroarsine (3), readily loses one of the benzyl
groups on heating to yield phenyl(2-phenylbenzyl)chloroarsine
(4). Cyclization of compound (4) gives 5,6-dihydro-5-phenyl-
dibenzo[bd]arsenin (5), methiodide, m.p. 195° (which gives a
picrate, m.p. 150-151°), and dichloropalladium derivative
m.p. 244-245°. 5,6-Dihydro-5-methyldibenzo[bd]arsenin,
methiodide, m.p. 212-215°, obtained from its hemihydrate,
m.p. 210-215° (G.J. Cookson and F.G. Mann, J. chem. Soc.,
1949, 2888).

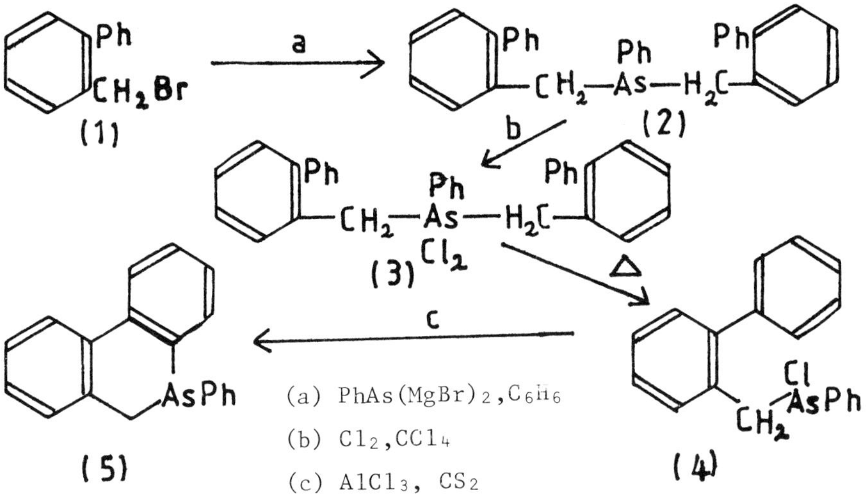
a
b
c
(1)
(2)
(3)
(4)
(5)
(a) PhAs(MgBr)2,C6H6
(b) Cl2,CCl4
(c) AlCl3, CS2

(d) Miscellaneous heterocycles containing an arsabenzene ring

Arsaphenols (1) on treatment with propargyl bromide yield the corresponding arsacyclohexadienones (2), which undergo thermal rearrangement to pyrano- and furano-arsenins, for example, compound (2; R^1=H, R^2=H) affords the pyranoarsenin (3), and (2; R^1=Ph, R^2=H) the furanoarsenin (4) (G Märkl and J.B. Rampal, Tetrahedron Letters, 1979, 1369).

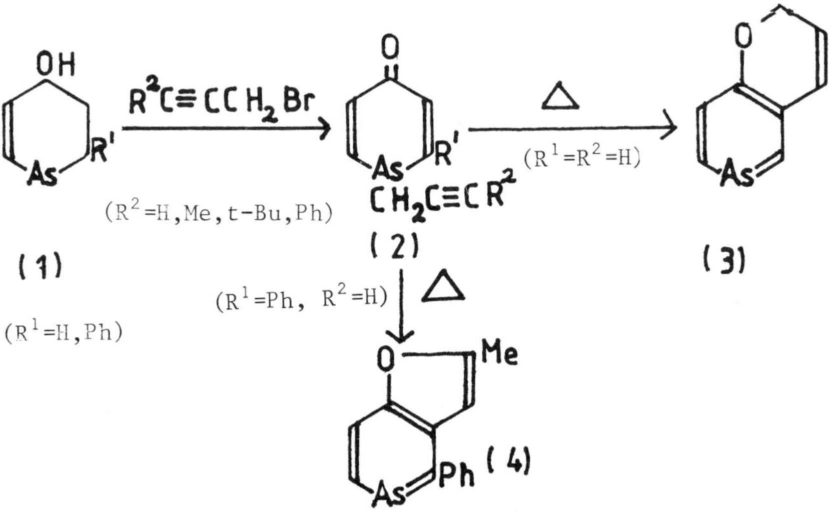
(R2=H,Me,t-Bu,Ph)
(R1=R2=H)
(R1=Ph, R2=H)
(R1=H,Ph)
(1)
(2)
(3)
(4)

3. *Antimony Compounds*

(a) *Mononuclear compounds*

The ^{1}H-nmr spectrum of stibabenzene (1) has been reported and it shows the characteristic signal pattern of the Group VB heteroaromatics, the α-protons giving a doublet at extremely low field, while the β- and γ-proton signals are in the normal aromatic region. Its ms (A.J. Ashe, J. Amer. chem. Soc., 1971, 93, 6691) and ^{13}C-nmr spectrum have also been reported (Ashe, R.R. Sharp, and J.W. Tolan, *ibid.*, 1976, 98, 5451).

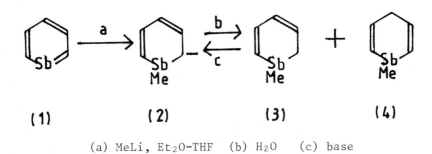

(1)   (2)   (3)   (4)

(a) MeLi, Et$_2$O-THF   (b) H$_2$O   (c) base

Stibabenzene (1) on treatment with methyllithium yields anion (2), which on quenching affords a mixture of 1,2-dihydro--1-methyl- (3) and 1,4-dihydro-1-methyl- (4) -stibabenzene. The anion (2) may be obtained from 1,4-dihydro-1-methyl--stibabenzene (4), prepared by the reaction between methyl-lithium and 1-chloro-1,4-dihydrostibabenzene. ^{1}H- and ^{13}C-nmr spectral data have been recorded (Ashe and T.W. Smith, Tetrahedron Letters, 1977, 407).

1,1-Dibutyl-1,4-dihydrostannaphenyllithium (5) on quenching with methyl iodide gives 1,1-dibutyl-1,4-dihydro-4--methylstannabenzene (6), which on treatment with antimony trichloride followed by 1,8-diazabicyclo[5.4.0]undec-7-ene yields 4-methylstibabenzene (7).

(5) (6)

(7) (R = Me)

(8) (R = t-Bu)

4-*tert*-Butylstibabenzene (8) may be obtained by a similar
route. The 4-methyl- and 4-*tert*-butyl- derivatives are
easily distillable liquids and at 25° the former shows no
change after 1h, but after 24h it has polymerised. Pure
stibabenzene (1) polymerises rapidly at 25°. Alkyl
derivatives (7) and (8) do not form any detectable quantities
of a Diel-Alder dimer (*cf* bismabenzene p. 57), although (7)
gives the expected Diels-Alder adduct with dimethyl acetylenedicarboxylate (Ashe, T.R. Diephouse, and M.Y. El-Sheikh,
J. Amer. chem. Soc., 1982, 104, 5693). Nmr studies relating to
the molecular structure of 4-methylstibabenzene have been
reported (T.C. Wong, M.G. Ferguson, and Ashe, J. mol. Struct.,
1979, 52, 231) and an investigation has been made of the
temporary anion states of stibabenzene (see p. 24). A
dissertation on some of the chemistry of stibabenzene has been
published (G.D. Fong, Diss. Abs. Int. B, 1979, 39, 4907).

1-Methylantimonane (1-methylstibacyclohexane) (9), b.p.
77-79°/17 mm., has been obtained as one of the products from
the reaction between 1,5-dibromopentane and dimethylstibyl-
sodium in liquid ammonia (H.A. Meinema, H.F. Martens, and
J.G. Noltes, J. organometallic. Chem., 1976, 110, 183). The
tetramethylammmonium salt (10) of the complex anion 1,1,1,1-
-tetrachloroantimonate has been synthesized and its ^{121}Sb
Mössbauer spectral data recorded (R. Barbieri *et al.*, J. chem.
Soc., Dalton, 1979, 1925).

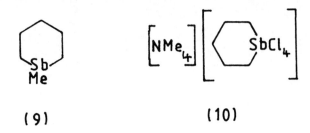

(9)   (10)

1-Chloro-, b.p. 58-59°/0.01 mm, m.p. 35°, 1,1,1-trichloro-
-antimonane, m.p. 103-105°; tetramethylammonium 1,1,1,1-
-tetrachloroantimonanate, m.p. 130° decomp. (Meinema *et al.*,
J. organometallic. Chem., 1977, <u>136</u>, 173).

(b) Fused ring compounds

Attempts have been made to synthesize 9-stibaanthracene
(dibenz[b,e]antimonin) (1) and 10-phenyl-9-stibaanthracene but
both compounds are too unstable for isolation. The reaction
between 9,10-dihydro-9,9-dimethyl-9-stannaanthracene (2) and
antimony trichloride gives 9-chloro-9,10-dihydro-9-
-stibaanthracene (3), which on treatment with 1,5-diaza-
bicyclo[5.4.0]undec-5-ene yields compound (4) (F. Bickelhaupt
et al., Rec. trav. chim., 1979, <u>98</u>, 3).

9-Chloro-9,10-dihydro-9-stibaanthracene (5-chloro-5,10-
-dihydrodibenz[b,e]antimonin) (3) on chlorination in
chloroform at 0° using an equimolar amount of sulphuryl
chloride affords 9,10-dihydro-9,9,9-trichloro-9-stibaanthracene
(5), m.p. 135-165° decomp., which with tetramethylammonium
chloride yields the tetramethylammonium 9,9,9,9-dihydro-9,9,9,9-
-tetrachloro-9-stibaanthracide, m.p. 250° decomp. (Meinema
et al., loc. cit.). 9,10-Dihydro-9,10-dimethyl-9-stiba-
anthracene, m.p. 95-96° (C. Jongsma et al., Tetrahedron,
1977, 33, 205).

Stibatriptycene (7), m.p. 177-178°, has been obtained by the
cyclization of 9-(2-chlorophenyl)-9,10-dihydro-9-stiba-
anthracene (6) using an excess of lithium piperidide. Its
structure was confirmed by ir, [1]H-nmr, and ^{13}C-nmr spectral
data (Jongsma et al., loc. cit.).

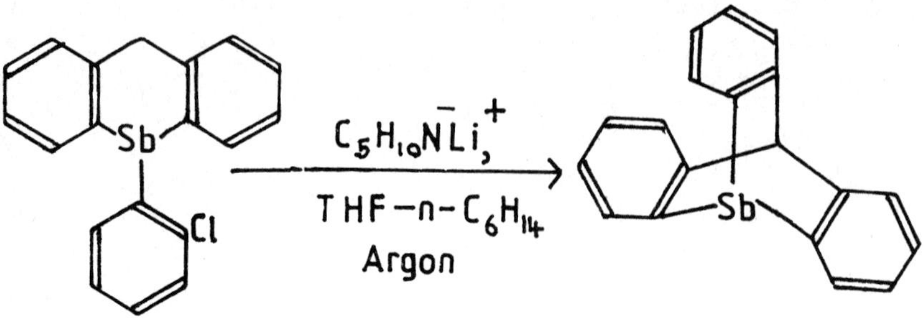

166

(6)   (7)

4. Bismuth Compounds

Attempts have been made to obtain bismabenzene (bismin)
(3) by reacting 1,4-dihydro-1,1-dibutylstannabenzene (1) with
bismuth trichloride to give 1,4-dihydro-1-chlorobismabenzene
(2), which on treatment with 1,5-diazabicyclo[4.3.0]non-5-ene
loses hydrogen chloride exothermically to yield polymeric
material. The addition of hexafluorobutyne at low
temperature, after precipitation of the hydrochloride (2)
affords a 1:1 adduct (4) (A.J. Ashe and M.D. Gordon, J. Amer.
chem. Soc., 1972, 94, 7596).

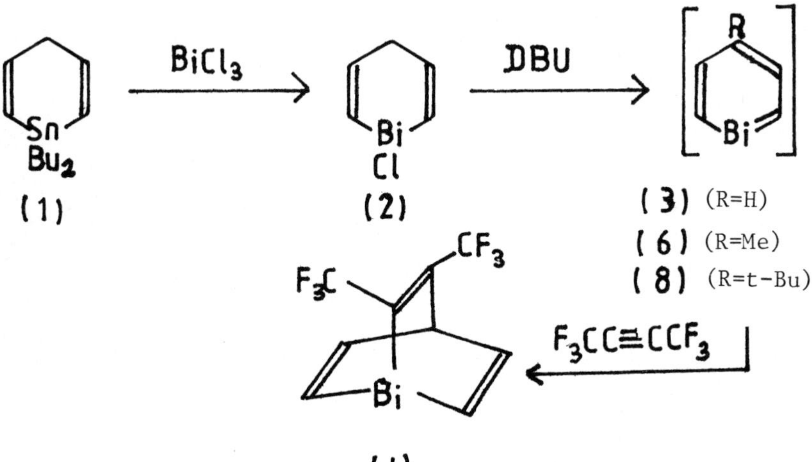

(1)   (2)   (3) (R=H)
                                     (6) (R=Me)
                                     (8) (R=t-Bu)

(4)

The [1]H-nmr spectroscopic detection at low temperatures
of bismabenzene (3) and its dimer (5) have been reported (Ashe,
Tetrahedron Letters, 1976, 415) along with photoelectron
spectral data (J. Bastide *et al.*, *ibid.*, p.411). The former
shows very low field signals for the α-protons due to the very
large diamagnetic anisotropy of the bismuth atom (Ashe, T.R.
Diephouse, and M.Y. El-Sheikh, J. Amer. chem. Soc., 1982, 104,
5693).

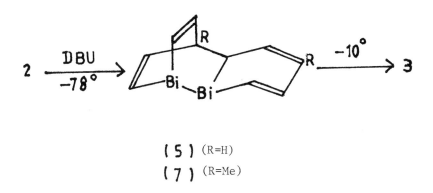

(5) (R=H)
(7) (R=Me)

4-Alkylbismabenzenes have been prepared and although 4-
-methylbismabenzene (6) is more stable towards polymerisation
than bismabenzene (3) it is still in mobile equlibrium in
tetrahydrofuran with its head to head Diels-Alder dimer (7).
On cooling solutions of 4-*tert*-butylbismabenzene (8) in
tetrahydrofuran no dimer could be detected and they were stable
for several hours at 0°C (Ashe, Diephouse, and El-Sheikh,
loc. cit.).

Attempts to form molybdenum-carbonyl complexes of
bismabenzene have been unsuccessful because of its extreme
lability (Ashe and J.C. Colburn, J. Amer. chem,. Soc., 1977,
99, 8099).

Chapter 30

PYRIDINE AND PIPERIDINE ALKALOIDS

MALCOLM SAINSBURY

Pyridine and piperidine bases occur widely in Nature, both
as alkaloids in plants and in the defence secretions of
certain insects and amphibians. They frequently co-occur
with tetrahydroquinolines and quinolizines with which they
share common biosynthetic origins.

1. *Alkaloids from the* Achillea *and* Piper *genera*

N-Acylpiperidines are common in plants of the unrelated *Achillea*
and *Piper* genera. Typically the flavour producing components
of pepper plants contain piperine, isopiperine, chavicine and
isochavicine which are geometrical isomerides of structure
(1) (R. Delleyn and M. Verzele, Bull.Soc.chim.Belges, 1975,
84, 435). Dihydro derivatives of these alkaloids occur in
P. novae hollandiae (T.R. Govindachari *et al.*, Ind.J.Chem,
1968, *7*, 308; J.W. Loder and G.B. Russell, Austral.J.Chem.,
1969, *22*, 1271) and in *P. guineese* (I.A.-Mensah, F.B. Torto
and I.Baxter, Tetrahedron Letters, 1976, 3049). This last
plant also produces 4,5-dihydro-2-methoxypiperine and wisane
(2'-methoxypiperine) (*idem,* Phytochem., 1977, *16*, 483).
An alkaloid from *P. peepuloides,* at first considered to
have *trans* stereochemistry , has now been shown to have the
cis-configuration (2) (O.P. Vig *et al.*, Ind.J.Chem., 1979,
17B, 427; 521). This plant also forms 1-(2-methoxy-4,5-
methylene-dioxycinnamoyl) piperidine and the 2-*trans*, 4-*cis*-
isomer of isowisanine (Mensah *et al.*, Planta Med., 1981, *41*,
200).

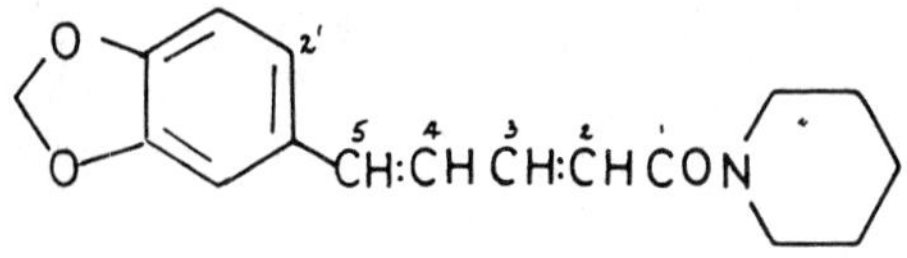

1

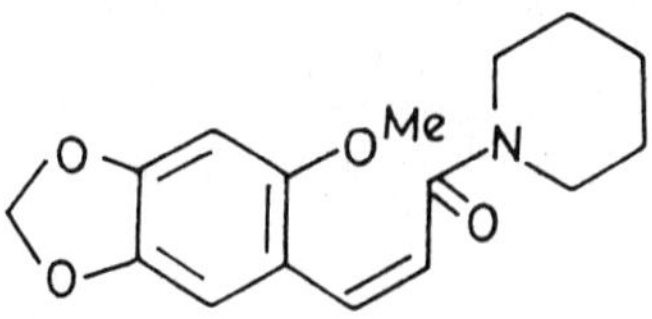

2

Piperine *(trans, trans)*

Isoperine (2,3-*cis*, 4,5-*trans*)

Chavicine *(cis, cis)*

Isochavicine (2,3-*trans*, 4,5-*cis*)

Piplartine (3), another pepper alkaloid, is dimerised on irradiation with ultraviolet light to the compound (4), which occurs as a natural product in its own right - as an extractive from *P. tuberculatum* (R.B.-Filho, M.P. DeSouza and M.E.O. Mattos, Phytochem., 1981, *20*, 345).

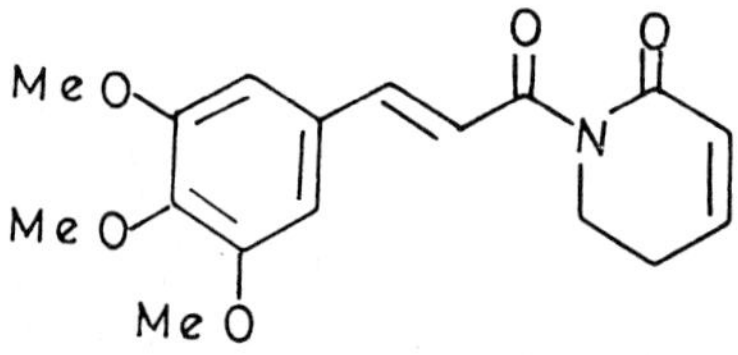

3

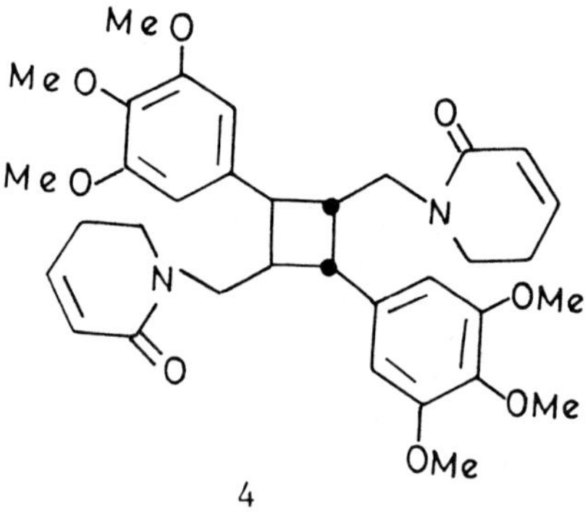

4

Pipermethystine isolated from *P.methysticum* is likely to have structure (5) since it is hydrolysed to 3-phenylpropanoic acid and the dihydropyridone (6) (R.M. Smith, Tetrahedron, 1979, *35*, 437).

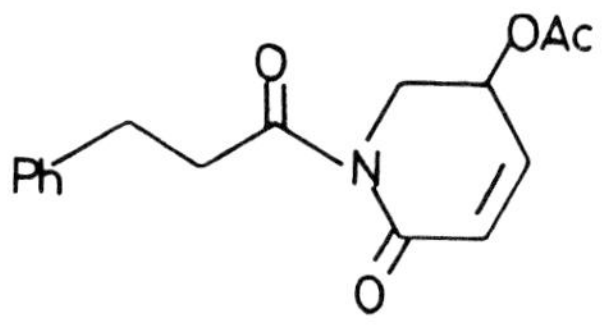

5 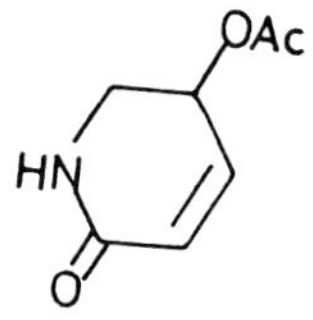   6

Alkaloids from *Achillea* species lack a aryl ring and often contain acetylenic alkyl chains linked through an amide unit to a piperidine ring. For example the following structures (7) and (8) occur in *A.spinulifolia* while (9) and (10) have been isolated from *A.grandifolia*. Similarly the amides (11) and (12) are elaborated by *A.biebersteinii* and *A. lycaonica* respectively (H. Greger, M. Genz and F. Bohlmann, Phytochem., 1981, *20*, 2579; 1982, *21*, 1071). Two other compounds (13) and (14) arise in *A. millefolium* (F. Bohlmann, C. Zdero and A. Suwita, Ber., 1974, 1038).

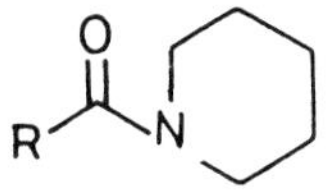

7   R = $Me(CH_2)_2CH\overset{c}{=}CH.C\equiv C(CH_2)_2CH\overset{t}{=}CH.CH\overset{t}{=}CH-$

8   R = $MeC\equiv C.C\equiv C.C\equiv C(CH_2)_2CH\overset{t}{=}CH.CH\overset{t}{=}CH-$

9   R = $MeCH\overset{t}{=}CH.C\equiv C.C\equiv C(CH_2)_2CH\overset{t}{=}CH.CH\overset{t}{=}CH-$

10   R = $MeCH\overset{t}{=}CH.C\equiv C.C\equiv C.CH\overset{t}{=}CH.CH\overset{t}{=}CH.CH\overset{t}{=}CH-$

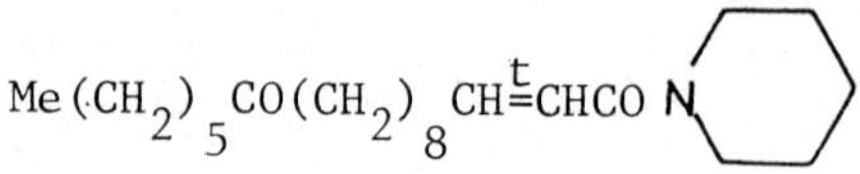

$$Me(CH_2)_5CO(CH_2)_8CH\overset{t}{=}CHCO N\langle\bigcirc\rangle$$

11

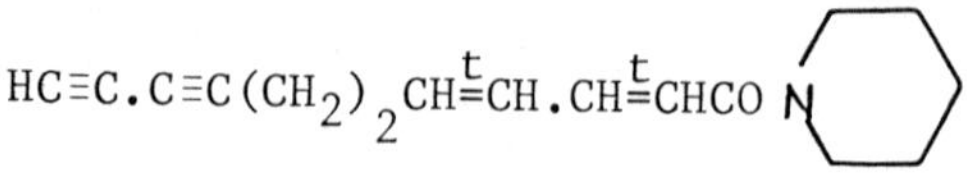

$$HC\equiv C.C\equiv C(CH_2)_2CH\overset{t}{=}CH.CH\overset{t}{=}CHCO N\langle\bigcirc\rangle$$

12

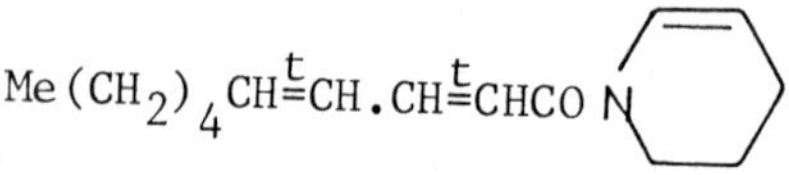

$$Me(CH_2)_4CH\overset{t}{=}CH.CH\overset{t}{=}CHCO N\langle\bigcirc\rangle$$

13

$$n-C_9H_{19}CH\overset{t}{=}CH.CH\overset{t}{=}CH.CO N\langle\bigcirc\rangle$$

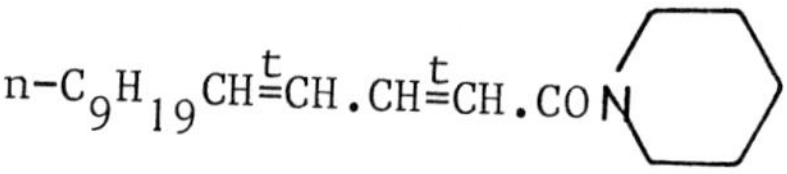

14

The two amides (15) and (16) occur in *A. falcata*; the first is unusal in having a hydroxyl group at *C*-4 in the piperidine ring (Greger, Zdero and Bohlmann, Ann., 1983, 1194.

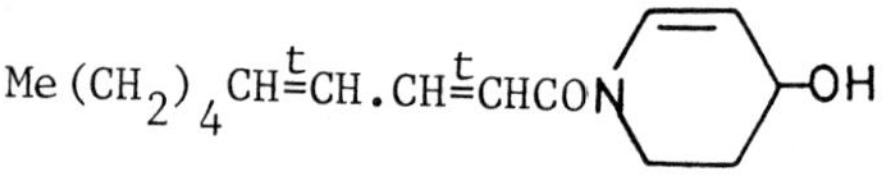

$$Me(CH_2)_4CH\overset{t}{=}CH.CH\overset{t}{=}CHCON\langle\bigcirc\rangle-OH$$

15

$$MeCH=CH(CH_2)_2CH\overset{t}{=}CHCH\overset{t}{=}CHCO N\langle\bigcirc\rangle$$

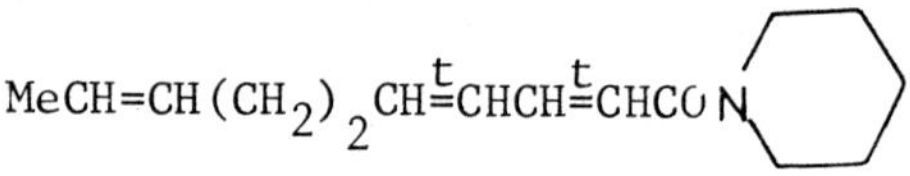

16

Interestingly *Otanthus maritimus*, another plant from the Anthemidae family, yields the thiophene derivatives (17) and (18) (Bohlmann, Zdero and Suwita *loc.cit.*). Piperoleine A (19,n = 6) and piperoleine B (19, n = 4) occur in black pepper *P.nigrum* (R. Grewe *et al*, Ber., 1970, *103*, 3752).

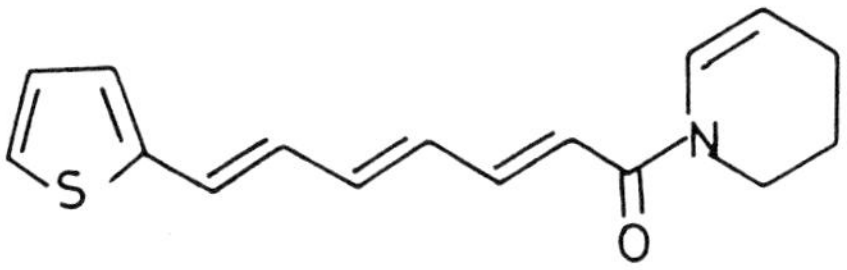

17

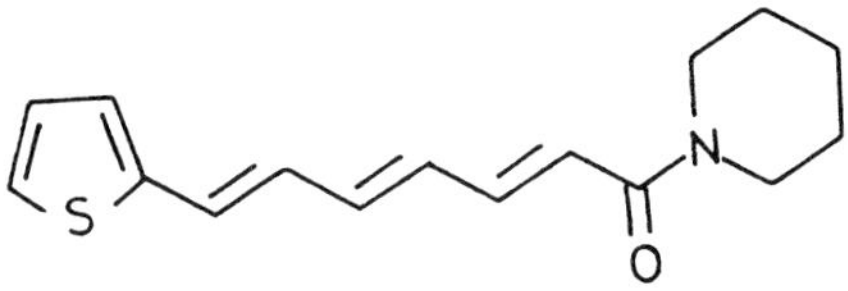

18

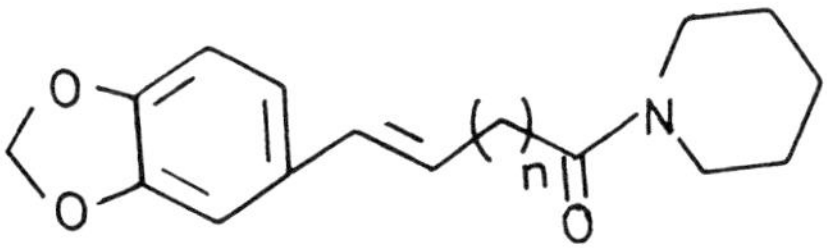

19

2. *Miscellaneous N-acylpiperidine alkaloids*

A complex dipiperidine alkaloid (-)-baptifoline (20) occurs in the Leguminous plant *Hovea longipes* indigenous to Australia (J.S. Fitzgerald *et al.*, Anales de Quim., 1972, *68*, 737) and the unnamed amide (21) is found in the stem bark of *Exoecharia sagallocha* (S. Prakash *et al*, Phytochem., 1983, *22*, 1836).

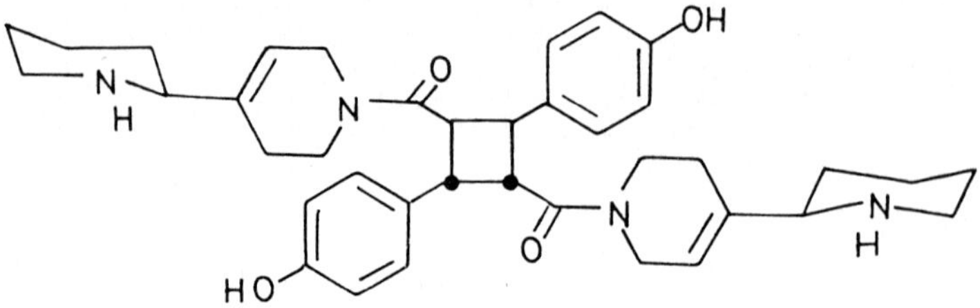

20

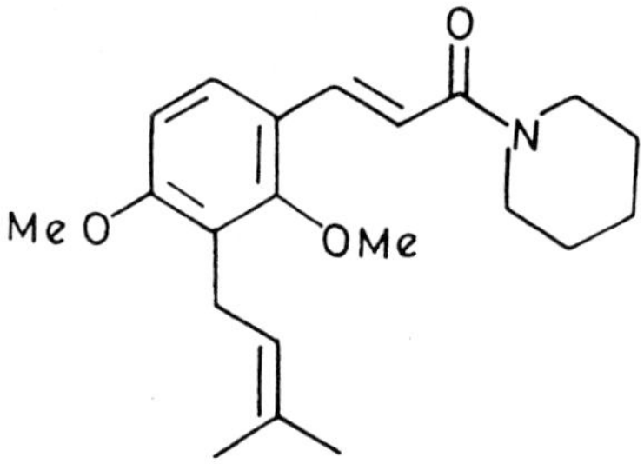

21

3. Sedum alkaloids

An X-ray diffraction analysis of (-)-sedinine, an alkaloid
present in several *Sedum* species, establishes the position of
the double bond in the heterocycle at $^3\Delta$ rather than at $^4\Delta$ as
had been suggested earlier.   The structure (22) for the
alkaloid represents its absolute stereochemistry (C. Hootele
et al., Tetrahedron Letters, 1980, *21*, 5063), and the (±)-
form has now been synthesised (M. Ogawa and M. Natsume,
Heterocycle, 1985, *23*, 831).   Sedacryptine, isolated from
S. acre, has the relative stereochemistry indicated in
formula (23) (Hootele *et al*, Tetrahedron Letters, 1980, *21*,
5061) and its racemate has also been synthesised
(A.P. Kozikowski and R.J. Schmiesing, J.org.Chem., 1983, *48*,
1000; M Ogawa and M. Natsume, Heterocycles, 1985, *23*, 831).

22

23

(±)-Sederine which co-occurs with sedacryptine in *S. acre* has structure (24) (Hootelé, J.P. Etienne and B. Colan, Bull.Soc. chim.Belges, 1979, *88*, 111). Other minor bases of this plant are sedinone (25) and dihydrosedinine (26), but the most abundant alkaloid is sedacrine which has the constitution (27) (B. Colan and Hootele, Canad. J.Chem., 1983, *61*, 470).

24

25

26

27

(+)-Sedridine (28) and (−)-allosedridine (29) have the absolute stereochemistries indicated in the respective formulae. These assignments are based on a von Braun type degradation of *O*,*N*-dibenzoylsedridine with phosphorus (v) bromide, followed by catalytic hydrogenolysis and hydrolysis, which gave (*S*)-(+)-2-octanol. This result sets the stereochemistry of the secondary alcohol centre at C-2' as (*S*) and since it was already known that the configuration at C-2 is also (*S*) the absolute stereochemistry of the sedridine is fully established. (−)-Allosedridine was converted into the racemic bicyclic derivative (30) the structure of which was determined by [1]H n.m.r. spectroscopy and by an X-ray crystallographic analysis. Assuming that no rearrangement occurred during derivatisation (-)-allosedridine must have (2*S*, 2'*R*).stereochemistry (D. Butruille *et al.*, Tetrahedron, 1971, *27*, 2055).

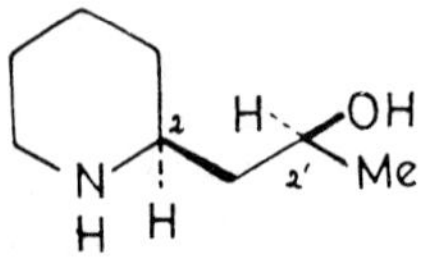

28

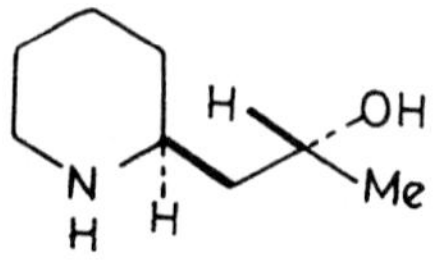

29

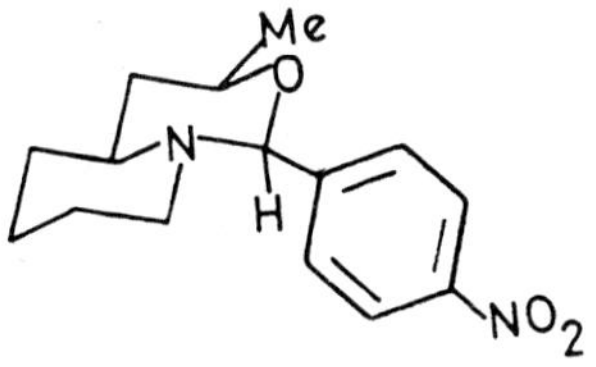

30

N-Methyl-*allo*-sedridine (31) occurs in *S. sarmentosum,* its stereochemistry was deduced by o.r.d. analyses and chemical correlations with the other alkaloids of this group (H.C. Beyerman *et al.*, Rec.Trav.chim., 1972, *91*, 1441).

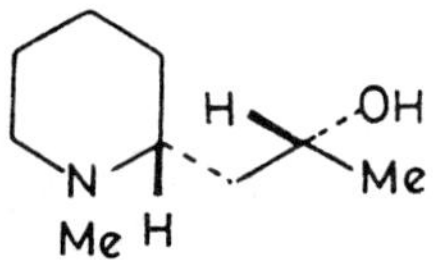

31

A mixture of $(S)-(-)$-sedamine (32, R = H) and $(S)-(-)$-
allosedamine (33, R = H), alkaloids of a number of *Sedum*
species, has been synthesised from $(S)-(+)$-piperid-2-one-
6-acetic acid (T. Wakabayashi *et al.*, Chem.Letters, 1977,
223). $(+)-$ 4-Hydroxysedamine (32, R = OH) and $(+)-4-$
hydroxyallosedamine (33, R = OH) occur naturally as minor
alkaloids of *S. acre* (F. Halin, P. Slosse and C. Hootele,
Tetrahedron, 1985, *41*, 2891). $(+)$-Sedamine has also been
synthesised from L-lysine as starting material (K. Irie
et al, Chem.Comm., 1985, 633).

32   33

Sesbanimide A (34), together with its two stereoisomers
sesbanimide B-1(35), sesbanimide B-2 (36) and sesbanimide
C (37) all occur in *Sesbania drummondii*, the structures
rest on spectral and X-ray diffraction analyses.   Each
compound is strongly cytotoxic (R.C. Powell *et al.*,
J.Amer.chem.Soc., 1983, *105*, 3739; Phytochem., 1984, *23*,
2789).

34   35   36

37

178

Sesbanine (43), also cytotoxic, is congeneric with the
sesbanimides (R.G. Powell *et al. ibid.*, 1979, *101*, 2784)
and has been the subject of several synthetic programmes
(J.C. Bottaro and G.A. Berchtold, J.org.Chem., 1980, *45*,
1176; M.J. Wanner, G.-J. Koomen and U.K. Pandit,
Heterocycles, 1981, *15*, 377; Tetrahedron, 1982, *38*, 2741;
M. Iwao and T. Kuraishi, Tetrahedron Letters, 1983, *24*,
2649).

   One approach (M. Wada, Y. Nishara and K.-Y. Akiba,
Tetrahedron Letters, 1985, *26*, 3267) involves the addition
of the *O*-silyl derivative (38) of methyl   pent-3-ene
carboxylate and the salt (39) of methyl nicotinate.
Oxidation of the product (40) with DDQ affords the diester
(41) which, after protonation, is treated successively
with mercury (II) acetate and sodium borohydride to yield
the diester (42).   Finally this compound when reacted with
ammonia in methanol affords (±)-sesbanine.   (Scheme 1)

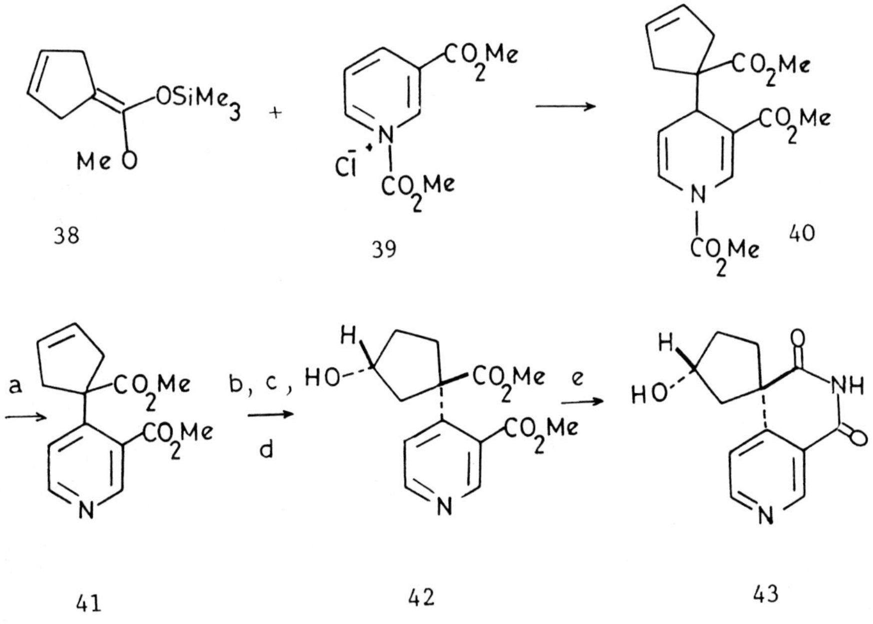

38   39   40

41   42   43

Scheme 1

      Reagents: a) DDQ, b) HBF$_4$, c) Hg(OAc)$_2$,

      d) NaBH$_4$, e) NH$_3$

4. *The alkaloids of* Prosopis *and* Cassia *species*

Prosopine (44a), prosopinine (44b), isoprosopinine A (44c),
isoprosopinine B (44d), prosophylline (44e), prosofrine
(44f), and prosofrinine (44g) are metabolites of *Prosopis
africana* (Q.K.-Huu *et al.*, Bull.Soc.chim. Belges, 1972, *81*,
425; 443). Two related alkaloids prosopinone and alkaloid
D occur in *Cassia carnaval* (D. Lythgoe *et al.*, Anales Asoc.
Quim.Argentina, 1972, *60*, 317), their structures are not
totally secure but are likely to be (44h) and 44ℓ)
respectively.

44

a; R^1 = OH,R^2 = H,R^3 = $(CH_2)_{10}CH(OH)Me$
b; R^1 = OH,R^2 = H,R^3 = $(CH_2)_9COCH_2Me$
c; R^1 = OH,R^2 = H,R^3 = $(CH_2)_6CO(CH_2)_4Me$
d; R^1 = OH,R^2 = H,R^3 = $(CH_2)_7CO(CH_2)_3Me$
e; R^1 = OH,R^2 = H,R^3 = $(CH_2)_9COCH_2Me$
f; R^1 = H, R^2 = H,R^3 = $(CH_2)_9CH(OH)CH_2Me$
g; R^1 = H, R^2 = H,R^3 = $(CH_2)_9COCH_2Me$
h; R^1 = OH,R^2 = H,R^3 = $(CH_2)_{10}COMe$
i; R^1 = H, R^2 = H,R^3 = $(CH_2)_3CH(OH)(CH_2)_{10}CH(OH)Me$

The above structural assignments owe much to chemical
degradations and to ^1H n.m.r. spectroscopy, but stereoselective synthesis of (-)-deoxoprosopinine (45, 6βH), and
(-)-deoxoprosophylline (45, 6α-H), derivatives of the
corresponding natural products, and also of prosafrinine
have now been reported (Y. Saitoh *et al.*, Tetrahedron
Letters, 1980, *21*, 75; Bull.chem.Soc.(Japan), 1981, *54*, 283;
M. Natsume and M. Ogawa, Heterocycles, 1980, *14*, 615). Other
syntheses include that of (±) isoprosopinine B (and also of
(±)-desoxoprospinine) (A.B. Holmes *et al.*, Chem.Comm., 1985,
37) Prosophylline has also been synthesised (*idem. ibid.*,
1981, *16*, 973).

45

Whereas the above alkaloids have a *trans* geometry between the hydroxyl group and the C-2 side chain, cassine (46) from *C. exdelsa* has a *cis*-orientation. An arrangement which is established by the synthesis summarised in Scheme 2 (E. Brown and A. Bonte, Tetrahedron Letters, 1975, 2881).

46

Scheme 2

Reagents: a) Br_2, b) $NaNH_2$, c) Hg^{2+}/H_2O
d) PBr_3, e) $Me_2C=CHCH_2CH_2COCH_2CO_2Et$,
f) $Ba(OH)_2/\Delta$, g) $O_3/Zn/HOAc$,
h) $EtNO_2/NaOEt$, i) $Pd/C/H_2$

N-Methylcassine occurs in several *Prosopis* species (I.B.
Giarinetto, J.L. Cabrera and H.R. Juliani, J.nat.Prod., 1980,
43, 632), while isocassine (47) spectalinine (48) and iso-6-
carnavaline (49) are obtained from *C. spectablis* (I.
Christofidis, A. Welter and J. Jadot, Tetrahedron, 1977, *33*,
977; 3005). Related compounds occur in *Prosopis julifora*
(V.U. Ahmad, A. Basha and W. Haque, Z. Naturforsch, 1978,
33b, 347), these are called juliforidine, juliflorine and
juliforicine. The first has been allocated structure (50)
without stereochemical qualification.

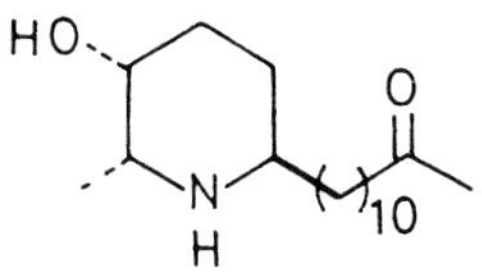

47

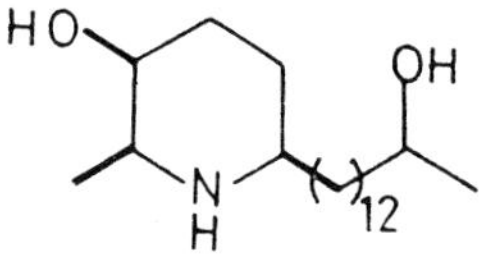

48

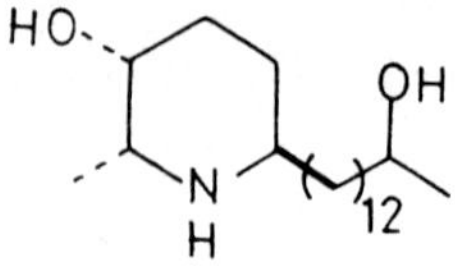

49

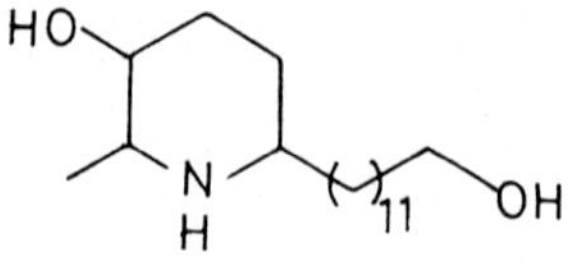

50

Melochinine (51,R=H) from the leaves of *Melochia pyramidata* has been converted into the same product (52) as that obtained from cassine by dehydrogenation, thus establishing that melochinine has the same pattern of substitution about the heterocyclic ring (E. Medina and G. Spiteller, Ber., 1981, 114, 814). A synthesis of this alkaloid has been described (G. Voss and H. Gerlach, Ann., 1982, 1466). Other alkaloids of *M. pyramidata* are the glucoside (51,R=glucosyl) of melochinine and melochininone (53) (idem., Ann., 1981, 538; 2096).

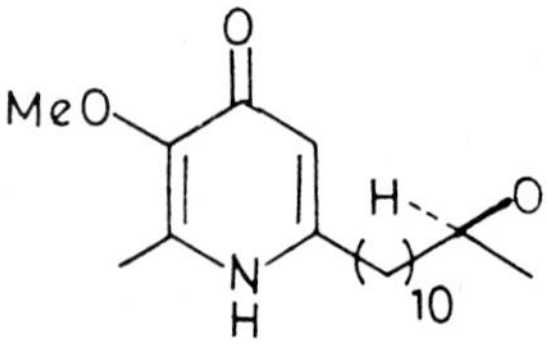

51

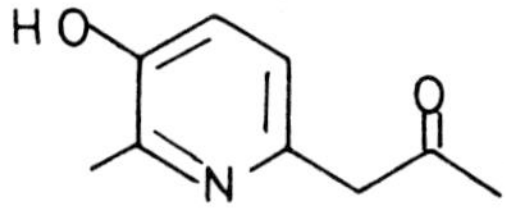

52

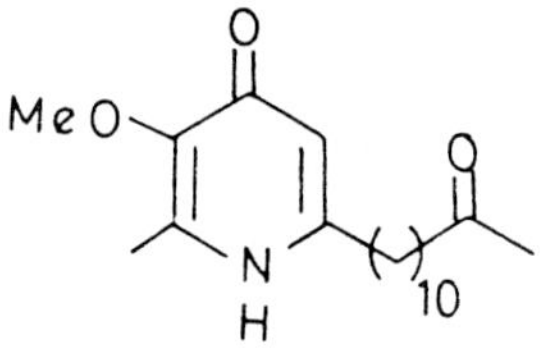

53

5. *Alkaloids of tobacco*

N-oxides of nicotine - both *cis*- and *trans*- isomers - are
widely distributed throughout the various parts of tobacco
plants (J.D. Phillipson and S.S. Hands, Phytochem., 1975,
14, 2683), and *N*-acylated derivatives of nornicotine have
been isolated from *Nicotiana tabacum.* These include the *N*-fo-
·rmyl and *N*-acetyl compounds,and others bearing n-hexanoyl
and n-octanoyl side chains (A.J.N. Bolt, Phytochem., 1972,
11, 2341; A.H. Warfield, W.D. Galloway and A.G. Kallianos,
Phytochem., 1972, *11*, 3371).

New alkaloids from *N.tabacum* include compounds (54) and
(55) (E. Demole and C. Demole, Helv., 1975, *58*, 523).
Additionally 5-methyl-2,3'-bipyridyl (56) is also present
in some varieties of *N.tabacum* (A.H. Warfield, W.D. Galloway
and A.G. Kallianos, Phytochem., 1972, *11*, 3371).

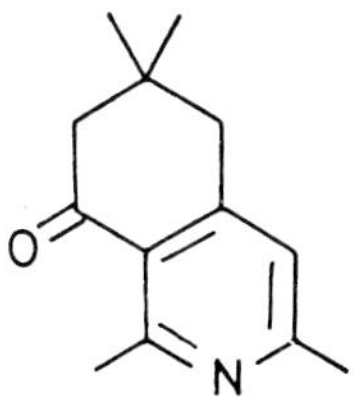

54

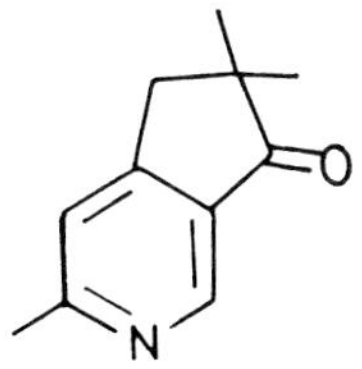

55

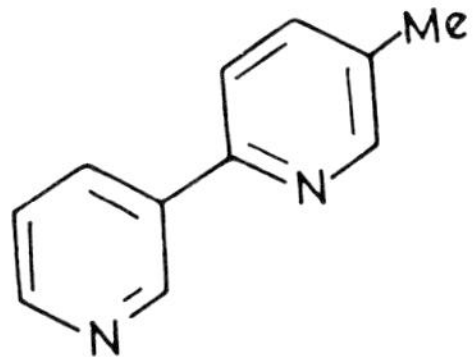

56

A versatile short synthesis of the tobacco alkaloids has been developed (G.F. Alberici *et al.*, Tetrahedron Letters, 1983, *24*, 1937). (Scheme 3)

3-Lithiopyridine with cyclobutanone gives the alcohol (57). This when treated with azoic acid and sulphuric acid undergoes the .Schmidt reaction to afford the azide (58) which re-arranges to myosmine (59). Reduction with sodium cyanoboro-hydride yields (±)-nornicotine and thence by N-methylation (±)-nicotine. Replacement of cyclobutanone by cyclopentanone allows the formation in turn of (±)-noranabastine and (±)-anabastine.

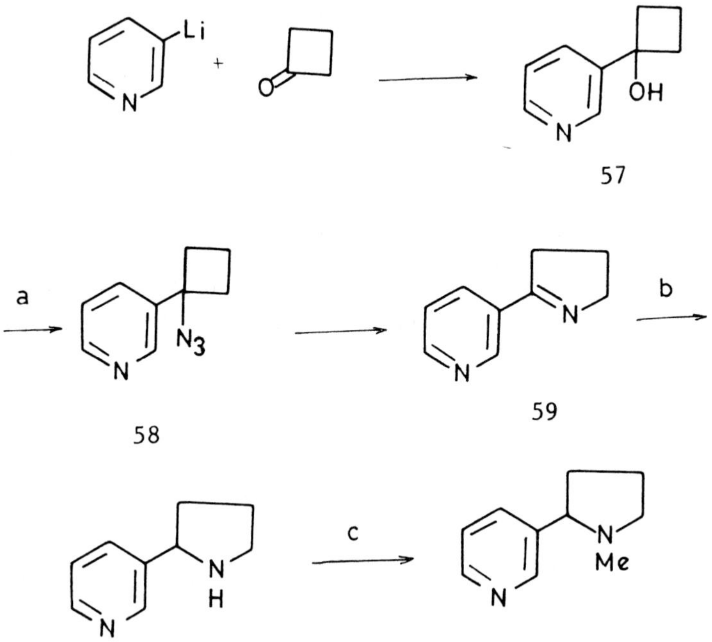

Scheme 3

Reagents: a) HN_3/H_2SO_4, b) $NaBH_3CN$, c) $HCHO/HCO_2H$

(+)-Nicotine of high optical purity is obtained by selective
utilisation of the (-)-enantiomer by the bacterium
Pseudomonas putida when fed the racemate.

A new synthesis of nornictyrine (61) requires as the final
steps reactions of the *N*-oxide (60) with phosphorus (III)
bromide and then with sodium hydroxide (S. Saeki, T. Takaaki
and M. Hamana, Heterocycles, 1984, *22*, 545).

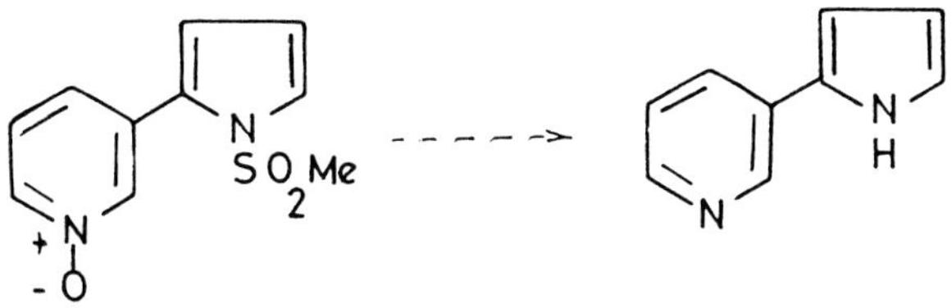

60   61

6. Alkaloids from papaya

The papaya plant *Carica papaya* is a source of the known
alkaloid (+)-carpaine (62, R=H) and also of the dehydro-
carpaines-I (63) and -II (64) (C.-S. Tang, Phytochem.,
1979, *18*, 651), both of which form carpaine on
hydrogenation.

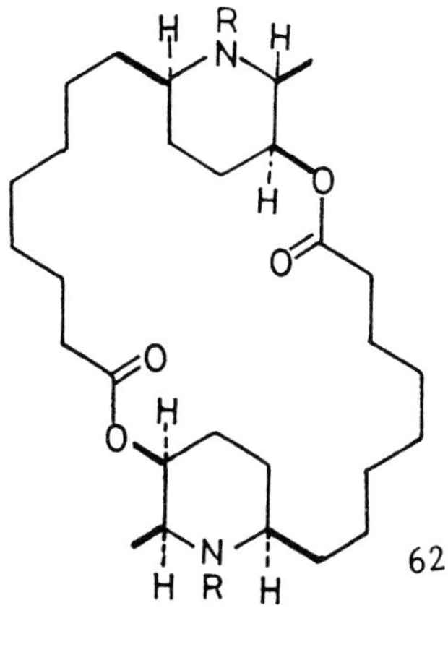

62 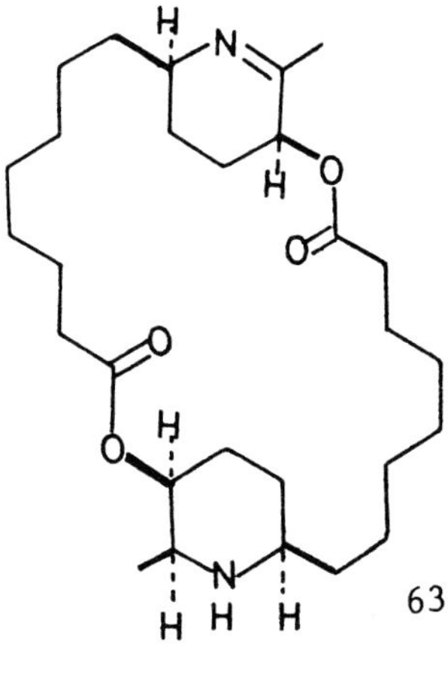   63

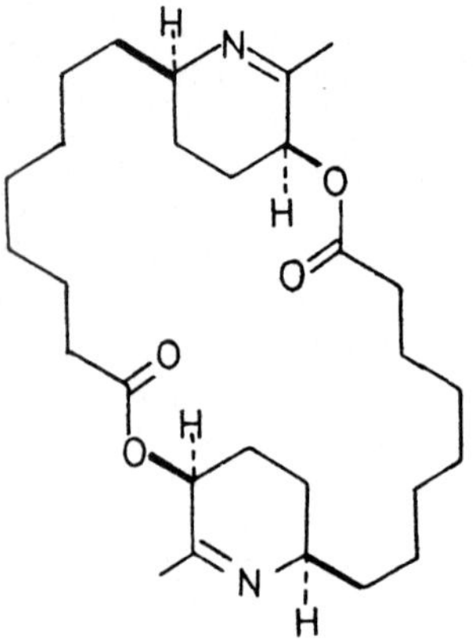

64

Hydrolysis of carpaine affords carpamic acid (65;n=7,R=H) the racemate of which has been synthesised (E. Brown and A. Bourgouin, Tetrahedron, 1975, *31*, 1047) N-Benzyloxycarbonyl carpamic acid (65,n=7;R=CO_2CH_2Ph), 2,2'-diphenyl disulphide and triphenylphosphine when heated together yield *N,N'*-bis-benzyloxycarbonylcarpaine (62,R=CO_2CH_2Ph) (E.J. Corey, K.C. Nicolaou and L.S. Melvin, J.Amer.chem.Soc., 1975, *97*, 654).

Azimine (66) from the leaves of *Azima tetracantha* on hydrolysis provides (+)-azimic acid (65,R=H;n=5), a total stereospecific synthesis of this compound, and also of (+)-carpamic acid, from (+)-glucose has been described (S. Hanessian and R. Frenette, Tetrahedron Letters, 1979, 3391).

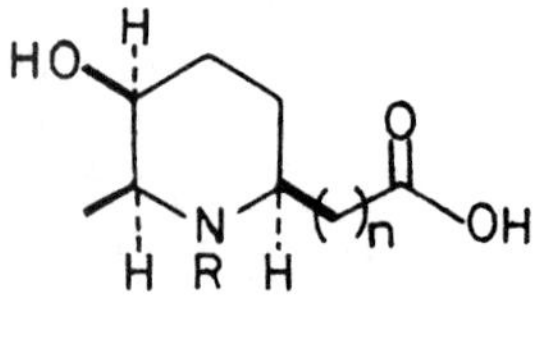

65

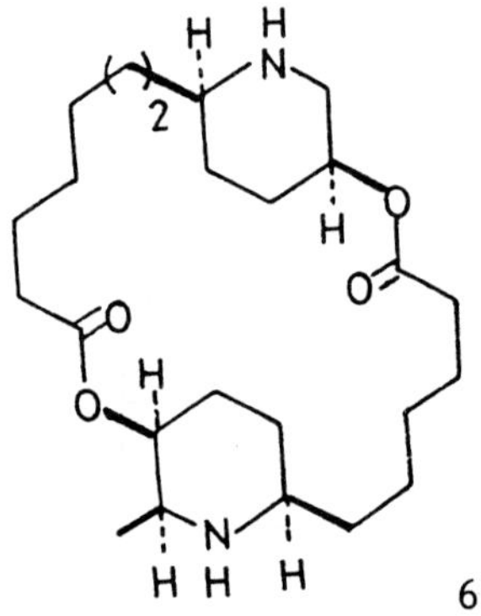

66

7. *Nuphar alkaloids*

(±)-Nupharamine (68, β-*3Me*) and (±)-3-epi-nupharamine
(68, α-3Me) are alkaloids of the water lilies *Rhiazama
nupharis* and *Nuphar luteus*. Racemic forms of both of
these bases have been synthesised by reduction of the
furanylpyridine (67), followed by hydration of the double
bond of the side chain with a mixture of formic and
perchloric acids. The two bases were separated by
chromatography (J. Szychowski, J.T. Wrobel and A.
Leniewski, Can.J.Chem., 1977, *55*, 3105). The absolute
configuration of 3-epinupharamine was confirmed by an X-
ray crystallographic examination of its hydrobromide
(M. Sabat *et al.*, *ibid.*, p.3111).

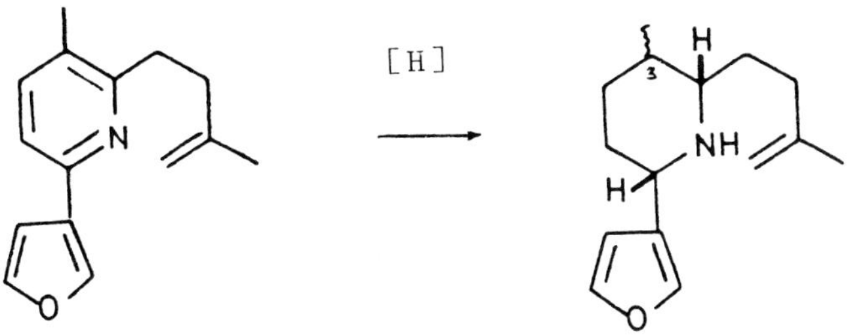

67

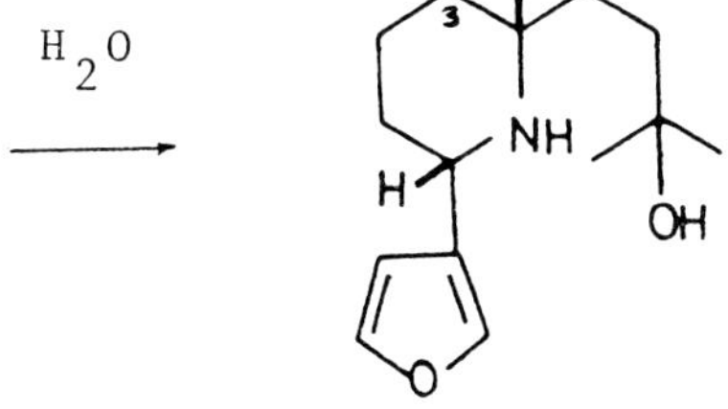

68

(±)-Anhydronupharamine (70;6-α-furanyl) and its 6-epimer
(70;6-β-furanyl) have also been prepared by reduction of
the tetrahydropyridine (69) with sodium borohydride (R.T.
LaLonde, N. Muhammad and C.F. Wong, J.org.Chem., 1977, 42,
2113).

69   70

The structure of nupharamine is obviously very close to that
of the quinolizidine alkaloids which also occur in *Nuphar*
species (see Chap.38) and, for example, it is possible to
degrade the quinazoline, alkaloid nupharidine (71) to (-)-
nupharamine *via* the enamine (72) (R.T. LaLonde *et al.*,
J.Amer.chem.Soc., 1971, 93, 2501).

71   72

8. Miscellaneous piperidines and pyridines from plant sources

Cryptophorine and crytophorinine, $C_{17}H_{29}NO_2$, are alkaloids of *Bathiorhamnus cryptophorus*. Structure (73) is allocated to cryptophorine on the basis of its ^1H n.m.r. spectrum and the fact that it absorbs four moles of hydrogen on catalytic reduction. Further degradative work indicates that cryptophorine is a 2,6-dialkylated 1-methylpiperidin -3-ol and the conclusion that a second methyl substituent is at position 2, rather than at position 6, is in line with other alkaloid structures from related species of plants. A possible structure for cryptophorinine is (74), although the compound does not readily dehydrate to cryptophorine which is puzzling (J. Bruneton *et al.*, Plant.Med.Phytother., 1975, *9*, 21; Tetrahedron Letters, 1975, 739).

73

74

A new alkaloid from *Conium* species is *N*-methylpseudoconhydrine (75) (M.F. Roberts and R.T. Brown, Phytochem., 1981, *20*, 447). It shows a strong similarity to some of the *Cassia*-type bases. Several stereoselective syntheses of pseudo-conhydrine have been reported (T. Shono *et al.*, Chem.Letters, 1984, 1101; 1129; G.W.J. Fleet, M.J. Gough and P.W. Smith, Tetrahedron Letters, 1984, *25*, 1853). (±)-α-Conhydrine (76) has also been prepared (S. Pilard and M. Vaultier, *ibid.*, p.1556).

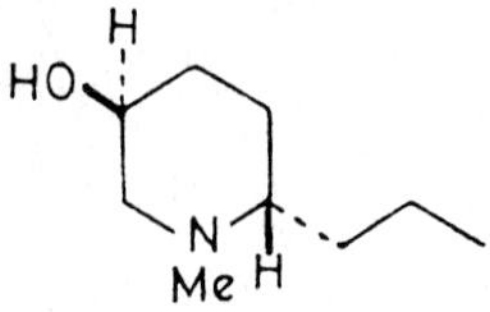

75

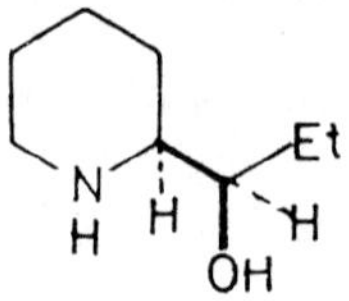

76

The spiro piperidine (77) is a constituent of the seeds of
the leguminous plants *Lonchocarpus sericeus* and *L.*
costaricensis (L.E. Fellows *et al.*, Chem.Comm., 1979, 977),
and the simple phenacylpiperidine (78) is an antifungal
agent from the plant *Boehermia cyclindrica* (W. Döpke *et al.*,
Z.Chem., 1981, *21*, 358). Other piperidines include
dumetorine (79), which is obtained from the yam *Pioscorea*
dumetorum (D.G. Corley, M.S. Tempesta and M.M. Iwu,
Tetrahedron Letters, 1985, *26*, 1615), and the simple
glucoside (80) an extractive of *Xanthocercis zambesiaea*
(S.V. Evans *et al.*, *ibid.*, p 1465).

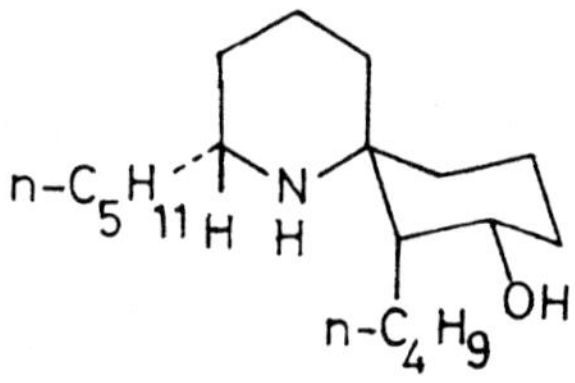

77

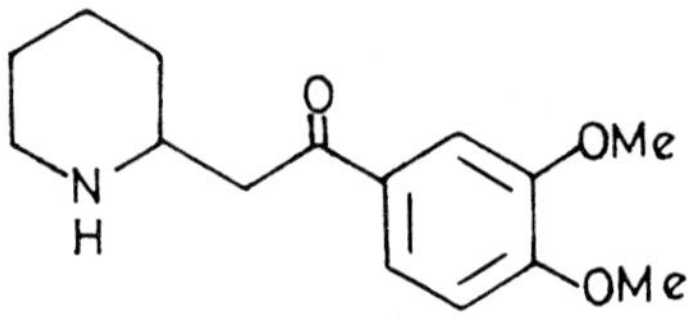

78

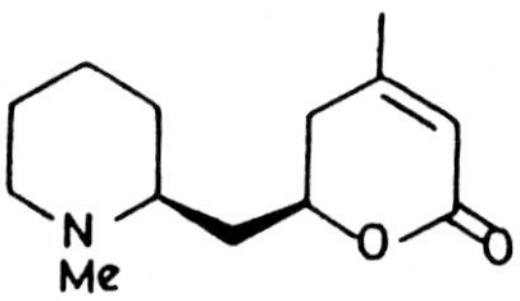

79

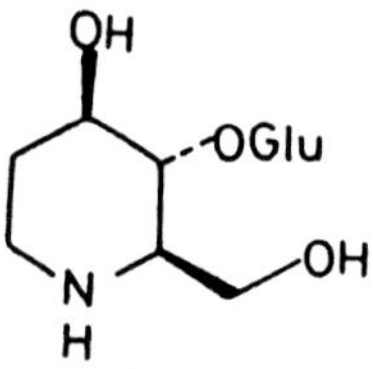

80

Schumanniophytine (81) is a chromonopyridine from the root
bark of *Schumanniophyton problematicum* which is also a source
of the piperidones (82, R=H) and (82, R=Me) (E. Schittler and
U. Spitaler, Tetrahedron Letters, 1978, 2911).

Related to these structures is rohitukine (83) the main
alkaloid of the leaves and stems of *Amoora rohituka* (syn.
Aphanamixis polystachya) Meliaceae (A.D. Harmon, V. Weiss
and J.V. Silverton, *ibid.*, 1979, 721).

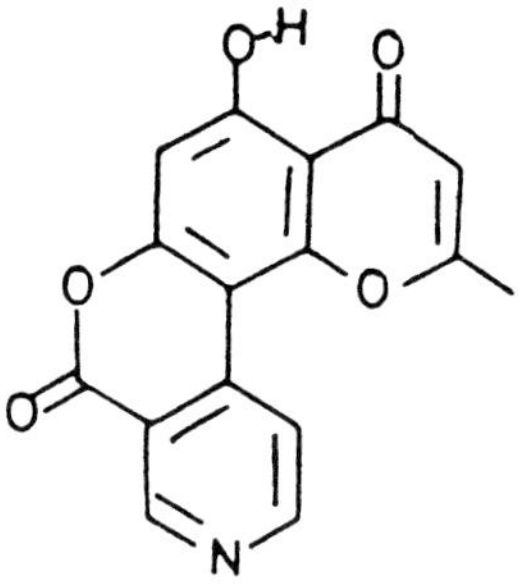

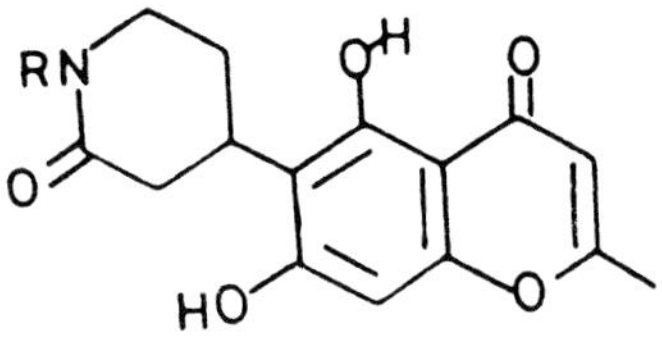

81

82

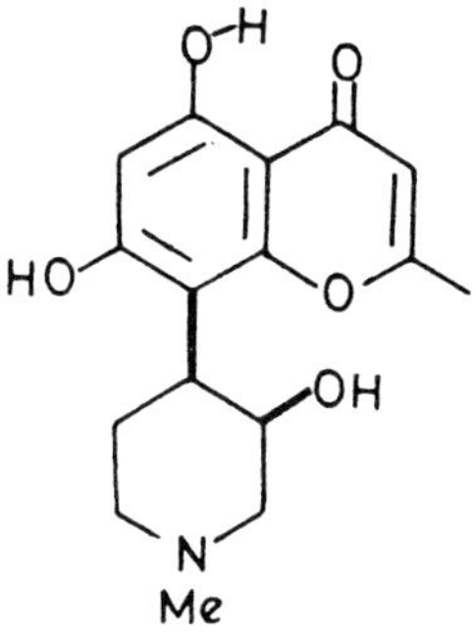

83

Acalyphin (84) is obtained from the weed *Acalypha indica* (A.
Narstedt, J.-D. Kant and V. Wray, Phytochem., 1982, *21*, 101);
whereas another 2-pyridinone, campedine (85), has been
isolated from *Campanula medium* (W. Döpke and G. Fritsch,
Pharmazie, 1970, *25*, 128). It should be noted however that
N-ethyl groups are most uncommon in nature, and the possibility
that this compoud is an artefact is very likely.

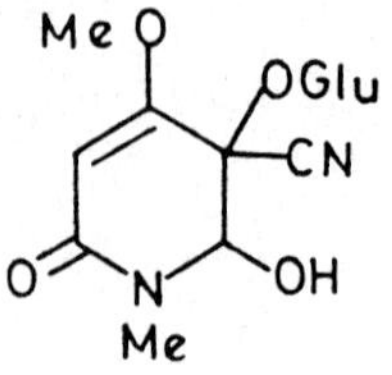

84

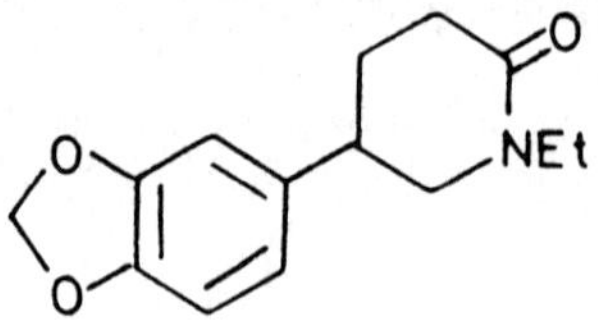

85

A more complex 2-pyridinone is xylostosidine (86) an unusual
sulphur bearing monoterpene glycoside from *Lonicera xylosteum*
(R.K. Chandhuri, O. Sticher and T. Winkler, Helv., 1980, *63*,
1045; Tetrahedron, 1981, *22*, 559). (+)-Kuraramine (87) is a
bispiperidine structure isolated from the flowers of *Sophora
flavescens* (I. Murakoshi *et al.*, Phytochem., 1981, *20*, 1407).
Dinglageine (88) and the corresponding secondary amide,
strychnovoline (89) occur as metabolites of *Strychnos dinklagei*
(S. Michel *et al.*, J.nat.Prod., 1985, *48*, 86).

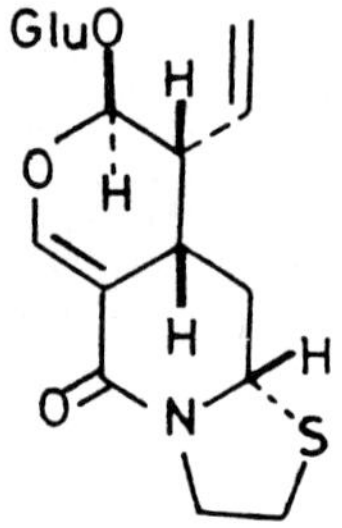

86

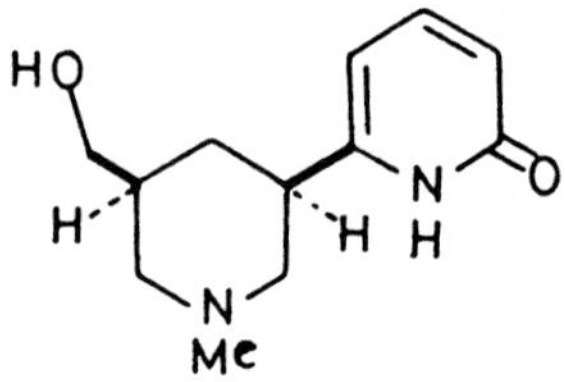

87

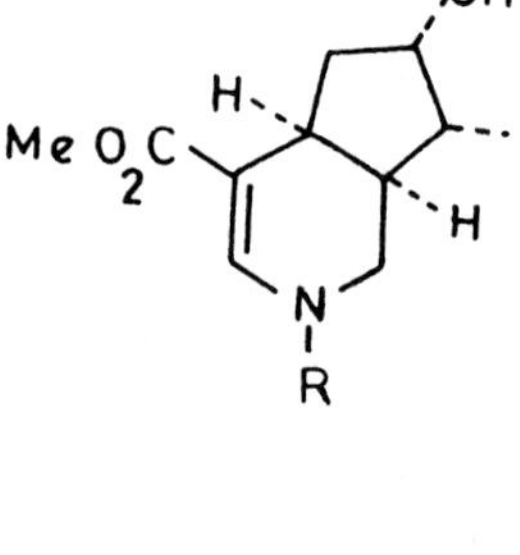

88 , R=Me

89 , R=H

(-)-Swainsonine (90) is a potent inhibitor of α-D-mannosidase.
It occurs in some plants e.g. *Swainsonia canescens* (S.M.
Colegate, P.R. Dorling and C.R. Huxtable, Austral.J.Chem.,
1979, *32*, 2257) and *Astralagus lentiginosus* (R.J. Molyneux and
L.F. James, Science, 1982, *216*, 190) and in the fungus
Rhizoctonia leguminicola (L.D. Hohenschulz *et al.*, Phytochem.,
1981, *20*, 811).
The alkaloid has been synthesised from two starting materials
(a) from methyl 3-acetamido-2,4,6-tri-*O*-acetyl-3-deoxy-α-*D*-
mannopyranoside (T. Suami, K. Tadano and Y. Iimura, Carbohydr.
Res., 1985, *136*, 67) and (b) from 3-amino-3-deoxy-*D*-mannose
(M.H. Ali, L. Hough and R.C. Richardson, *ibid.*, p.225).

90

(-)-Pinidine (91) is an alkaloid of established structure
from *Pinus jeffreyi*. It has been synthesised from 2,6-
lutidine in three simple steps (Scheme 4), followed by a
separation of the isomers and a resolution of the
appropriate racemate using (-)-6,6'-dinitrodiphenic acid
(E. Leete and R.A. Carver, J.org.Chem., 1975, *40*, 2151).
An alternative approach to this alkaloid has been reported
by S. Arseniyadis and J. Sartoretti (Tetrahedron Letters,
1985, *26*, 729).

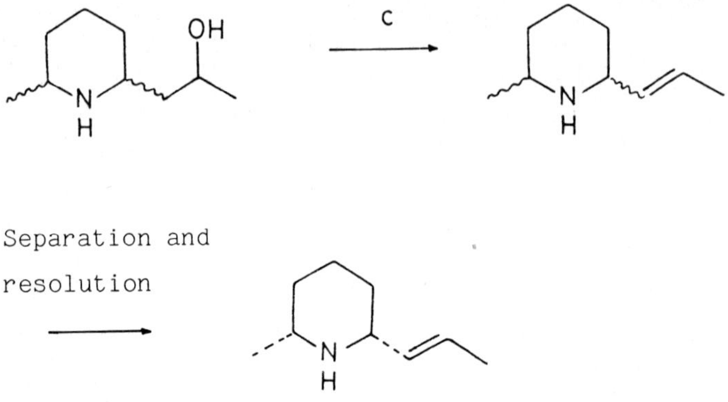

Separation and

resolution

91

Scheme 4

Reagents: a) MeCHO/BuLi, b) H_2/Pt/HCl,

(c) $KHSO_4$

(+)-Actinidine (92) previously found in *Actinidia polygama*, (main work, p. 158; see also section 10 below), also occurs in the roots and rhizomes of *Valeriana officinalls* (M.M. Janot *et al.*, Ann.Pharm.Fr., 1979, *37*, 413), and boschiakine (93) is found in *Plantago sempervirens* (H. Ripperger, Pharmazie, 1979, *34*, 577). Venoterpine (alkaloid RW-47) (94) occurs in *Stiga hermoteca* (M. Baona *et al.*, Phytochem., 1980, *19*, 718). This last compound has previously been isolated from other sources, its stereo-chemistry has been deduced by chemical correlation with alkaloids of known structure (T. Ravao *et al.*, Tetrahedron Letters, 1985, *26*, 837).

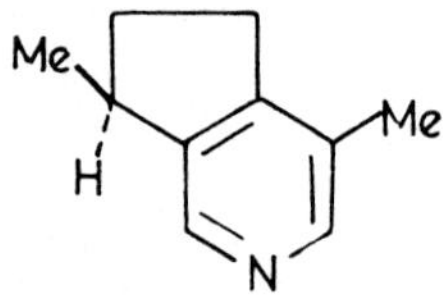

92

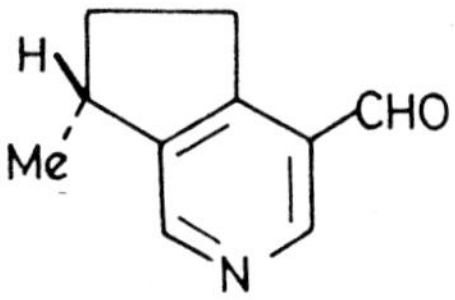

93

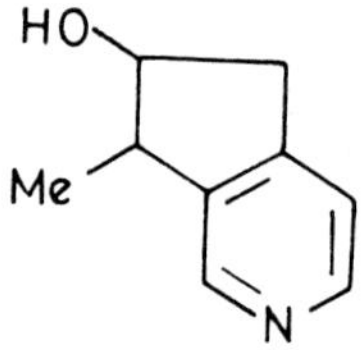

94

Reduced analogues of venoterpine include tecomanine (95), an
alkaloid from *Tecoma stans*. The structure of this base is
known from X-ray studies and a stereoselective synthesis of
its racemic modification has been published (T. Imanishi,
N. Yagi and M. Hanaoka, Tetrahedron Letters, 1981, *22*, 667).

95

1,2,3,6-Tetrahydropyridine-2-carboxylic acid (96) occurs in
the sea-weed *Corallina officinalis* (J.C. Madgwick *et al.*,
Arch.Biochem.Biophys., 1970, *141*, 766), and pegaline from
Peganum harmala has been shown to be identical with L-(-)-4-
hydroxypipecolic acid (97) (V.U. Ahmad and M.A. Khan,
Phytochem., 1971, *10*, 3339). The related base (98) is
present in *Pongamia glabra* (P.S.S. Kumar, V.V.S. Murti and
T.R. Seshadri, Tetrahedron Letters, 1971, 4451).

96

97   $R^1=R^2=H$

98   $R^1=Me; R^2=OH$

The medicinal Indian plant *Abrus precatorius* elaborates
the ester precaterine (99), and a number of simple
pyridine Zwitterionic alkaloids of established structure,
such as trigonelline (100) (S. Ghosal and S.K. Dutta,
Phytochem., 1971, *10*, 195).

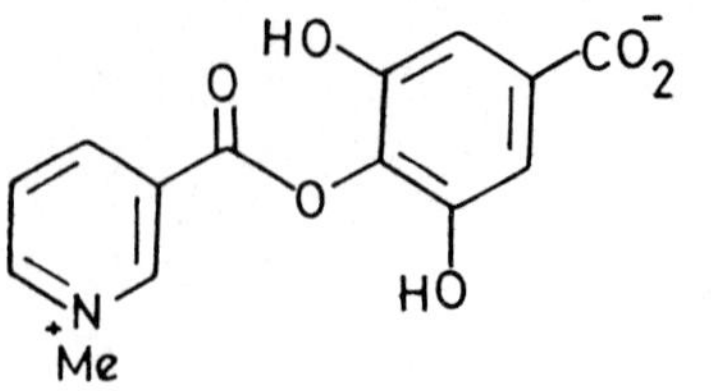

99

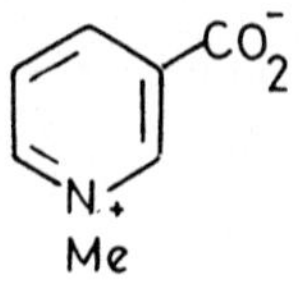

100

Plants of *Amaryi plumerieri* from Jamaica,metabolise the
nicotinamides (101) and (102) (B.A. Burke and H. Parkins,
Tetrahedron Letters, 1978, 2723), whereas onychine (103)
from *Onychopetalum amazonicum* is based on a 2- benzoyl -
pyridine sub-structure (J. Koyama *et al*., Heterocycles,
1979, *12*, 1017).

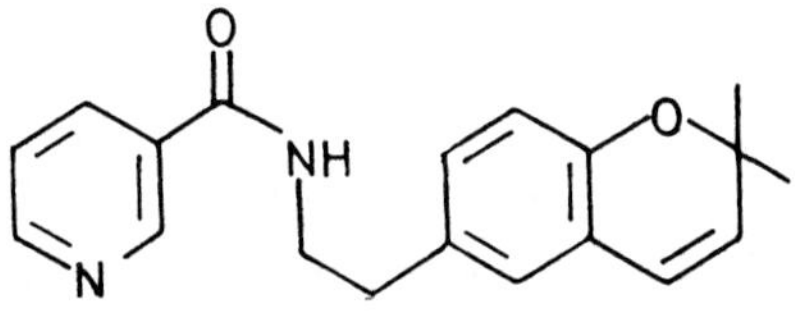

101

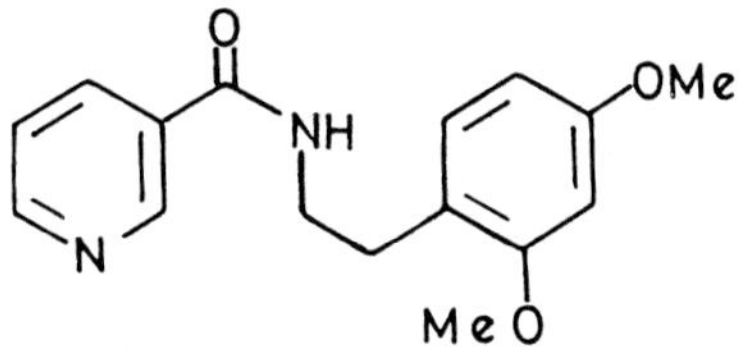

102

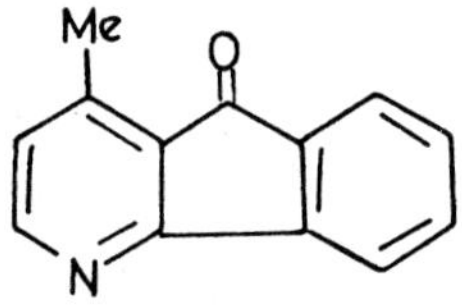

103

Nitramine and isonitramine alkaloids of *Nitraria* spp.,
previously described as decahydroquinolines, are now
considered to be the epimeric spiropiperidines (104 and
(105) respectively (A.A. Ibragimov *et al*., Khim.prirod.
Soedin., 1981, 623; Z. Osmanov *et al*., *ibid*., 1982, 126).

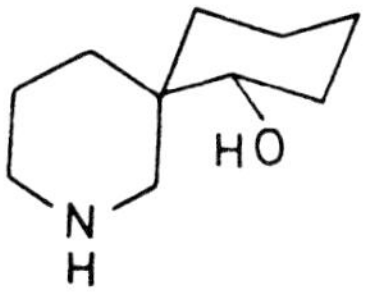

104

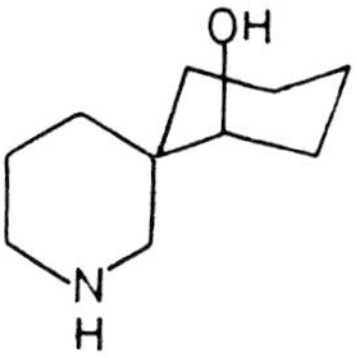

105

9. Fungal metabolites.

Originally there was some debate of the structure of
flavipucine (*syn.* glutamicine) an alkaloid from *Aspergillus
flavipes,* however the constitution (106) has been deduced by
X-ray crystallography (P.S. White, J.A. Findlay and W.H.J.
Tam, Canad.J.Chem., 1978, *56,* 1904). A probable co-metabolite
of the fungus is isoflavipucine which has structure (107)
(Findlay *et al.,* Canad.J.Chem., 1977, *55,* 600). Flavipucine
has been synthesised (N.N. Girotra, Z.S. Zelawski and N.L.
Wendler, Chem.Comm., 1976, 566; Girotra and Wendler,
Heterocycles, 1978, *9,* 417; but also see Findlay, *ibid.,*
1979, *12,* 389). Flavipucine is readily converted into its
isomer by heat or by base treatment.

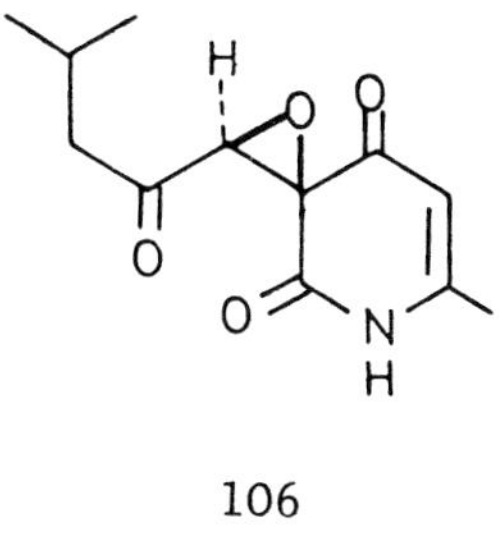

106

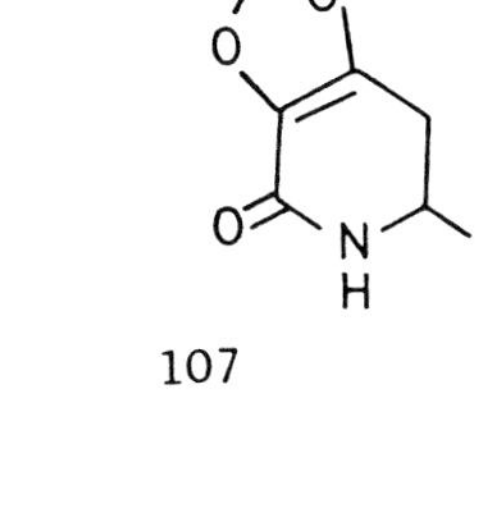

107

Streptomyces species yield the bicyclic base abikoviromycin
(108) (M. Onda *et al.*, Chem.pharm.Bull., 1974, *22*, 2916; 1975,
23, 2462).
S. tendae is a source of a number of relative complex
antibiotics containing a pyridine nucleus.   Some examples
are nikkomycin Qx (109, R=R^1) nikkomycin Qz (109, R=R^2),
nikkomycin Ox (110, R=R^1) and nikkomycin Oz (110, R=R^2).
Others have similar structures to the O series but lack the
hydroxyl group at C-5 in the pyridine ring (C.Bormann *et al.*,
J.Antiobiotics, 1985, *38*, 9).

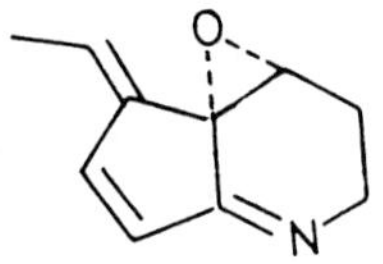

108

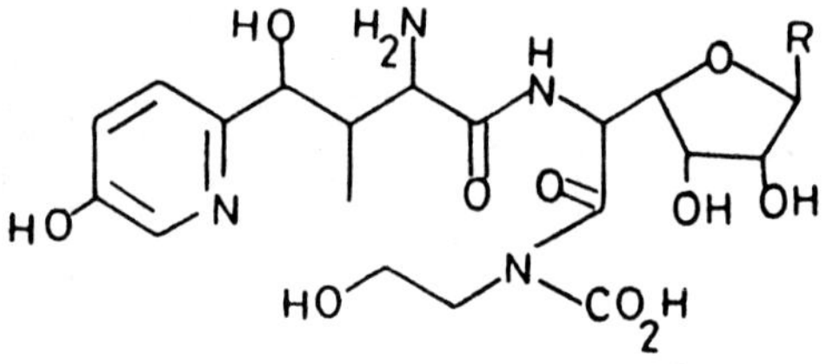

109

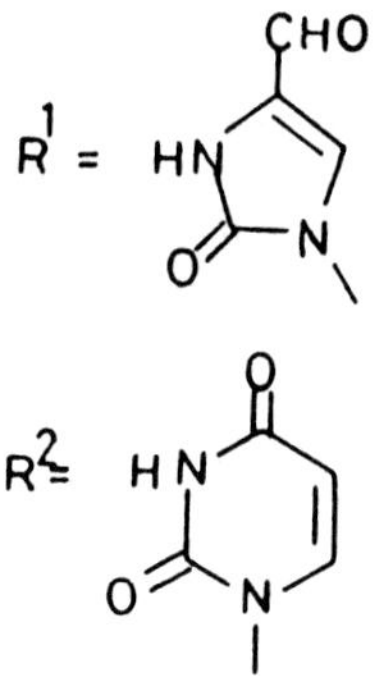

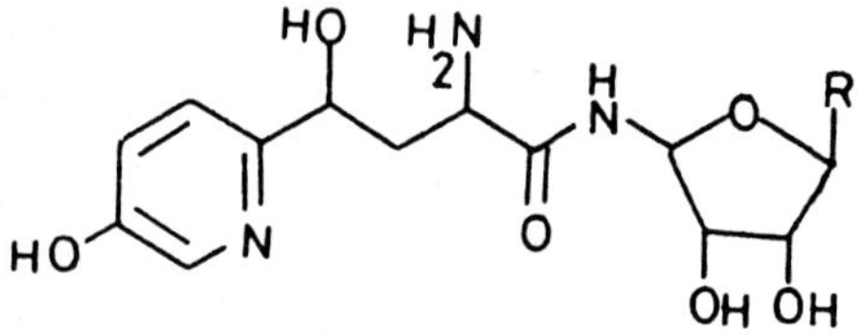

$R^1 =$

$R^2 =$

110

The mushrooms *Cortinarius orellanus* and *C. speciossimus* are sources of three bipyridyl bases which are named orelline orellanine and orellinine (W.Z. Antkcwiak and W.P. Gessner, Tetrahedron Letters, 1979, 1931; Experientia, 1985, *41*, 769). Orellanine is the bis-*N*-oxide of orelline (111), which exists in tautomeric equilibrium with the pyridone form (112). Orellinine is the mono-*N*-oxide of orelline.
Another highly oxygenated natural product is rubrifacine (113) a red pigment of the bacterium *Erwinia rubrifaciens* (G. Feistner and H. Budzikiewicz, Canad.J. 1985, *63*, 495.

111   112   113

10. *Piperidine and pyridine bases from insects*

(a) *Ant venoms*

The veom of fire ants (genus *Solenopsis*) contains 2,6-dialkylpiperidines. Typically these bear a methyl group at C-2 and a long alkyl or alkenyl chain at C-6. Some examples from *S. saevissima* are structures (114) and (115) {where n=10, 12 and 14} and (116) and (117) {where n=3 and 5} (J.G. MacConnell, M.S. Blum and H.M. Fales, Tetrahedron, 1971, *27*, 1129). In addition some of the bases occur naturally as the *N*-nitroso derivatives.

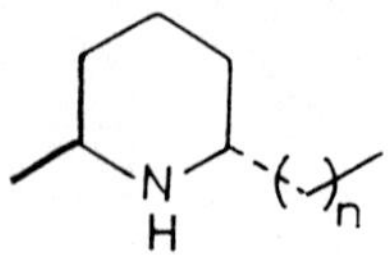

114

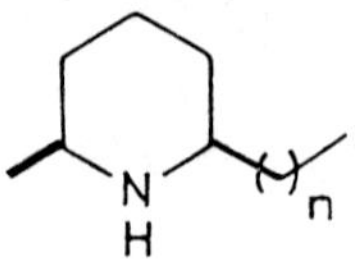

115

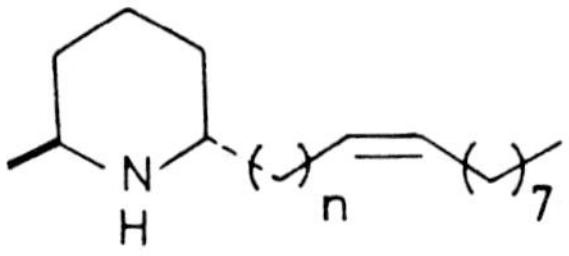

116

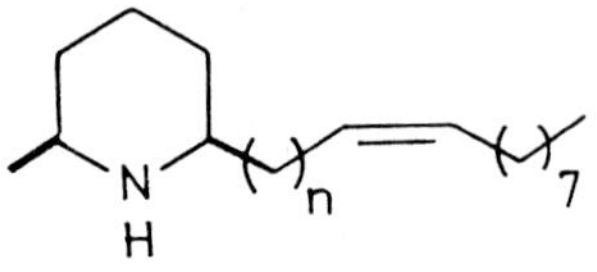

117

An interesting investigation was carried out to ascertain
if there are differences in the alkaloidal composition of
the venoms from worker and soldier ants and also between
those of red and black races, and it was found that in
some cases there is indeed a variation. Thus in red forms
of *S. saevissima trans*-isomers predominate, whereas in the
venom of black ants mixed *cis* and *trans*-isomers are present,
although structures (114,n=14) (115,n=14), (114,n=5) and
(115,n=5) are only minor components (J.M. Brand *et al.*,
Toxicon, 1972, *10*, 259; Insect Biochem., 1973, *3*, 45).

A number of groups have reported syntheses of various ant
toxins; see, for example, Y. Moriyama *et al.*, Tetrahedron
Letters, 1977, 825; R.K. Hill and T. Yuri, Tetrahedron, 1977,
33, 1569). A total synthesis of solenopsin-A is
illustrative (K. Fuji, K. Ichikawa and E. Fujita, Chem.pharm.
Bull., 1979, *27*, 3183) (Scheme 5).

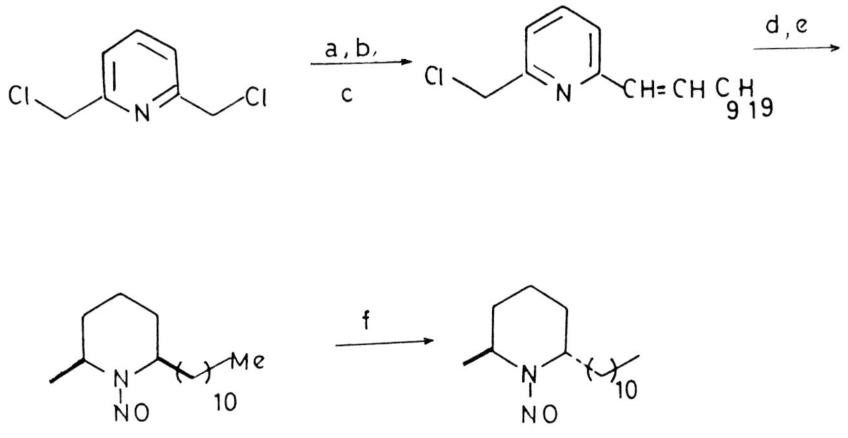

selenopsin-A

Scheme 5

Reagents: a) Ph$_3$P, b) NaH, c) decanal,
d) Ni/Pt/H$_2$, e) isoamylnitrite,
f) KO^{t}Bu

Actinidine (92), a plant alkaloid (see section 8), is also a minor component of the defence secretion of the Australian cock-tail ant *Iridomyrmex nitidceps* (G.W.K. Cavill *et al.*, Tetrahedron, 1982, *38*, 1931), a fact which indicates that ants may obtain toxins (or at least their precursors) from dietary sources. Actinidine has been synthesised (M. Nitta, A. Sekiguchi and H. Koba, Chem.Letters, 1981, 933). Anabaseine (118), a dihydro derivative of anabasine a well known tobacco alkaloid, is present in the poison glands of *Aphaenogaster* ants for which it also an attractant (J.W. Wheeler *et al.*, Science, 1981, *211*, 1051). Ants from Puerto Rico produce the simple tetrahydropyridine (119) (T.H. Jones, M.S. Blum and H.M. Fales, Tetrahedron, 1982, *38*, 1949).

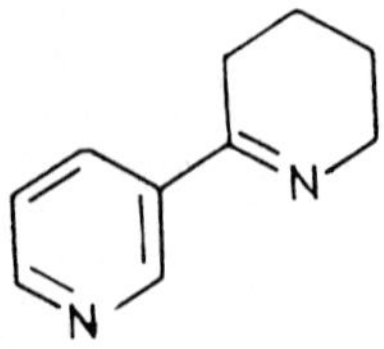

118

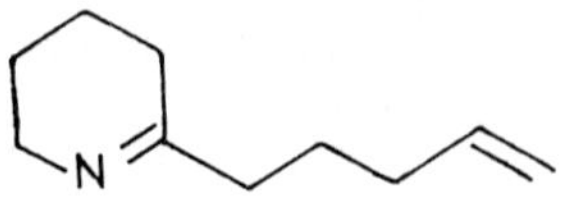

119

(b)   Bases from bugs and beetles

The staphylinid beetle *Stenus comma* secretes the alkaloid
stenusine (120) which has water spreading properties and
thus allows the insect to move easily over the surface of
the water.   The absolute stereochemistry of the base has
not been established, although the gross structure (120)
has been synthesised (H. Schildknecht *et al.*, Angew Chem.
Intern.Ed., 1975, *14*, 427).   Australian mealy bugs
Cryptolaemus montrouzieri excrete the dialkylpiperidine
(121), and probably also the *trans*-isomer as well (W.V.
Brown and B.P. Moore, Austral.J.Chem., 1982, *35*, 1255).   The
resemblance between this structure and those metabolised by
fire ants is obvious (see (a) above).

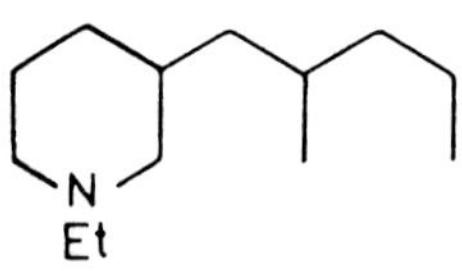

120

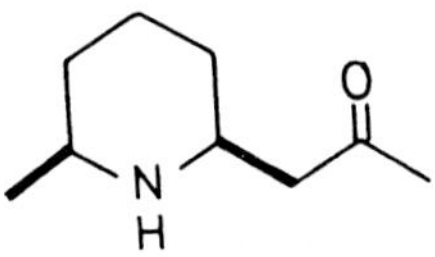

121

11. *The pumilotoxins*

The skin of the Panamanian frog *Dendrobates pumilo* is used by
natives as an arrow poison. Two of the alkaloids
responsible for this property are pumiliotoxin-A (122) and
pumiliotoxin-B (123) (J.W. Daly and C. Myers, Science, 1967,
156, 970; B. Witkop and E. Gossinger, in "The Alkaloids",
ed. A. Brossi, Academic Press, New York, 1983, Vol.21, Ch.5).
A less complex compound pumiliotoxin-251D (124) is metabolised
by *D.tricolor* (Daly *et al.*, J.Amer.chem.Soc., 1980, *102*, 830)
and the establishment of its structure by X-ray crystallography,
greatly aided the elucidation of the constitutions of the A
and B toxins (T. Tokuyama *et al.*, Tetrahedron Letters, 1982,
23, 2121; L.E. Overman and R.J. McCready, *ibid.*, p.2355;
M. Vemura *et al.*, *ibid.*, p.4369), although the stereochemistry of the indolizidine unit remained uncertain for some
time.

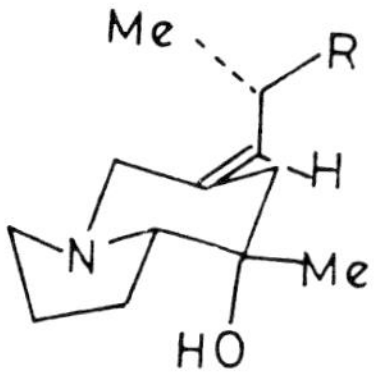

122   R =

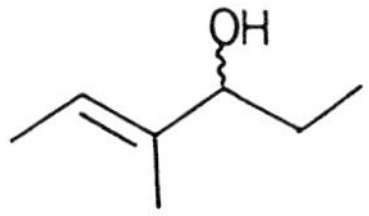

123   R =

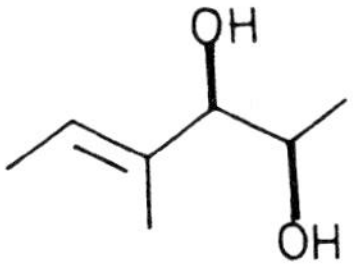

124   R = n-C_3H_7

The question of the configuration of the heterocyclic system
was solved, however, by an enantioselective total synthesis
(Overman, K.L. Bell and F. Ito, J.Amer.chem.Soc., 1984, *106*,
4192; Overman and N.H. Lin, J.org.Chem., 1985, *50*, 3669).
In this work the (R)-silyalkyne (125) was treated with
lithium di-isopropylamide and methyl lithium and then the
epoxide (126) was added. This gave the lactone (127) which
with potassium hydroxide in ethanol produced the protected
amino alcohol (128). Reaction of this compound with formalin
afforded the cyclopentaoxazolidine (129) and this when heated
with one molar equivalent of camphorsulphonic acid and
chromatography yielded the indolizidine (130, R=Bn).
Deprotection and oxidation under Swern conditions gave the
aldehyde (131). Finally a Wittig reaction between this
aldehyde and the ylide (132) produced the enone (133) which
was reduced with lithium aluminium hydride to yield (+)-
pumiliotoxin-B, together with a small amount (-6%) of its
erythro-isomer (Scheme 6).

125

126

a

127   Bn=CH$_2$Ph

b

128

c

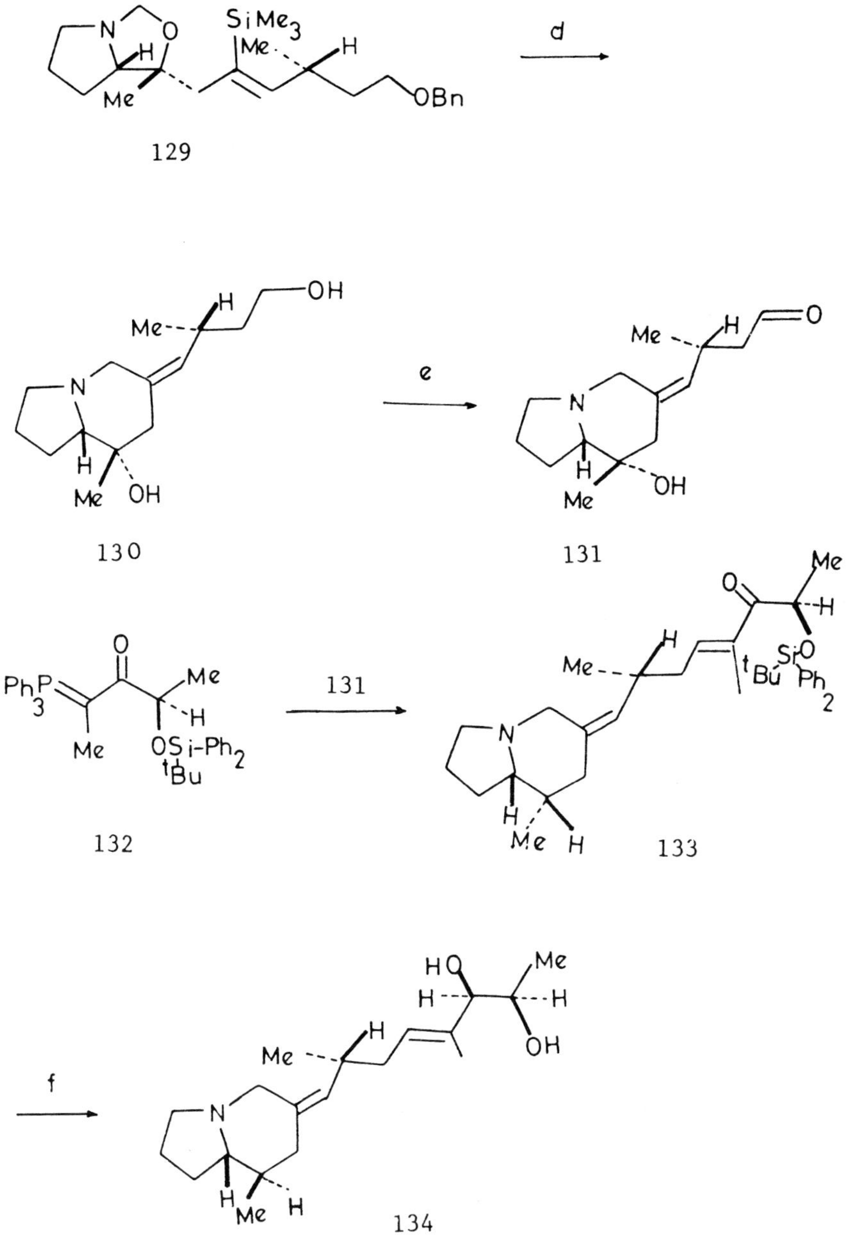

129

130

e

131

132

131

133

f

134

Scheme 6

Reagents: a) MeLi, b) KOH/EtOH, c) HCHO,
d) H$^+$, e) (COCl)$_2$/DMSO, f) LAH

The same methodology was adapted to synthesise pumiliotoxin-251D and subsequently modified further to obtain the structurally related allopumiliotoxin A alkaloids 267A (134) and 339B (135) (Overman and S.W. Goldstein, J.Amer.chem.Soc., 1984, *106*, 5360).

These last structures and others of the same type are also present in the defence secretions of neotropical frogs of the Dendrobatidae family.

Histrionicotoxin (137) and dihydroisohistrionicotoxin (138) are the main toxins produced by *D. histrionicus* from Columbia (Daly *et al.*, Proc.Nat.Acad.Sci.USA., 1971, *68*, 1870). The structure of the latter compound has been confirmed by X-ray crystallographic studies (I.L. Karle, J.Amer.chem.Soc., 1973, *95, 4036*).

The same frogs metabolise several other bases including allodihydrohistrionicotoxin (139), but at lower concentrations, these compounds are closely related to histrionicotoxin (T. Tokuyama *et al.*, Helv., 1974, *57*, 2597; Daly *et al.*, *ibid.*, 1977, *60*, 1128; Tetrahedron, 1983, *39*, 49).

137 R =

138 R =

139 R =

Stereochemically controlled syntheses of fully reduced
histrionicotoxin and its octahydro derivative (a natural
product) have been announced, and preliminary approaches
to histrionicotoxin itself (M. Aratani *et al.*, J.org.Chem.,
1975, *40*, 2009; 2011; E.J. Corey, J.F. Arnett and G.N.
Widiger, J.Amer.chem.Soc., 1975, *97*, 430; J.J. Tufariello
and E.J. Trybulski, J.org.Chem., 1974, *39*, 3378; E.J. Corey,
M. Petzilka and Y. Ueda, Tetrahedron Letters, 1975, 4343;
S.A. Godleski and D.J. Heacock, J.org.Chem., 1982, *47*,
4822; A.J. Pearson and P. Ham, J.chem.Soc.Perkin I, 1983,
1421; Godleski *et al.*, J.org.Chem., 1983, *48*, 2101) have
culminated in a total synthesis of (±)-histrionicotoxin
(S.C. Carey, M. Aratani and Y. Kishi, Tetrahedron Letters,
1985, *26*, 5887).

12. *Pyridines from marine organisms*

The sex attractant of the sea slug *Navanax inermis* contains
pyridine bases, one of which is the 3-pyridyltetraenone (140)
(H.L. Sleeper and W. Fenical, J.Amer.chem.Soc., 1977, *99*,
2367). Another structure, perhaps of vegetable origin, is
pulo'upone (153) which has been isolated as a trace
constituent of the Hawaiian mollusc *Philinopsis speciosa*
(S.J. Coval *et al.*, Tetrahedron Letters, 1985, *26*, 5359).

140

141

Chapter 31

THE QUINOLINE ALKALOIDS

MALCOLM SAINSBURY

1.   Known quinolines from micro-organisms and higher plants

Since the main work (C.C.C. 2nd edn., Vol. 1VG, Ch.31, pp 171-255) was written interest in this group of natural products has continued apace. Many new alkaloids have been isolated and their structures determined principally by ^{1}H-and ^{13}C-n.m.r. spectroscopy, and by part syntheses from known alkaloids. In addition "new" plants of the Rutacea family have been examined and found to contain familiar structures. A measure of this latter activity is provided by the Tables 1 - 6 in which extractives of established structure present in these plants are listed.

RUTACEOUS ALKALOIDS OF ESTABLISHED STRUCTURE
(see the main work for details)

TABLE 1

Simple quinolines and 4-quinolinones

Alkaloid name	Botanical source (genus)
Graveolinine	Ruta[26]
N-Methyl-2-phenyl-4-quinolinone	Flindersia,[60] Haplophyllum[17]
Graveoline	Haplophyllum[13,17,61,66]
Isodictamnine	Dictamnus[46]
Isopteleine	Dictamnus[46]
Norgraveoline	Haplophyllum[80]
Ifflaiamine	Flindersia[104]

TABLE 2

2-Quinolinones

Alkaloid name	Botanical source (genus)

4-Methoxy-1-methyl-2-quinolinone	Myrtopsis,[21] Zanthoxyllum[43,91,98]
Flindersine	Atlantia,[25] Fagara,[107] Haplophyllum,[18] Micromelum,[94] Zanthoxyllum[90]
N-Methylflindersine	Almeidea,[101] Euxylophora,[76] Fagara,[48] Myrtopsis,[20] Ptelea[6]
Edulinine	Dutaillyea,[75] Fagara,[28] Haplophyllum Zanthoxyllum[72]
Folimine	Haplophyllum[17,109]
Oricine	Oricia[85]
Atanine	Afraegle[73]
Preskimmianine	Dictamnus,[46,110] Citrus[99]
Ptelefoline methyl ether	Ptelea[6]
Foliosidine	Haplophyllum[17]
N-Methylatanine	Almeidea,[101] Citrus,[99] Melicope[83]

TABLE 3

Furoquinolines

Alkaloid name	Botanical source (genus)

Dictamnine	Afraegle,[73] Boronella,[74] Dutaillyea,[75] Esenbeckia,[62] Flindersia,[63] Glycosmis,[64,65] Haplophyllum,[17] Melicope,[83] Myrtopsis,[20,21] Toddalia,[86] Zanthoxylum[1,2,42,43,88,91]
Skimmianine	Aegle,[100] Araliopsis,[14] Datura,[16] Dictamnus,[26] Esenbeckia,[62] Euxylophora,[76]
Halfordinine	Araliopsis,[14] Diphasia,[47] Oricia,[56] Teclea[23]
Haplophydine	Haplophyllum[12]
Platydesmine	Haplophyllum,[17] Flindersia,[60] Zanthoxylum[91]
Myrtopsine	Haplophyllum[80,91]
7-(-3-Methylbut-2-enyloxy)-γ-fagarine	Haplophyllum[102]

	Flindersia,[63] Glycosmis,[58,65] Haplophyllum,[53,79,109] Melicope,[32,103] Monnieria,[106] Murraya,[55] Myrtopsis,[20,21] Oricia,[85] Teclea,[23,35] Toddalia,[68,86] Tylophora,[41] Vespris,[87] Zanthoxylum[1,3,4,37,39,43,57,70,71,72,89,91,92,93]
γ-Fagarine	Aegle,[100] Dictamnus,[46] Erythrochiton,[27] Flindersia,[63]Haplophyllum,[53,66,79,102,105] Monnieria,[106] Myrtopsis,[20] Toddalia,[68,86] Tylophora,[41] Vespris,[87] Zanthoxylum[42,43]
7-Isopentenyloxy-γ-fagarine	Haplophyllum,[11,30]
Kokusaginine	Acronychia,[44] Baurella,[59] Dutaillyea,[75] Esenbeckia,[62,108] Glycosmis,[58] Haplophyllum,[105]Melicope,[103]Monnieria,[106] Oricia,[56,85] Sargentia,[22] Teclea,[36,96] Vespris[37]
Pteleine	Dutaillyea,[75] Pteleine[5]
Haplopine	Afraegle,[73] Melicope,[103] Monnieria,[84] Haplophyllum,[30]Zanthoxylum[2,43,91,102,103]
Maculine	Esenbeckia,[62,108] Sargentia,[22] Teclea,[35]
Maculosidine	Esenbeckia,[62] Oricia[85]
Evolitrine	Acronychia,[44] Dutaillyea,[75] Esenbeckia,[62] Glycosmis,[50] Melicope[83,103]
Robustine	Thamnosma,[24] Zanthoxylum,[2,43,91]
Anhydroevoxine	Haplophyllum[102]
Evodine	Haplophyllum[19,102]
Dubinidine	Haplophyllum[17]
Evoxine	Haplophyllum,[53,78,102] Monnieria,[84,106] Orixa,[69] Teclea[34]
Evoxine acetate	Haplophyllum[102]
Glycoperine	Haplophyllum[10,51,102]
Flindersiamine	Esenbeckia,[62,108] Oriciopsis,[95] Teclea[35,36,96]
Acronycidine	Baurella[49]
Confusameline	Melicope,[103] Myrtopsis[21]

TABLE 4

Furoquinolones

Alkaloid name	Botanical source (genus)
Lemobiline (Spectabiline)	Euxylophora[76]
Isodictamnine	Dictamnus[46]
Isomaculosidine	Pteleine[5]
Isoplatydesmine	Araliopsis,[8,14] Ptelea[9]

TABLE 5

Furoquinolinium salts

Alkaloid name	Botanical source (genus)
Pteleatinium salt	Ptelea[7]
•Methylplatydesminium salt	Choisya[15]
•Methylhydroxyluninium salt	Choisya[45]

TABLE 6

Pyranoquinolones

Alkaloid name	Botanical source (genus)
Khaplofoline	
Ribalinine	Araliopsis,[8,14] Fagara,[28] Ruta,[26] Zanthoxyllum

References to Tables 1-6

1. S. Najjar, G.A. Cordell and N.R. Farnsworth, Phytochem., 1975, *14*, 2309.
2. H. Ishii et al., Yakugaku Zasshi, 1974, *94*, 322; C.A., 1974, *81*, 132753.
3. F. Fish, I.A. Meshab and P.G. Waterman, Phytochem., 1975, *14*, 2094.
4. E.F. Nesmelova, I.A. Bessonova and S.Yu. Yunusov, Khim.prir.Soedin., 1975, 666; C.A., 1976, *84*, 105863.
5. L.A. Mitscher et al., Lloydia, 1975, *38*, 117.
6. J. Reisch et al., Phytochem., 1975, *14*, 1678.
7. L.A. Mitscher et al., Lloydia, 1975, *38*, 109.
8. J. Vaquette et al., Phytochem., 1976, *15*, 743.
9. I. Ya Isaev and I.A. Bessonova, Khim.prir.Soedin., 1974, 815; C.A., 1975, *82*, 121677.

10. V.I. Akhedzhanova, I.A. Bessonova and S.Yu. Yunusov, Khim.prir.Soedin., 1974, 680; C.A., 1975, *82*, 73261.
11. I.A. Bessonova, V.I. Akhedzhanova, S.Yu. Yunusov, Khim.prir.Soedin., 1974, 677; C.A., 1975, *82*, 86462.
12. Kh.A. Abdullaeva, I.A. Bessonova and S.Yu. Yunusov, Khim.prir.Soedin., 1974, 684; C.A., 1975, *82*, 73260.
13. A.G. Gonzalez, R.E. Reyes and E.D. Chico, Anales de Quim., 1974, *70*, 281; C.A., 1974, *81*, 117048.
14. F. Fish, I.A. Meshal and P.G. Waterman, Planta Med., 1976, *29*, 310.
15. R. Garestier and M. Rideau, C.R.Congr.Natl.Soc.Savantes Sect.Sci., 1973, *98*, 183; C.A., 1976, *85*, 156534.
16. E.G. Sharova, S.Yu. Aripova and A.U. Abdibalimov, Khim.prir.Soedin., 1977, 127; C.A., 1977, *87*, 50201.
17. D.M. Kazakova, I.A. Bessonova and S.Yu. Yunusov, Khim.prir.Soedin., 1976, 682; C.A., 1977, *86*, 136315.
18. V.I. Akmedzhanova, I.A. Bessonova and S.Yu. Yunusov, Khim.prir.Soedin., 1976, 320; C.A., 1977, *86*, 43861.
19. V.I. Akmedzhanova, I.A. Bessonova and S.Yu. Yunusov, Khim.prir.Soedin., 1977, 289; C.A., 1977, *87*, 98863.
20. M.S. Hifnawy et al., Phytochem., 1977, *16*, 1035.
21. M.S. Hifnawy et al., Planta Med., 1976, *29*, 346.
22. X.A. Dominguez et al., Rev.Latinamer.Quim., 1977, *8*, 47; C.A., 1977, *86*, 117636.
23. F. Fish, I.A. Meshal and P.G. Waterman, J.Pharm. Pharmacol., 1976, *28*, suppl. p.72P.
24. P.T.O. Chang et al., Lloydia, 1976, *39*, 134.
25. I.H. Bowen and J.R. Lewis, Lloydia, 1978, *41*, 184.
26. A.G. Gonzalez et al., Anales de Quim., 1977, *73*, 430; C.A., 1977, *87*, 148668.
27. S. Johne and S. Haerlting, Pharmazie, 1977, *32*, 415.
28. R. Torres and B.K. Cassels, Phytochem., 1978, *17*, 838.
29. B.P. Das and D.N. Chowdhury, Chem. and Ind., 1978, 272.
30. E.F. Nesmelova, I.A. Bessonova and S.Yu. Yunusov, Khim.prir.Soedin., 1977, 289; C.A., 1977, *87*, 81276.
31. Kh.A. Abdullaeva, I.A. Bessonova and S.Yu. Yunusov, Khim.prir.Soedin., 1977, 425; C.A., 1977, *87*, 148, 666.
32. A. Ahond et al., Phytochem., 1978, *17*, 166.
33. G.J. Kapadia, Y.N. Shukla and S.P. Basak, Phytochem., 1978, *17*, 1443.
34. J. Vaquette et al., Planta Med., 1978, *33*, 78.

214

35. F. Fish, I.A. Meshal and P.G. Waterman, Fisoterapia, 1977, *48*, 170; C.A., 1978, *88*, 166747.
36. J.I. Okogun and J.F. Ayafor, J.chem.Soc.Chem.Comm., 1977, 652.
37. R. Hamsel and E.-M. Cykulski, Arch.Pharm. 1978, 311 135.
38. H. Ishii, T. Ishikawa and J. Haginiwa, Yakugaku Zasshi, 1977, *97*, 890; C.A., 1977, *87*, 197 250.
39. F.R. Stermitz and I.A. Sharifi, Phytochem., 1977, *16*, 2003.
40. N. Decandain, N. Kunesch and J. Poisson, Ann.Pharm.Fr., 1977, *35*, 521; C.A., 1978, *89*, 43872.
41. T. Etherington, R.B. Herbert and F.B. Jackson, Phytochem., 1977, *16*, 1125.
42. V.H. Deshpande and R.K. Shastri, Indian J.Chem., 1977, 15B, 95.
43. H. Ishii et al., Yakugaku Zasshi, 1976, *96*, 1458; C.A., 1977, *86*,136 297.
44. L.B. de Silva et al., Phytochem., 1979, *18*, 1255.
45. M. Rideau et al., Phytochem., 1979, *18*, 155.
46. V.I. Akhmedzhanov, I.A. Bessonova and S.Yu. Yunusov, Khim.prir.Soedin., 1978, 476; C.A., 1978, *89*, 176 371.
47. P.G. Waterman et al., Biochem.Syst.Ecol., 1978, *6*, 239.
48. F.Y. Chou et al., Heterocycles, 1977, *7*, 969.
49. B. Couge et al., Plant.Med.Phytother., 1980, *14*, 208.
50. M. Sarker, S. Kundu and D.P. Chakraborty, Phytochem., 1978, *17*, 2145.
51. E.F. Nesmelova, I.A. Bessonova and S.Yu. Yunusov, Khim.prir.Soedin., 1978, 758; C.A., 1979, *91*, 20 830.

52. Kh.A. Abdullaeva, I.A. Bessonova and S.Yu. Yunusov, Khim.prir.Soedin., 1978, 219; C.A., 1978, *89*, 103 710.
53. A. Al-Shamma, N.A. El-Douri and J.D. Phillipson, Phytochem., 1979, *18*, 1417.
54. J. Mendez, Planta Med., 1978, *34*, 218.
55. M.T. Fauvel et al., Plant.Med.Phytother., 1978, 12, 207.
56. F. Fish, I.A. Meshal and P.G. Waterman, Planta Med., 1978, *33*, 228.
57. J. Vaquette, A. Cave and P.G. Waterman, Plant.Med.Phytother., 1978, *12*, 235.

58. I.H. Bowden, K.P.W.C. Perera and J.R. Lewis, Phytochem., 1978, *17*, 2125.
59. F. Tillequin et al., J.nat.Prod., 1980, *43*, 498.
60. F. Tillequin, M. Koch and T. Sevenet, Plant Med., 1980, *39*, 383.
61. V.I. Akmedzhanova, I.A. Bessonova and S.Yu. Yunusov, Khim.prir.Soedin., 1980, 803; C.A., 1981, *94*, 136157.
62. D.L. Dreyer, Phytochem., 1980, *19*, 941.
63. F. Tillequin, M. Koch and T. Sevenet, Plant.Med.Phytother., 1980, *14*, 4.
64. L.H. Bowen, K.P.W.C. Perera and J.R. Lewis, Phytochem., 1980, *19*, 1566.
65. K. Rastogi, R.S. Kapil and S.P. Popli, Phytochem., 1980, *19*, 945.
66. D.M. Razakova, I.A. Bessonova and S.Yu. Yunusov, Khim.prir.Soedin., 1979, 810; C.A., 1980, *93*, 22 586.
67. Kh.A. Abdullaeva, I.A. Bessonova and S.Yu. Yunusov, Khim.prir.Soedin., 1979, 873; C.A., 1980, *93*, 41 504.
68. P.N. Sharma et al., Indian J.Chem.Sect.B, 1979, *17*, 299.
69. T. Yajima, N. Kato and K. Munakata, Agric.biol.Chem., 1977, *41*, 1263.
70. D.L.Dreyer and R.C. Brenner, Phytochem., 1980, *19*, 935.
71. J.A. Swinehart and F.R. Stermitz, Phytochem., 1980, *19*, 1219.
72. F.R. Stermitz, M.A. Caolo and J.A. Swinehart, Phytochem., 1980, *19*, 1469.
73. J. Reisch, M. Muller and I. Mester, Planta Med., 1981, *43*, 285.
74. F. Bevalot, J. Vaquette and P. Cabalion, Plant.Med.Phytother., 1980, *14*, 218.
75. G. Baudouin et al., J.nat.Prod., 1981, *44*, 546.

76. L. Jurd and R.Y. Wong, Austral.J.Chem., 1981, *34*, 1625.
77. P. Wulff, J.S. Carle and C. Christopherson, Comp.Biochem.Physiol.B, 1982, *71*, 525.
78. D.M. Razakova and I.A. Bessonova, Khim.prir.Soedin., 1981, 673; C.A., 1982, *96*, 31 680.
79. D. Batsuren, E.Kh. Batirov and V.M. Malikov, Khim.prir.Soedin., 1981, 659; C.A., 1982, *96*, 48 968.
80. V.I. Akhmedzhanova and I.A. Bessanova, Khim.prir.Soedin., 1981, 613; C.A., 1982, *96*, 31 670.

216

81. D.M. Razakova and I.A. Bessonova, Khim.prir.Soedin., 1981, 528; C.A., 1982, *96*, 100 871.
82. S.A. Khalid and P.G. Waterman, Planta Med., 1981, *43*, 148.
83. M.Th. Fauvel et al., Phytochem., 1981, *20*, 2059.
84. G. Moulis et al., Planta Med., 1981, *42*, 400.
85. S.A. Khalid and P.G. Waterman, Phytochem., 1981, *20*, 2761.
86. P.N. Sharma et al., Indian J.Chem., Sect.B, 1981, *20*, 936.
87. S.A. Khalid and P.G. Waterman, J.nat.Prod., 1982, *45*, 343.
88. N. Ruanguengsi et al., J.sci.Soc.Thailand, 1981, *7*, 123; C.A., 1982, *96*, 31 659.
89. L. Ren and F. Xie, Yaoxue Xuebao, 1981, *16*, 672; C.A., 1982, *96*, 48 976.
90. M.R. Torres and B.K. Cassels, Bol.Soc.Chil.Quim., 1982, *27*, 260; C.A., 1982, *96*, 196 571.
91. H. Ishii et al., Yakugaku Zasshi, 1981, *101*, 504; 1982, *102*, 182; C.A., 1981, *95*, 111 726; 1982, *97*, 69 240.
92. M.-H. Wang, Yao Hsueh T'ung Pao, 1981, *16*, 48; C.A., 1981, *95*, 192 260.
93. Z. Chang et al., Yaoxue Xuebao, 1981, *16*, 394; C.A., 1982, *97*, 20 735.
94. P. Tantivantana et al., J.org.Chem., 1983, *48*, 268.
95. J.F. Ayafor et al., Phytochem., 1982, *21*, 2603.
96. J.F. Ayafor et al., J.nat.Prod., 1982, *45*, 714.
97. J.F. Ayafor and J.I. Okogun, J.nat.Prod., 1982, *45*, 182.
98. H. Ishii et al., Yakugaku Zasshi, 1983, *103*, 279; C.A., 1983, *98*, 50 251.
99. T.-S. Wu, C.-S. Kuoh and H. Furukawa, Phytochem., 1983, *22*, 1493.

100. M.S. Karawya, Y.W. Mirhom and I.A. Shehata, Egypt.J.pharm.Sci., 1982, *21*, 239; C.A., 1983, *98*, 68 855.
101. C. Moulis et al., Phytochem., 1983, *22*, 2095.
102. I.A. Bessonova and S.Yu. Yunusov, Khim.prir.Soedin., 1982, 530; C.A., 1983, *98*, 176 118.
103. F. Tillequin et al., J.nat.Prod., 1982, *45*, 486.
104. M.F. Grundon in The Alkaloids ed. R.H.F. Manske and R.G.A. Roderigo, Academic Press, 1979, Vol.XVII, pp.169-177.

105. A. Ulubelen, Phytochem., 1985, *24*, 372.
106. J. Bhattacharyya and L.M. Serur, J.nat.Prod., 1984, *47*, 379.
107. S. Ahmed, J.nat.Prod., 1984, *47*, 391.
108. F. Bevalot et al., Planta Med., 1984, *50*, 523.
109. D.M. Razakova, I.A. Bessonova and S.Yu. Yunusov, Khim.prir.Soedin., 1984, 635; C.A., 1985, *102*, 75690h.
110. (a) J.F. Frank et al., Acta Crystallographia, 1978, *13 B*,2316;
(b) A.H.J. Wang ibid., p.2319.

2. *New alkaloids*

(a) *Simple quinolines and 4-quinolinones*

Simple quinolines and quinolin-4-ones often occur outside of the Rutacae plant family, thus the brominated quinoline (1) is metabolised by the marine bryozoan *Flustra foliacea* (P. Wulff, J.S. Carle and C. Christophersen, Comp.Biochem.Physiol.B, 1982, *71*, 525) and an unusual alkaloid the bisquinolyl lactone, broussonetine (2), is found in a plant belonging to the Moracea family (A.A.L. Gunatilaka et al., Phytochem., 1984, 929).

Ephedralone (3) is claimed as an alkaloid from the Egyptian plant *Ephedra alata* (M.A.M. Nawar et al., Phytochem., 1985, *24*, 878).

4-Quinolones (4) with long alkyl chains attached to position 2 are metabolised by bacteria and have been given the name pseudanes to denote their initial discovery in the extractives of *Pseudomonas* species (see C.C.C. 2nd edn., Vol. 1VG, p.221). Such compounds also occur in higher plants(J. Reisch et al., Phytochem., 1975, *14*, 840). Malatyamine (5), in the form of its ethyl ester, has been obtained from *Haplophyllum cappadocicum* indigenous to Turkey (G. Arar et al., J.nat.Prod., 1985, *48*, 642). Since ethyl esters are very unusual in Nature it is probable that the free acid is the true alkaloid. Some N-methylated structures (4, R = Me; n = 10, 12, and 14) have been extracted from the fruit and leaves of *Evodia rutaecarpa* (T. Kamikado et al., Agric.biol.Chem.(Japan), 1976, *40*, 605; C.A., 1976, *84*, 180446), and the analogue (4, R = Me; n = 8) occurs in *Ruta graveolens* (M.F. Grundon and H.M. Okely, Phytochem., 1979, *18*, 1768). Not surprisingly the similar structures (6) and (7), incorporating unsaturated side chains, are also found in *Pseudomonas aeruginosa* (A.G. Kozlovski, et al., Izvest.Akad.Nauk.S.S.S.R.Ser.Khim., 1976, 1146; C.A., 1977, *86*, 29964; H. Budzikiewicz, Monatsch., 1979, *110*, 947). Hapovine (8) is yet another variant on this theme and is a metabolite of *Haplophyllum popovii* (D.M. Razakova and I.A. Bessonova, Khim.prir.Soedin., 1981, 528; C.A., 1982, *96*, 100871).

3

4

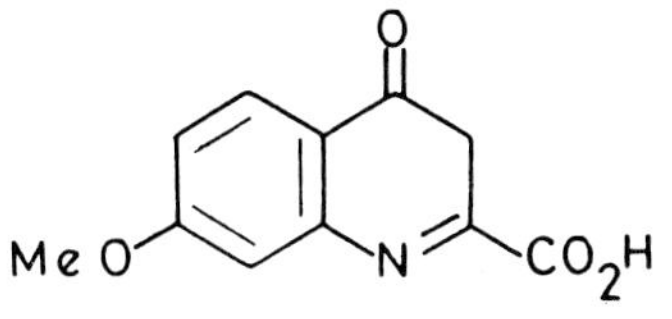

5

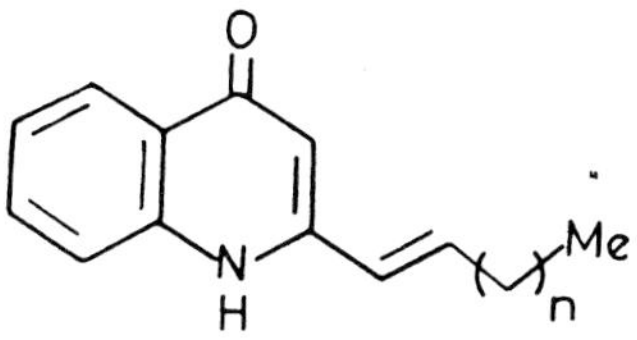

6

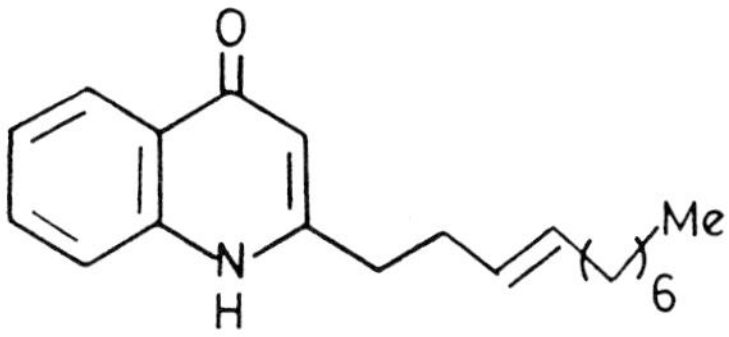

7

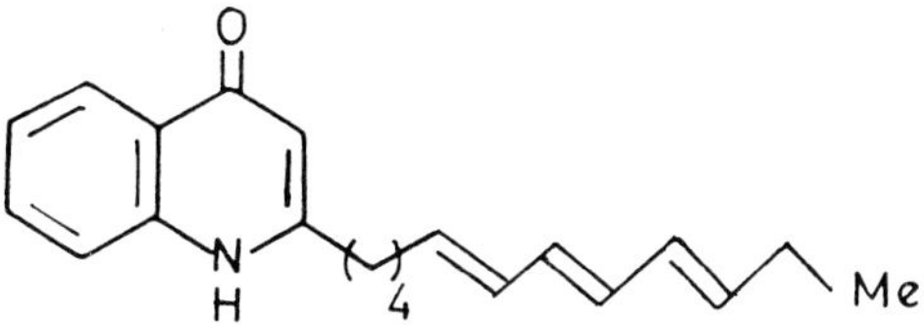

8

Another new quinolinone norgraveoline (9,R = H), is accompanied by graveoline (9, R = Me) in *Haplophyllum dubium* (Razakova, Bessonova and S.Yu.Yunusov, ibid., 1979, 810; C.A., 1980, *93*, 22586). The latter alkaloid is a common constituent of Rutaceous plants. Unusual structures are provided by melochinone (10) (G.J. Kapadia et al., J.Amer.chem.Soc., 1975, *97*, 6814) and melovine (11) (Kapadia, Y.N. Shukla and S.P. Basak, Phytochem., 1978, *17*, 1443) both from *Melochia tomentosa* , a plant of the Sterculiaceae family.

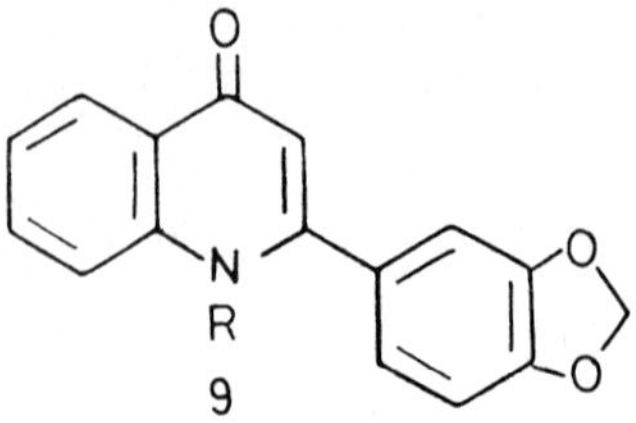

9

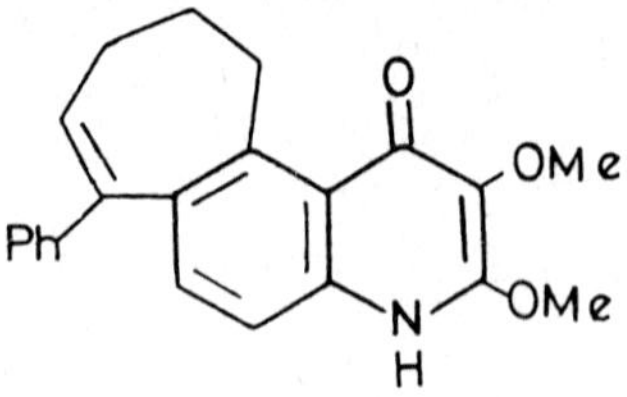

10

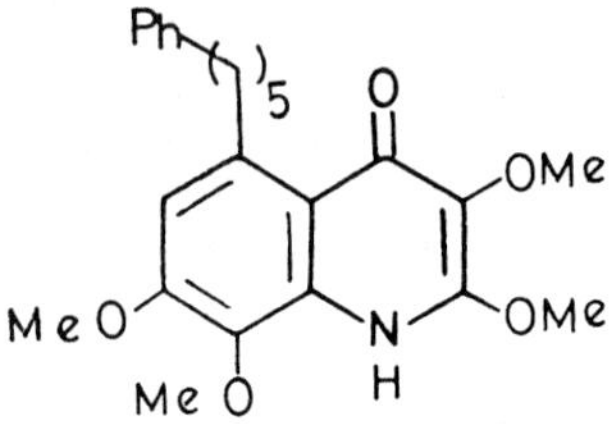

11

N- Methylkhapofoline (12,R = H), a new alkaloid from *Almeidea guyanensis*, was in fact also previously known as a synthetic product (C. Moulis et al., Phytochem., 1983, *22*, 2095). *Balfourodenron riedelianum* is a source of numerous alkaloids including the known compound (+)-ribalinidine (12, R = OH), and its isomer riedelianine (13), (L. Jurd and R.Y. Wong, Austral.J.Chem., 1983, *36*, 1615).

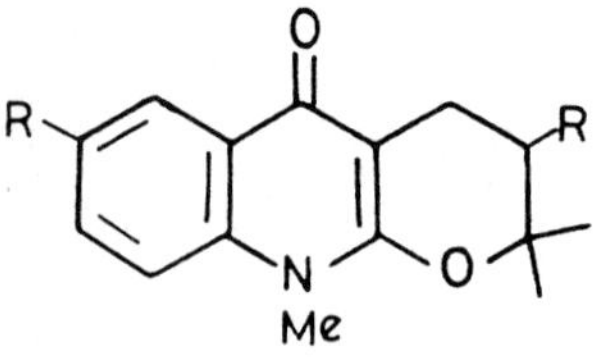

12

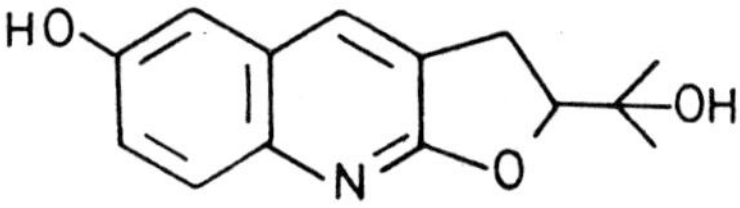

13

Geibalansine (14, R = H) and its O-acetyl derivative (14, R = OAc) (amorphous) are pyranoquinolines from *Geijera balansae*. Proof of structure for geibalansine was provided by a synthesis of O-acetylgeibalansine (M. Ramesh, P. Rajamanickan and P. Shanmugan, Heterocycles, 1984, *22*, 125), but in fact this compound had also been prepared earlier, as an intermediate in a synthesis of ribalinine (15) (R.M. Bowman and M.F. Grundon, J.chem.Soc.(C), 1966, 1504).

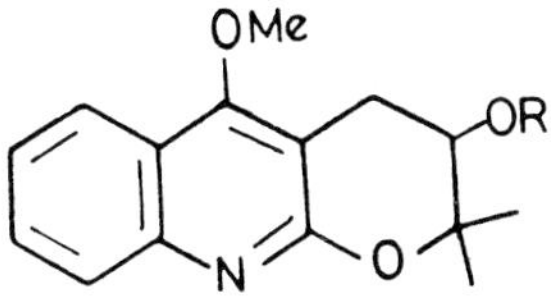

14

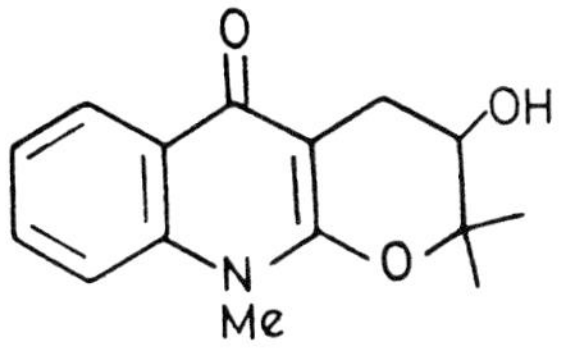

15

Glycarpine, isolated from *Glycosmis cyanocarpa*,was initially allocated structure (16); however, this compound has now been synthesised (as shown below) and shown to have m.p. 171°C. Since there is a difference between this figure and that quoted for the alkaloid, it is obvious that the original assignment is untenable (S.-C. Kuo et al., J.nat.Prod., 1984, *47*, 47).

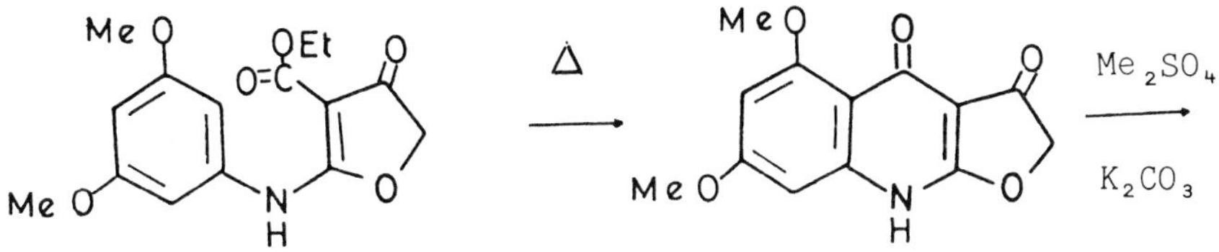

(b) 2-Quinolinones

Most often naturally occurring quinolin-2-ones are found in the Rutacea and all the alkaloids listed below arise in plants of this family.

The swietenidins A and B, which are the first naturally occurring 2-quinolinones to bear a methoxyl group at C-3, are present in the bark of the East Indian satin wood *Chloroxylon swietenia*.   Their structures, (17) and (18) respectively, were deduced from spectroscopic data (K.S. Bhide, R.B. Mujumdar and A.V.R. Rao, Indian J.Chem.Sect.B, 1977, *15B*, 440), but swietenidin A methyl ether has now been synthesised (Bhide and Mujumdar ibid., 1983, *22B*, 1254).   Swietenidin B is another example of a alkaloid prepared as a synthetic compound prior to its eventual isolation as a natural product.

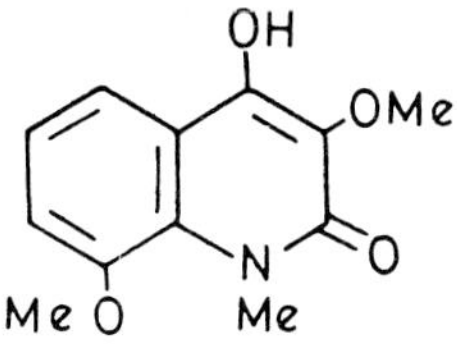

17

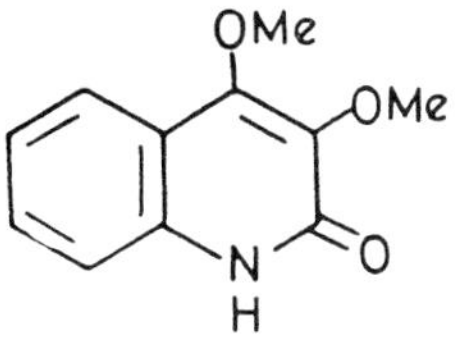

18

The simple 2-quinolone, integriquinolone (19), m.p. 257-260°C, is a constituent of *Zanthoxylum integrifolium* (H. Ishii et al., Yakugaku Zasshi, 1981, *101*, 504; C.A., 1982, *97*, 69240), while daurine (20) occurs in *Haplophyllum dauricum* (I.A. Bessonova et al., Khim. prir. Soedin., 1983, 116; C.A., 1983, *99*, 71039). Structure (20) for daurine is supported by the fact that it may be degraded to the known alkaloid folifidine (21).

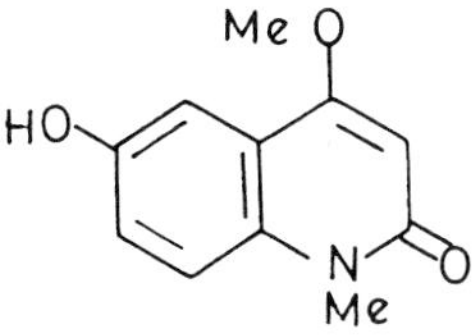

19

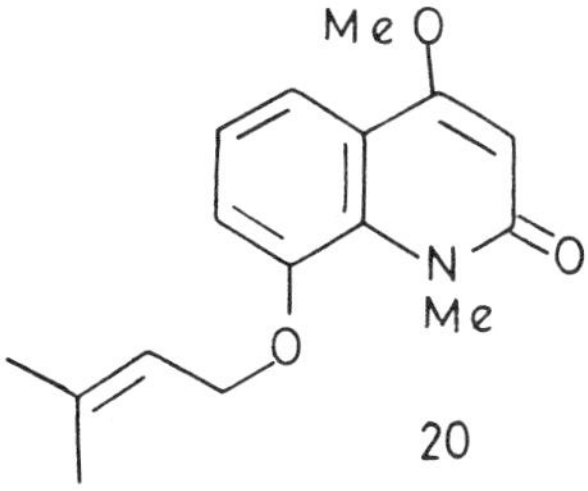

20

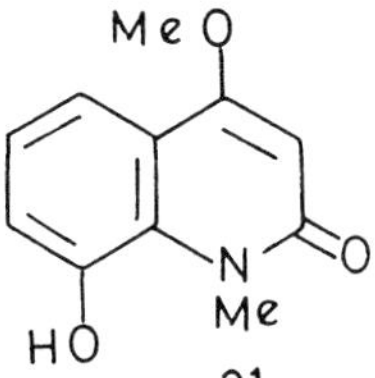

21

Glycophylone (22) (P. Bhattacharyya and B.K. Chowdhury, Chem.Ind., 1984, 352), glycosolone (23) (B.P. Das and D.N. Chowdhury, ibid., 1978, 272) and glycolone (24) (Bhattacharyya and Chowdhury, Phytochem., 1985, *24*, 634)are three more alkaloids from *Glycomis pentaphylla*-a prodigious source of quinoline derivatives.

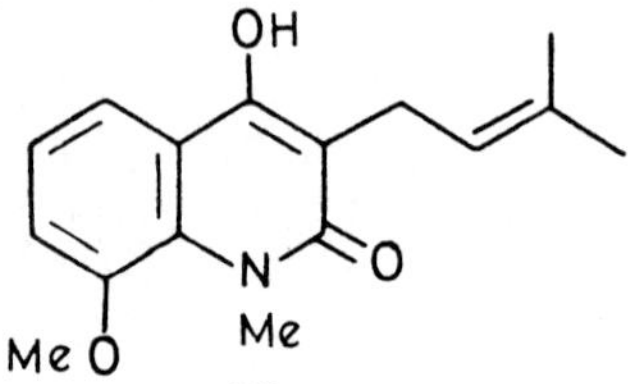

22

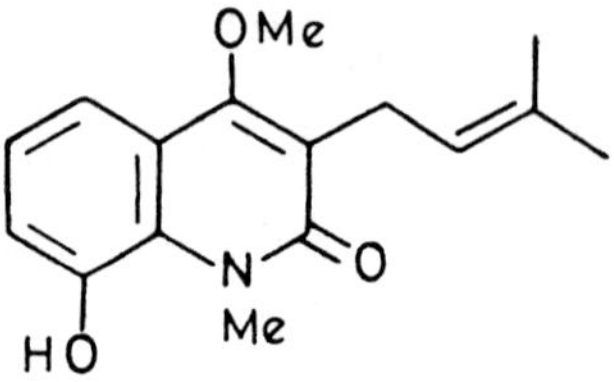

23

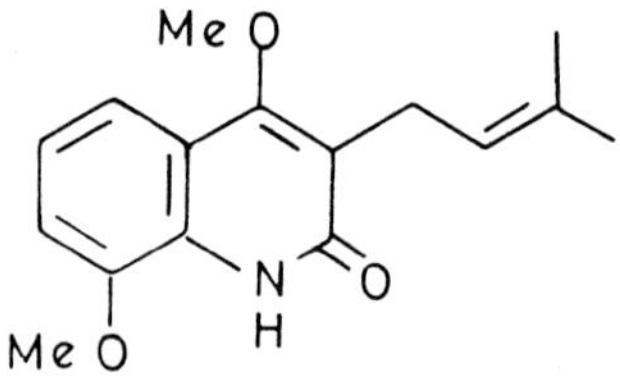

24

The roots of another plant in this genus - *G. mauritiana* - metabolise the related but unnamed structure (25) (K. Rastogi, R.S. Kapil and S.P. Popli, Phytochem., 1980, *19*, 945).

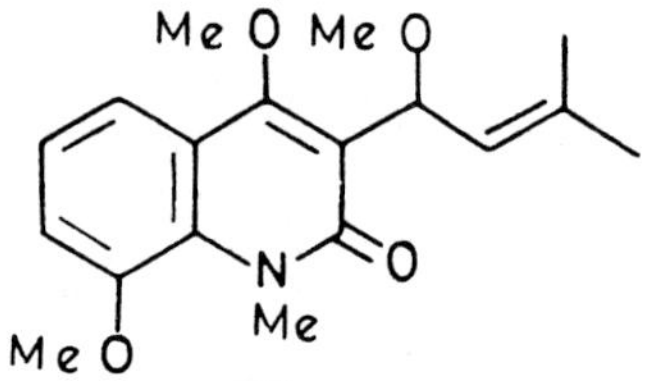

25

Almeine (26), amorphous, and 4-desmethyl-N-methylatanine
(27), m.p. 162-163°C, are present in the stem bark of
Almeidea guyanensis (C. Moulis et al., Phytochem., 1983, *22*,
2095). Treatment of the latter compound with DDQ gives
rise to N-methylflindersine (28), m.p. 84°C, which occurs in
the root bark of *Ptelea trifoliata* (J.Reisch et al., ibid.,
1975, *14*, 1678), and which is also the insect antifeedant
principle of *Fagara chalybaea* , *F. holstii* and
Xylocarpus granotum (F.Y. Chou et al., Heterocycles, 1977,
7, 969).

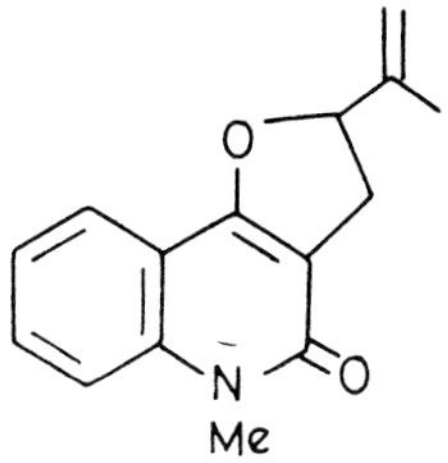

26

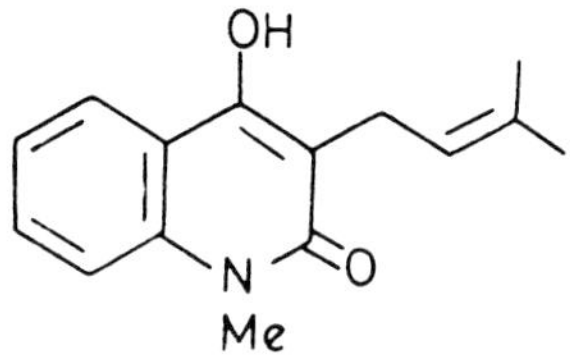

27

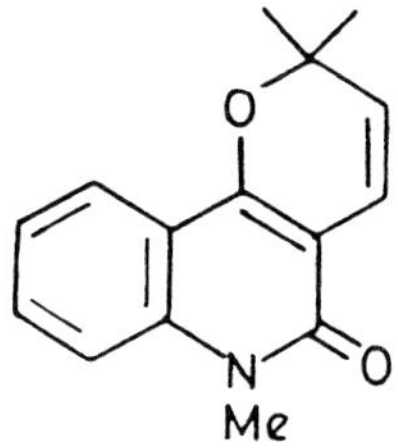

28

Structures based on the flindersine model are very common in
Rutaceous plants and the methoxy-N-methyl derivative (29),
is obtained from *Oricia renieri* (S.A. Khalid and
P.G. Waterman, Phytochem., 1981, *20*, 2761), while the three
alkaloids (30), (31) and (32) come from *Vespris stolzii*
(Khalid and Waterman, J.nat.Prod., 1982, *45*, 343).

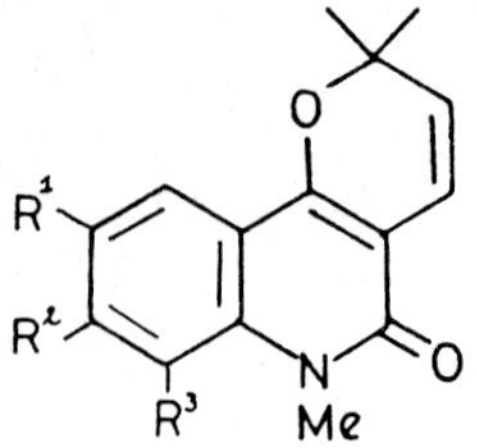

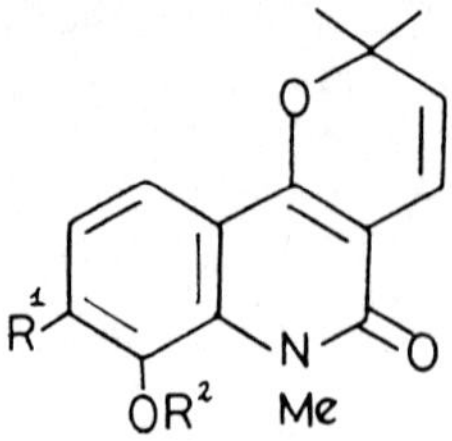

29; $R^1=R^3=H$, $R^2=OMe$, $R^4=Me$

30; $R^1=H$, $R^2=$

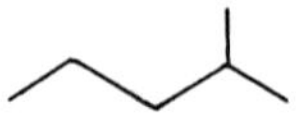

31; $R^1=Me$, $R^2=$

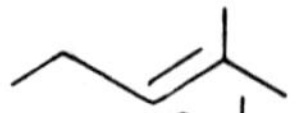

32; $R^1=Me$, $R^2=$

Ravesilone (33), another compound of this type, has been isolated from *Ravenia spectablis* (P. Bhattacharyya and B.K. Chowdhury, Phytochem., 1984, *23*, 1825).

Yet another related alkaloid is the hydrate (34) of N-methylflindersine; this occurs in *Euxylophora paraensis* (L. Jurd and E. Wong, Austral.J.Chem., 1981, *34*, 1625).

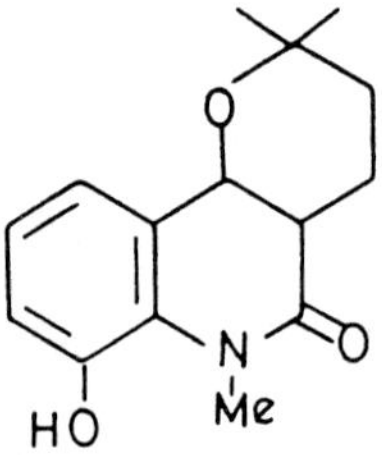

33

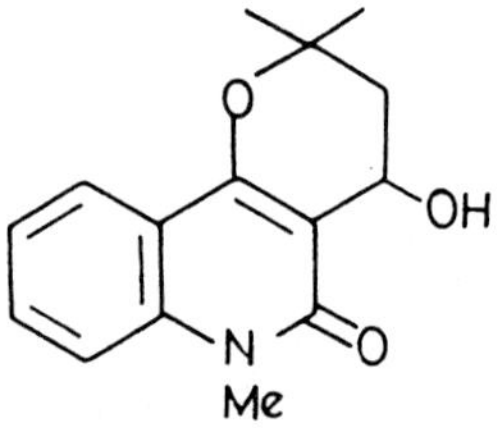

34

Zanthobungeanine (35) occurs in *Zanthoxylum bungeaum* (L. Ren and F. Xie, C.A., 1982, *96*, 48976) while zanthophylline (36) and desmethylzanthophylline (37) are present in *Z. monophyllum* (F.R. Stermitz and I.A. Sharifi, Phytochem., 1977, *16*, 2003). Zanthophylline, the more abundant alkaloid, has been synthesised by N-alkylation of 8-methoxyflindersine with acetoxymethyl chloride.

35

36 , R = Me
37 , R = H

Vespris louisii generates the alkaloids vesprisine (38,), N-methylpreskimmianine (39) and vesprisilone (40) (J.F. Ayafor, B.L. Sodengam and B. Ngadjui, Tetrahedron Letters, 1980, *21*, 3293; Phytochem., 1982, *21*, 955). The structure of vesprisilone was deduced by spectroscopy, and by its reduction with sodium borohydride to give the same diol as that obtained by treatment of preskimmianine first with 3-chloroperbenzoic acid and then with alkali. Vesprisine also occurs in *V. stolzii* (S.A. Khalid and P.G. Waterman, J.nat.Prod., 1982, *45*, 343) and it may be synthesised by treating preskimmianine with hydrogen chloride, and oxidising the product pyrano derivative with DDQ.

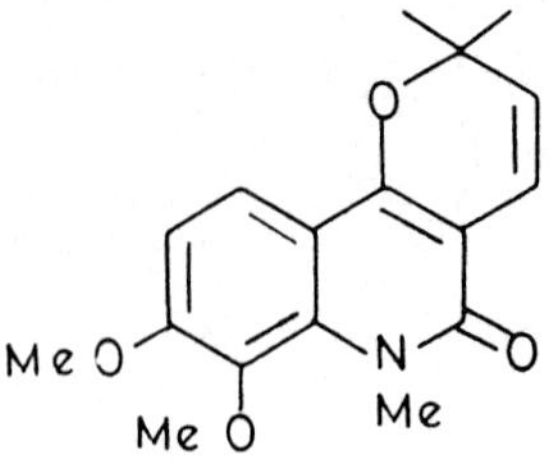

38

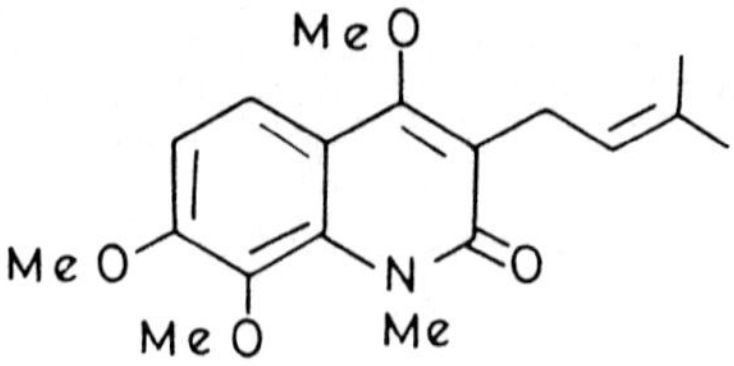

39

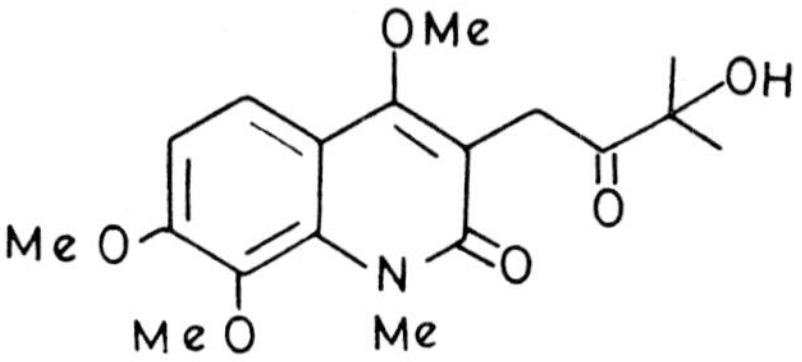

40

Hydroxylunidonine (41) occurs both in the stems and flowers of *Lunasia amara*, whereas 6-methoxylunidonine (42), 6-methoxylunidine (43) and 6-methoxyhydroxylunidine (44) have only been found in the stem of this plant (L.A. Mitscher et al., Lloydia, 1975, *38*, 117; J.A. Reisch et al., Phytochem., 1975, *14*, 1678).

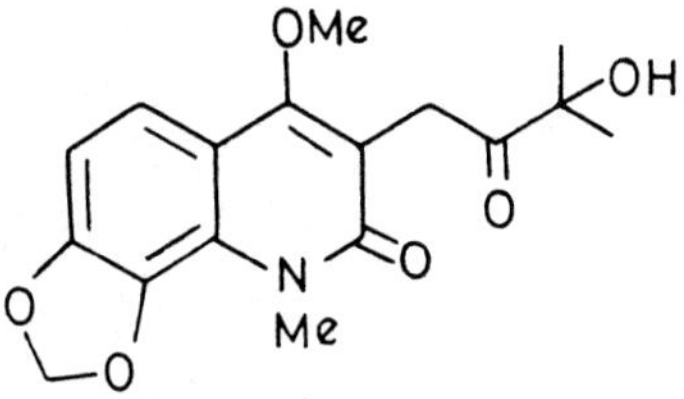

41

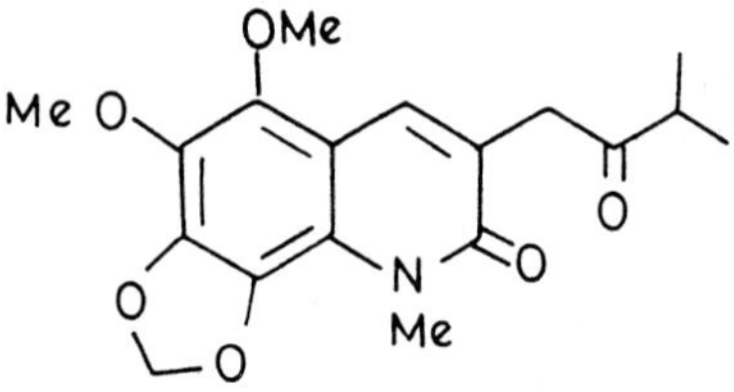

42

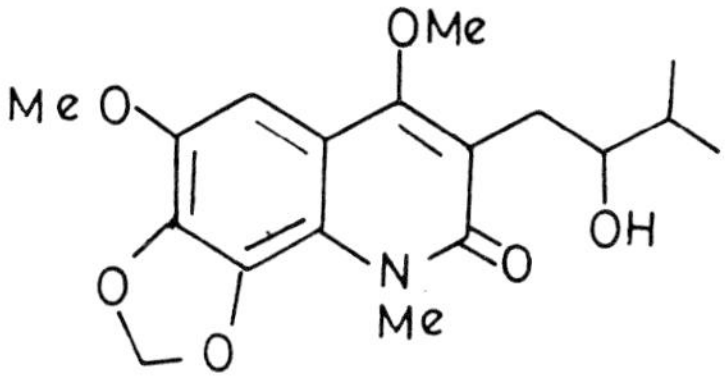

43

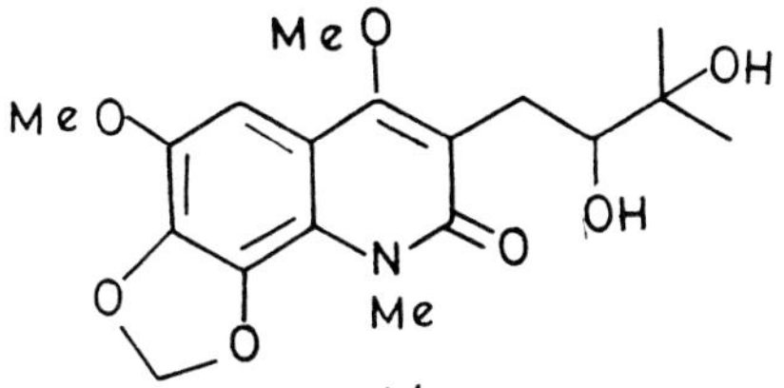

44

N-Desmethyllunidonine (45) is yet another metabolite of
P. trifoliata (K. Szendrei et al., Herba Hung., 1974, *13*,
49; C.A., 1975, *83*, 40 169).

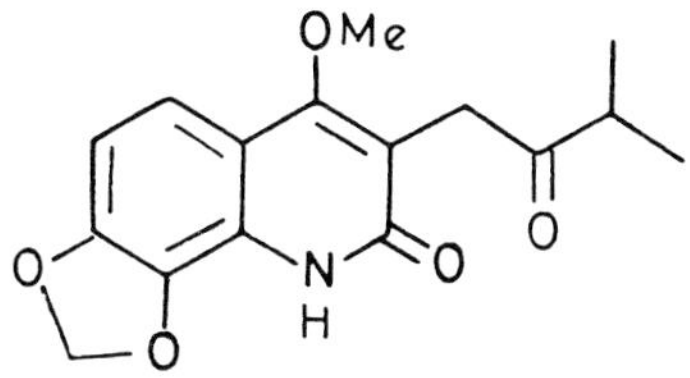

45

A general synthetic procedure to alkaloids of this type, as
well as to furanoquinoline structures, has been announced
(M. Ramesh, P.S. Mohan and P. Shanmugan, Tetrahedron, 1984,
40, 4041: see also p.3431). The versatility of this
approach is illustrated by syntheses of atanine,
flindersine, orcine, preskimmianine, O-methylglycosolone and
zanthobungeanine.

Glycarpine (46) is a simple furoquinolinone alkaloid
obtained from *Glycosmis cyanocarpa* (M. Sarker, S. Kundu and
D.P. Chakraborty, Phytochem., 1978, *17*, 2145), while
melineurine (47) has been isolated from *Melicope lasioneura*
(F. Tillequin et al., J.nat.Prod., 1982, *45*, 486).
Melineurine has been synthesised (Tillequin, G. Baudonin and
M. Koch, ibid., 1983, *46*, 132).

Taifine (48), isotaifine (49) and 8-methoxytaifine (50)are
claimed as alkaloids from *Ruta chalepensis* (B.A.H. El-Tawil
et al., Z. Naturforsch., Teil.B, 1981, *36*, 1169). However,
since N-ethyl groups are unusual it seems probable that
these compounds were formed during their isolation by
treatment of the corresponding NH compounds with hot
ethanolic potassium hydroxide. The secondary amines are
thus the true alkaloids of the plant.

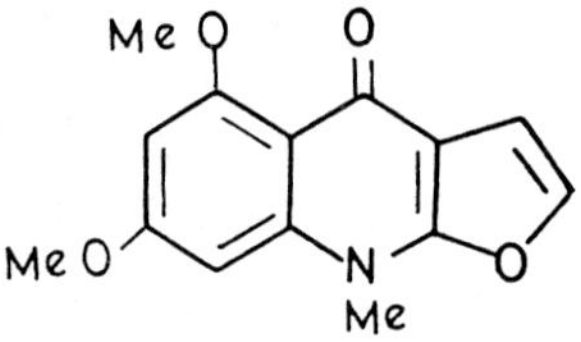

46

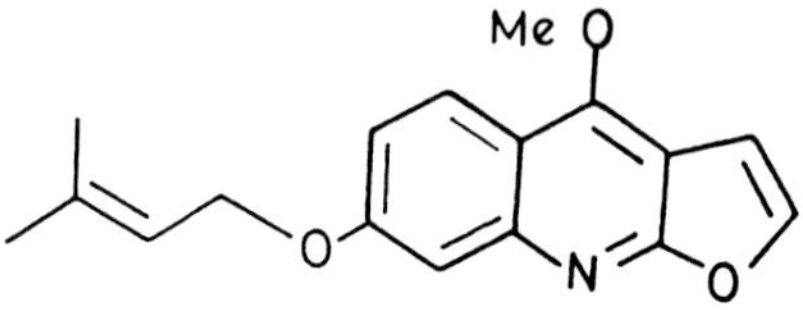

47

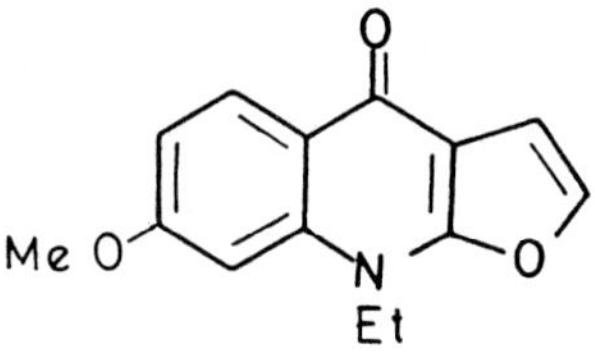

48

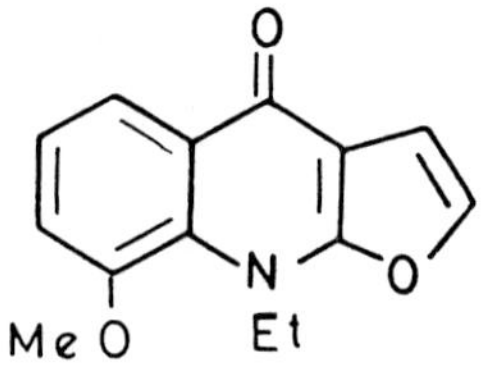

49

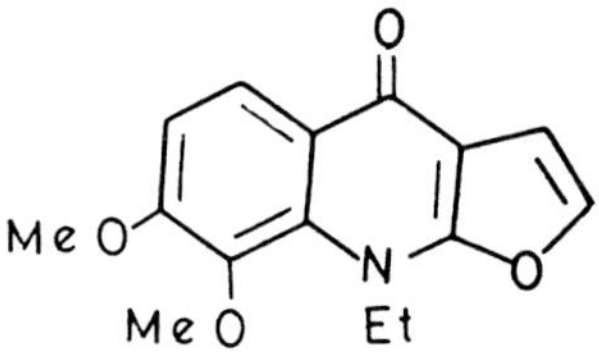

50

Buchapsine (51, R=H) and an unnamed alkaloid (52) both occur in *Haplophyllum bucharicum* (E.F. Nesmelova, I.A. Bessonova and S. Yu Yunusov, Khim.prir.Soedin., 1982, 532).

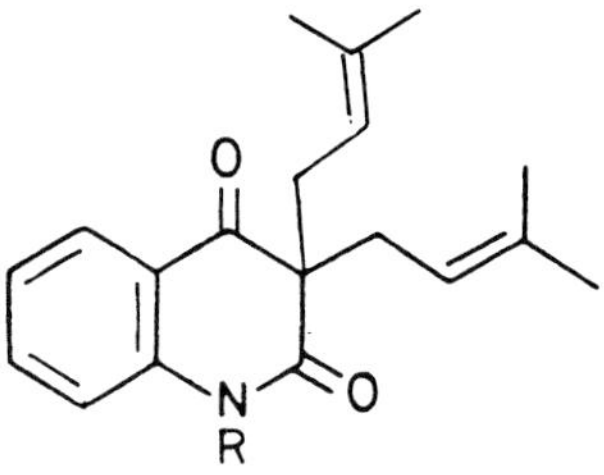

51

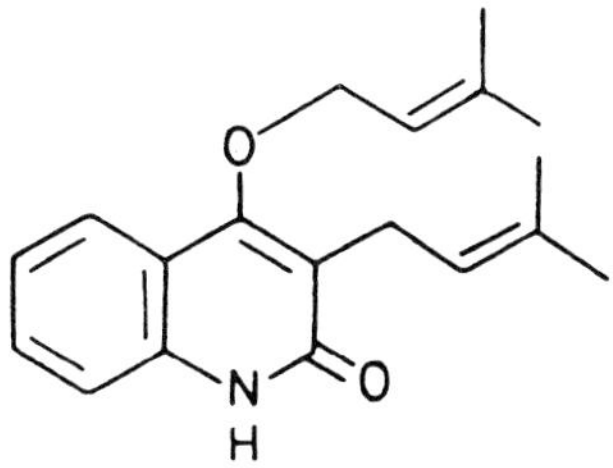

52

N-Methylbuchapsine (51, R≠Me) accompanies a number of established alkaloids in *Esenbeckia flava* (D.L. Dreyer, Phytochem., 1980, *19*, 941). The diol (53), together with the tricyclic compounds (54, R=H) and (54, R=Me), and the alkaloid praraensine (55) is present in *Euxylophora paraensis* (L. Jurd and M. Benson, Chem.Comm., 1983, 92; Jurd, Benson and R.Y. Wong, Austral.J.Chem., 1983, *36*, 759). These compounds are accompanied by the "dimeric" paraensidimerins A (57; α-Hd, α-He), B (56; R=CH$_2$C(OH)Me$_2$); C (57; α-Hd, β-He); D (56;R=CH=CMe$_2$); E (57; β-Hd, β-He); F (57; β-Hd, β-He), and G (58)(Jurd and Wong, ibid., 1981, *34*, 1625; Jurd, Wong and Benson, ibid., 1982, *35*, 2505).

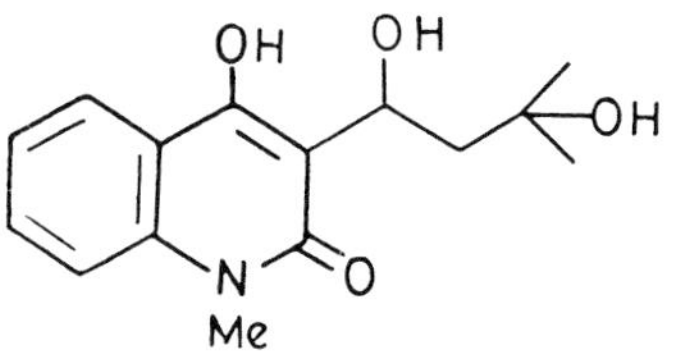

53

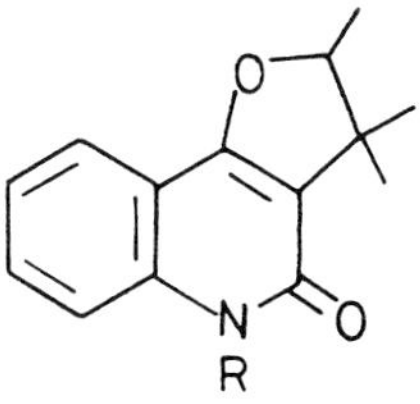

54

55

56

57

58

The methoxylated derivatives vepridimerines A (59; α-Hd, α-He) and B (59; α -Hd, β-He) and C (60; α-Hd, α -He) are present in *Vespris louisii* (T.B. Ngadjui et al., Tetrahedron Letters, 1982, *23*, 2041).

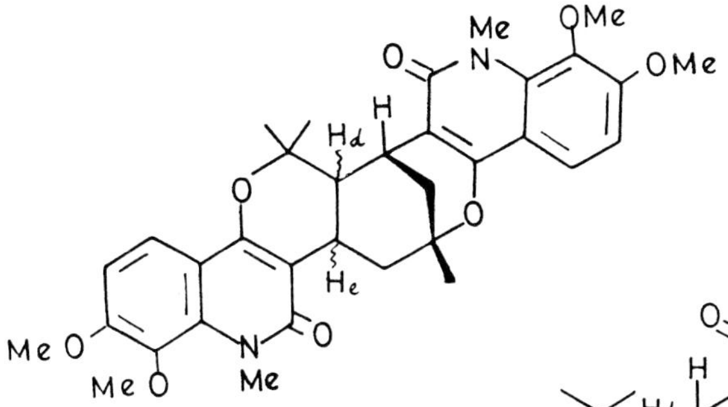

59

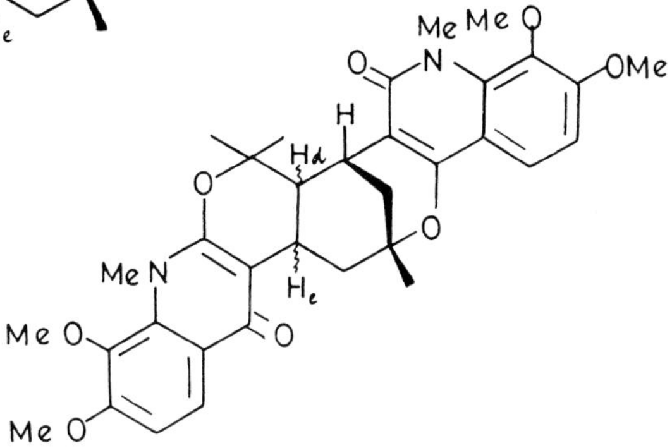

60

The structure determination of each of these compounds relies heavily upon spectroscopic analyses and single crystal X-ray diffraction studies. Vespridimerin D(60; α-Hd, β-He) occurs in *Oricia renieri* along with its analogues vespridimerins B and C (S.A. Khalid and P.G. Waterman, Phytochem., 1981, *20*, 2761).

Araliopsine (62), which is present in *Araliopsis soyauxii* (J. Vaquette et al., Phytochem., 1976, *15*, 743) and *Zanthoxyllum simulans* (Z. Chang et al., Yaoxuc Xuebao, 1981, *16*, 394; C.A., 1982, *97*, 20735) has been synthesised by G.M. Coppola (J. heterocyclic Chem.,1983, 1589) by reacting N-methylisatoic anhydride with the lithium enolate (61) and treatment of the intermediate epoxide with acid.

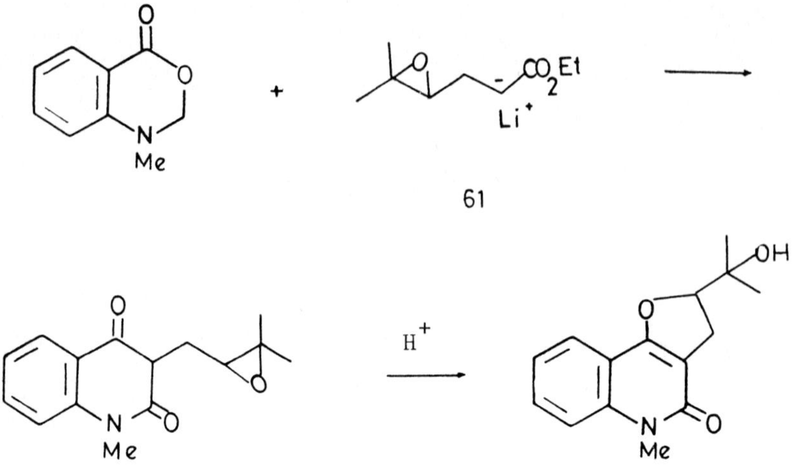

61

62

Pteledimeridine (63) and pteledimerine (64) from the root bark of *Ptelea trifoliata* (J. Reisch et al., Tetrahedron Letters, 1978, 3681; Meister et al., Ann., 1979, 1785) represent a half way stage in the "dimerisation" process of quinolone precursors in higher plants. The two alkaloids are isomeric, and differ only in the furoquinolinone unit which is based on a 2-quinolinone in pteledimeridine and a 4-quinolinone system in pteledimerine.

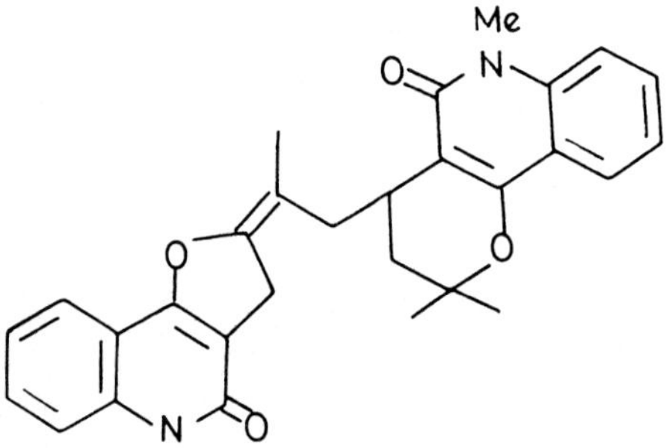

63

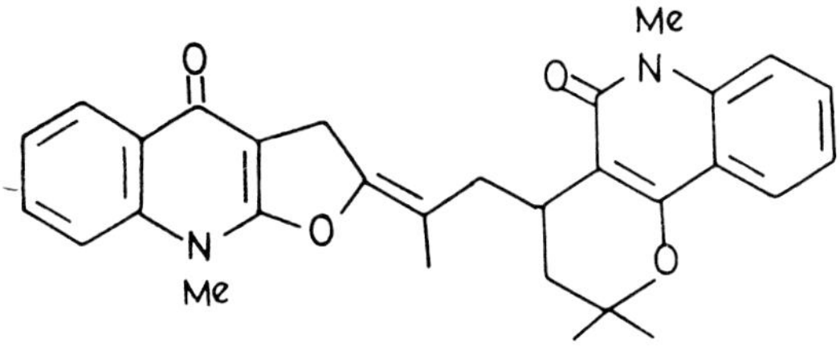

64

(c)   New Furoquinoline and dihydrofuroquinoline alkaloids

Tecleaverdoorine,   which   is   found   in   the   stem   bark   of
Teclea verdoorniana, has the structure (65).   The failure
of this alkaloid to afford a chroman derivative on acid
treatment eliminates other alternative formulations in which
the prenyl group and the phenolic hydroxyl group are *ortho*
to one another (J.I. Okogun and J.F. Ayafor, J.chem.Soc.,
Chem.Comm.,   1977,   652).   This   alkaloid   may   be
dihydroxylated in the side chain to give another extractive
of this plant, tecleaverdine, which must therefore have
structure (66) (Ayafor and Okogun, J.chem.Soc. Perkin I,
1982, 909).

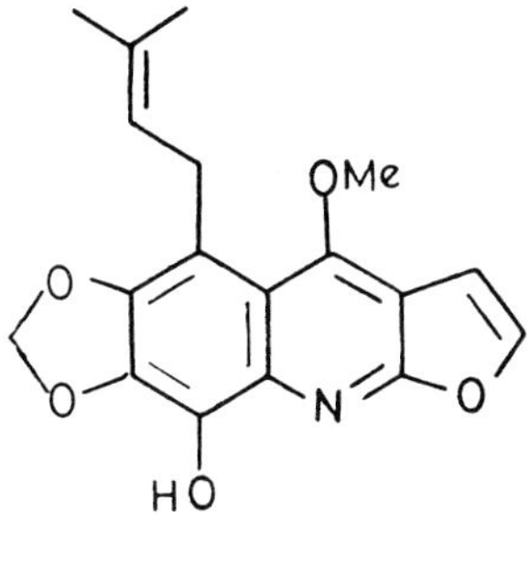

65

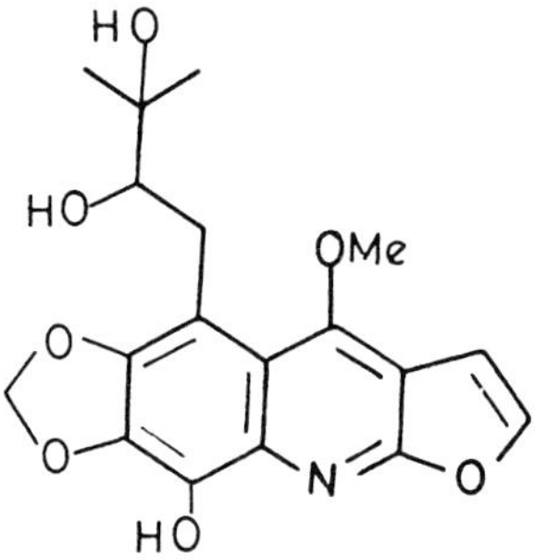

66

Tecleine, first obtained from *T. sudanica* (R.R. Paris and
A. Stambouli, C.rend.Med.Sci., 1959, *248*, 3736), is likely
to have the constitution (67) since it gives the known
alkaloid flindersiamine (68) on methylation. It is also
present in *T. verdooniana* (Ayafor and Okogun, loc.cit.).
Another alkaloid from this species is tecleamine (69)
(Ayafor et al., Phytochem., 1982, *21*, 2603).

67, R = H
68, R = Me
69, R =

Delbine, m.p. 229-231°C and montrifoline, m.p.245-247°C are
new furoquinoline alkaloids from *Monnieria trifoliata*. The
former yields kokusaginine (70) on methylation with
diazomethane, and is thus a hydroxymethoxydictamnine. Since
it is not identical with helipavifoline (71) delbine is
considered to be 6-hydroxy-7-methoxydictamnine (72).

Montrifoline is converted into delbine by fusion with alkali
and, as it is not the same as evolatine (73), it is assigned
structure (74) (J. Battacharyya and L.M. Serur,
Heterocycles, 1981, *16*, 371).

70

71

Heliparvifoline, m.p. 245-247°C, is metabolised by *Haplophyllum parvifolia* (P.T.O. Chang et al, J.pharm.Sci., 1976, *65*, 561). On O-methylation it too forms kokusaginine and was considered to be 7-hydroxy-6-methoxydictamnine. This conclusion is now confirmed by a synthesis (T. Sekiba, Bull.chem.Soc.Japan, 1978, *51*, 325).

Evolatine and heliparvifoline were claimed as natural products from *Monnieria sp.* (G. Moulis et al., Planta Med., 1981, *42*, 400), however, it is now clear that these compounds are really montrifoline and delbine (Battacharyya and Serur, Heterocycles, 1983, *20*, 1063).

Montrifoline (also called nkobisine) is also present in *Teclea ouabanguinensis* (J.F. Ayafor et al., Phytochem., 1982, *21*, 2603), *T. verdooniana* (Ayafor and J.I. Okogun, J.nat.Prod., 1982, *45*, 182) and in *Haplophyllum vulcanicum* (A. Patra et al., Heterocycles, 1984, *22*, 2821). This last plant also affords a new alkaloid-(+)-nigdenine (75).

72

73

74

75

Confusameline (76) is an extractive of the plant *Melicope confusa* (T.-H. Yang et al., J.pharm.Soc.Japan, 1971, *91*, 782; C.A., 1971, *75*, 95382). Confirmation of this structure is provided by a synthesis (scheme 1), which has been modified and applied to the formation of the known alkaloids robustine and haplophine (T. Sekiba, J.Sci. Hiroshima Univ.Ser.A.phys.Chem., 1976, *40*, 143).

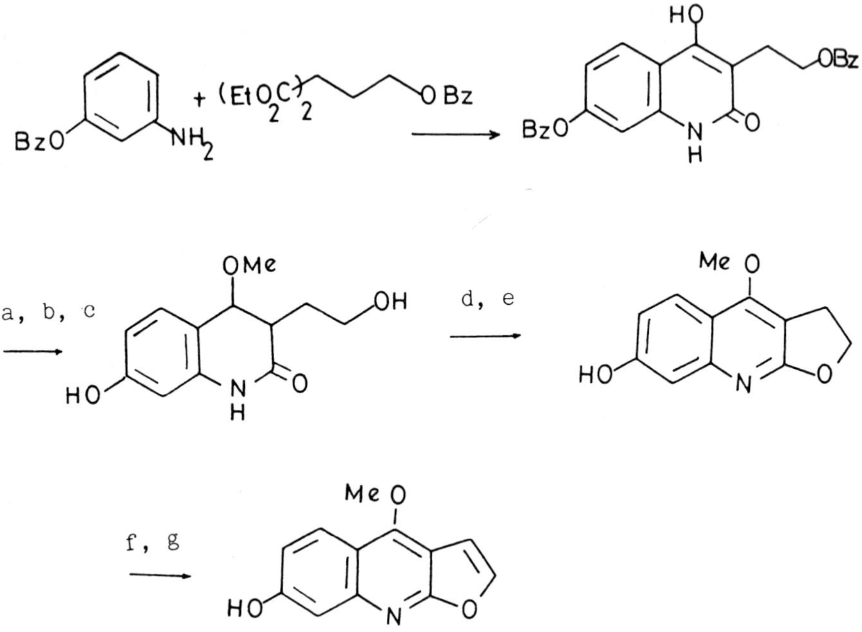

Scheme 1

Reagents: (a) Δ ; (b) CH_2N_2; (c) H_2-Pd;
(d) polyphosphoric ester; (e) $PhCH_2Cl$, NaOEt;
(f) DDQ; (g) HCl-EtOH

Dutadrupine, from *Dutaillyea* *drupacea*, contains a 2,2-dimethylpyrene ring and it has been assigned structure (77). This is confirmed by a synthesis of the alkaloid from confusamelin(76)(F. Tillequin, G. Baudouin and M. Koch, Heterocycles, 1982, *19*, 507) (scheme 2).

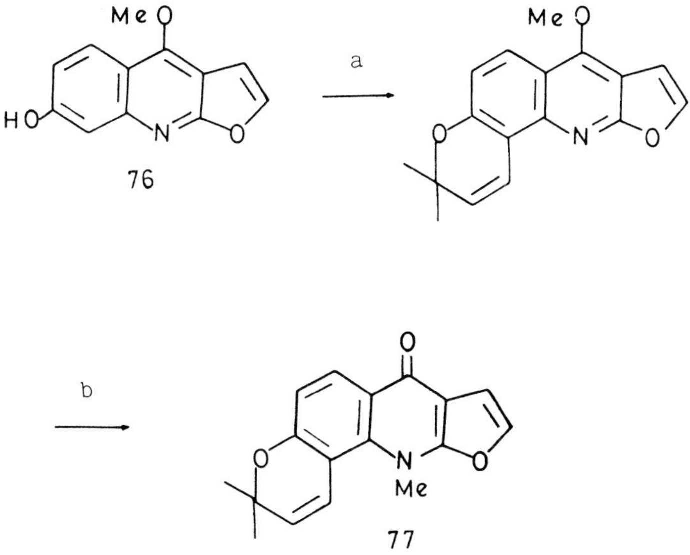

76

a

b

77

Scheme 2
Reagents: (a) Me$_2$C(Cl)C≡CH, K$_2$CO$_3$,KI; (b) MeI, $\triangle$.

8-Hydroxy-7-methoxydictamnine (78) is present in *Zanthoxylum arborescens*, together with its O-dimethylallyl derivative (79) (J.A. Grina, M.R. Ratcliff and F.R. Stermitz, J.org.Chem., 1982, *47*, 2648). Formerly structure (79) was allocated to the alkaloid perfamine, but since these compounds are not the same a new constitution for perfamine (80) has been proposed. This proposal is supported by some spectroscopic and degradative studies (D.M. Razakova, I.A. Bessonova and S. Yu Yunusov, Khim.prir.Soedin., 1983, 246; C.A., 1983, *99*, 22734).
Isomaculine (81), known already as a synthetic product, has been isolated from *Esenbeckia pilocarpoides* (F. Bevalot et al., Plant Med., 1984, *50*, 522).

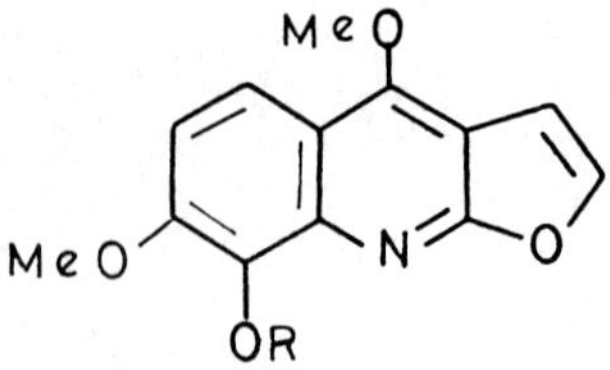

78, R=H

79, R =

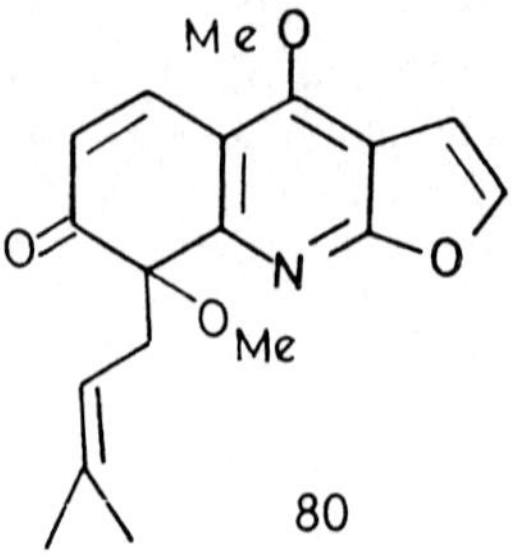

80

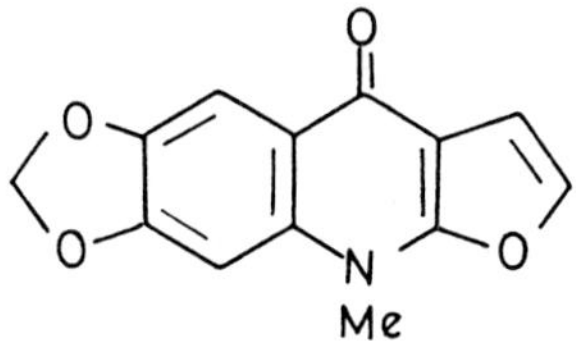

81

7-Methoxy-8-(3-methylbut-2-enyloxyl)-dictamnine (82) is obtained from *Zathoxylum arborescens* (J.A. Grina, M.R. Ratcliff and F.R. Stermitz, J.org.Chem., 1982, *47*, 2648).

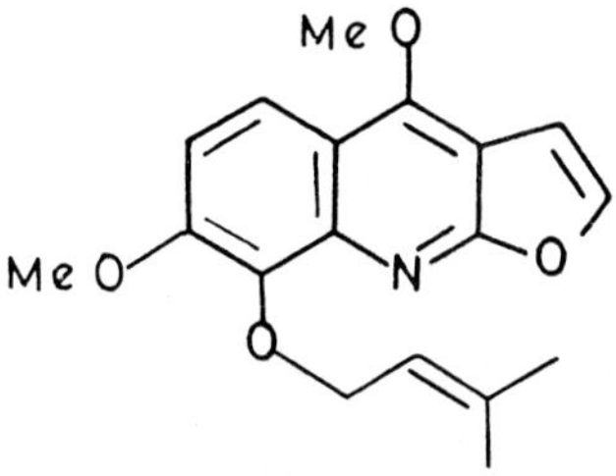

82

Four furoquinoline alkaloids containing a monosaccharide
unit are glycohaplopine (83) (Kh.A. Abdullaeva
I.A. Bessonova and S. Yu Yunusov, Khim.prir.Soedin., 1979,
873; C.A., 1980, *93*, 41504), triacetylglycoperine
(84,R=OAc), its diacetyl and monoacetyl derivatives
(idem.ibid., 1978, 219; C.A., 1978, *89*, 103 710). All are
obtained from the plant *Haplophyllum perforatum.* More
familiar. structures, from *H. buxaumii*, are the amorphous
alkaloids 4,5,6-trimethoxyfuroquinoline (85) and
4,5,7-trimethoxyfuroquinoline (86) (A. Ulubelen, Phytochem.,
1985, *24*,372).

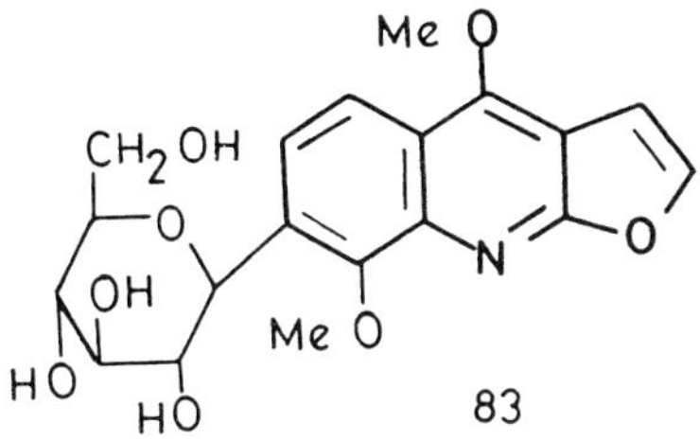

83

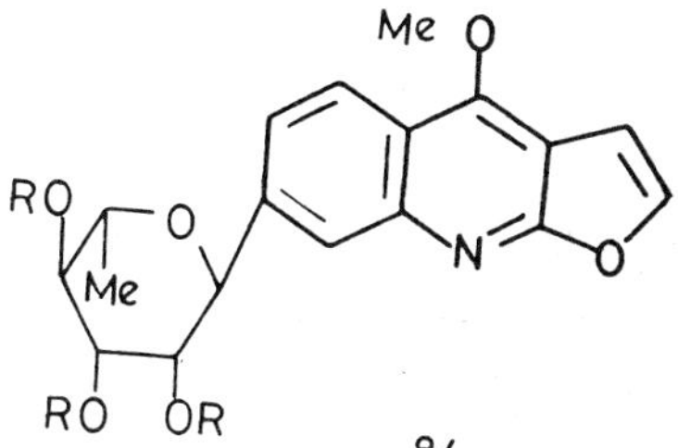

84

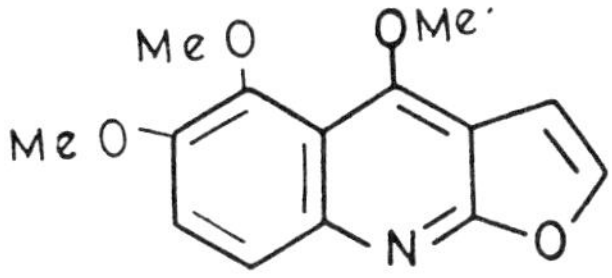

85

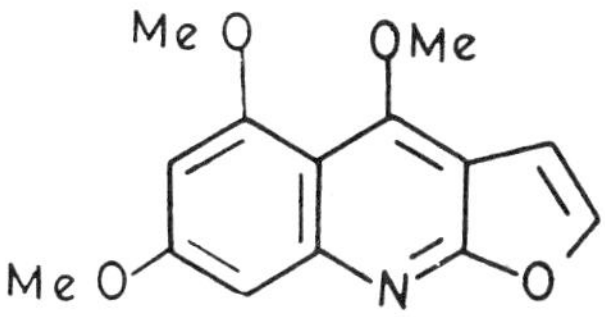

86

Dihydrofuroquinoline alkaloids are represented by
(-)-myrtopsine (87), from *Myrtopsis sellingi*, (M.S. Hifnawy,
et al., Plant Med., 1976, *29*, 346), vesprisinium salt (88),
from *Vespris louisii* (J.F. Ayafor, B.L. Sondengam and
B.T. Ngadjui, Tetrahedron Letters, 1981, *22*, 2685; Planta
Med., 1982, *44*, 139), and isoptelefolonium salt (89), from
Ptelea trifoliata (M. Rideau et al., Phytochem., 1979, *18*,
155). N-Methylplatdesminium salt (90) is also present in
Araliopsis tabonensis (P.G. Waterman, Biochemical
Systematics, 1973, *1*, 153).

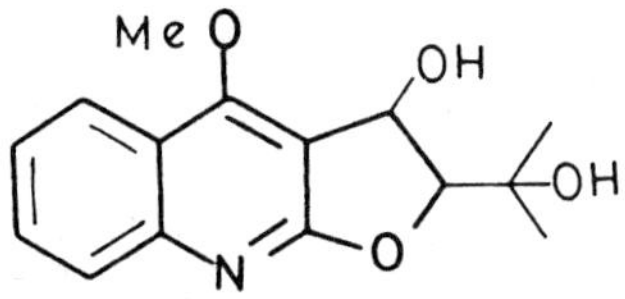

87

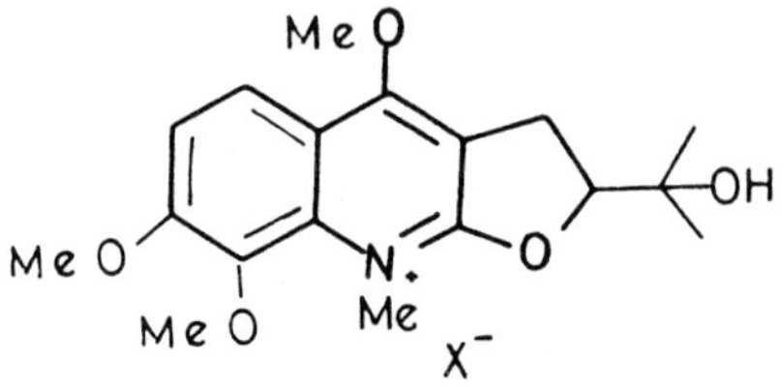

88

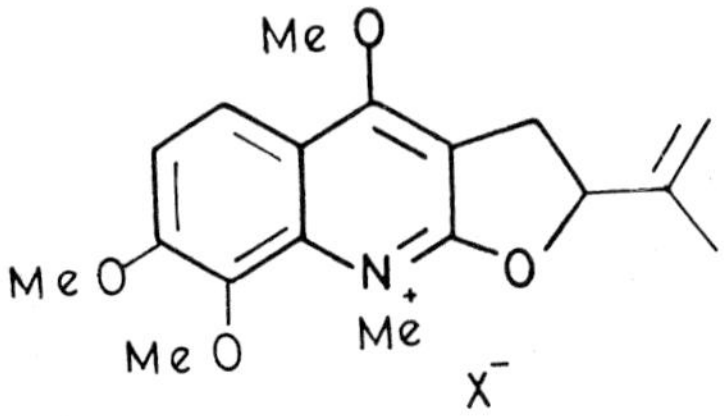

89

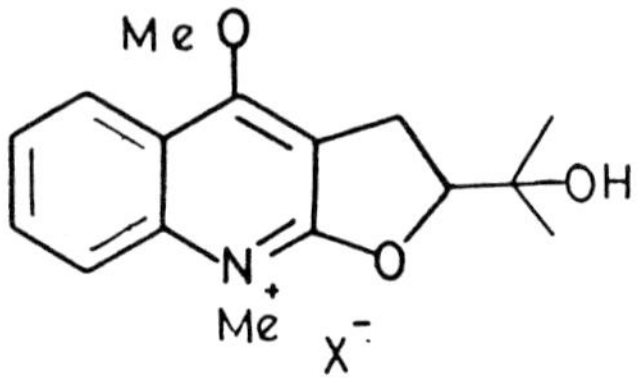

90

3. Non-Rutaceous miscellaneous alkaloids

Isokomarovine (91), a mixed β-carboline-quinoline, is obtained from *Nitraria komariovii* (T.S. Tulyaganov, A.A. Igragimov and S. Yu Yunusov, Khim.prir.Soedin., 1982, 635).

3-Methyltryptophan (92) has been shown to be a biosynthetic precursor of the antibiotic streptonigrin (93) in *Streptomyces flocculus* (J.K. Allen, K.D. Barrow and A.J. Jones, J.chem.Soc.Chem.Comm., 1979, 280). The antibiotic itself has been studied by other groups of chemists (S.J. Gould and S.M. Weinreb, Fortschr.chem. Org.Naturst., 1982, *41*, *cf*.77). Another antibiotic, streptonigrone (95), is metabolised by *Streptomyces sp.* (A.J. Helt, R.W. Rickards and J.-P. Wu, Antibiotics, 1985, *38*, 516), but lavendamycin (94), a related alkaloid from *S. lavendulae* (T.W. Doyle et al., Tetrahedron Letters, 1981, 4595), does not appear to be its natural precursor. Lavendamycin has been synthesised (A.S. Kende and F.H. Ebetino, Tetrahedron Letters, 1984, 923).

91

92

93

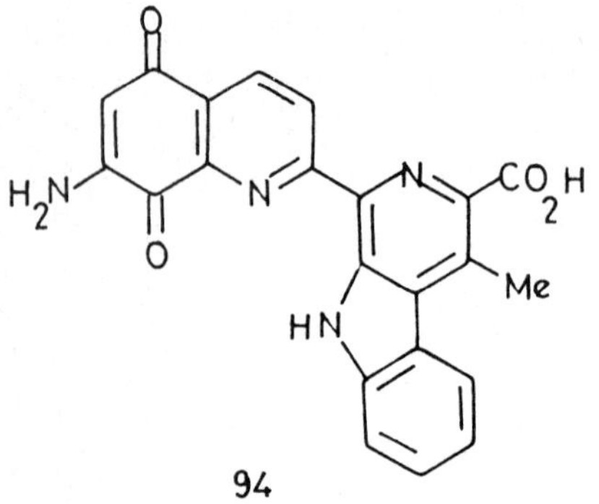

244

94

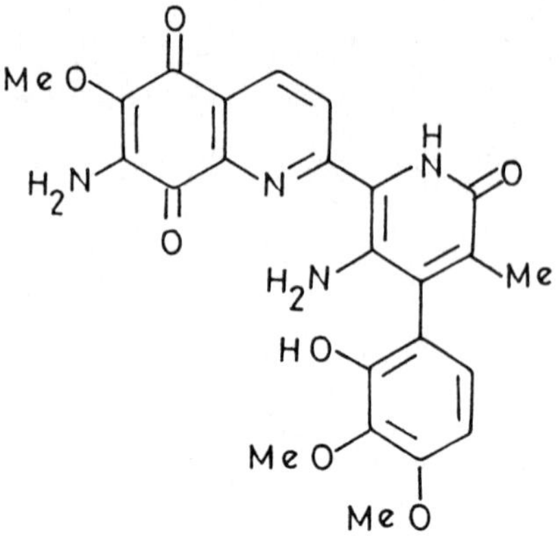

95

Decahydroquinoline "alkaloids" occur in the skins of Neotropical frogs of the *Dendrobatidae* genus (for a review see J.W. Daly, Fortschr.Chem.Org.Naturst., 1982, *41*, 205). These structures, of which pumiliotoxin-C (96) is a typical example, are obviously related to the pyridine bases described in Chapter 30. Gephyrotoxin (97) another natural product of this type has also been synthesised (Overman, D. Lesuisse and M. Hashimoto, J.Amer.chem.Soc., 1983, *105*, 5373; J. Royer and H.P. Husson, Tetrahedron Letters, 1985, *26*, 1515).

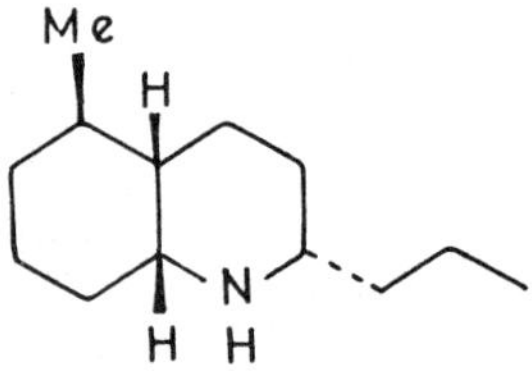

96

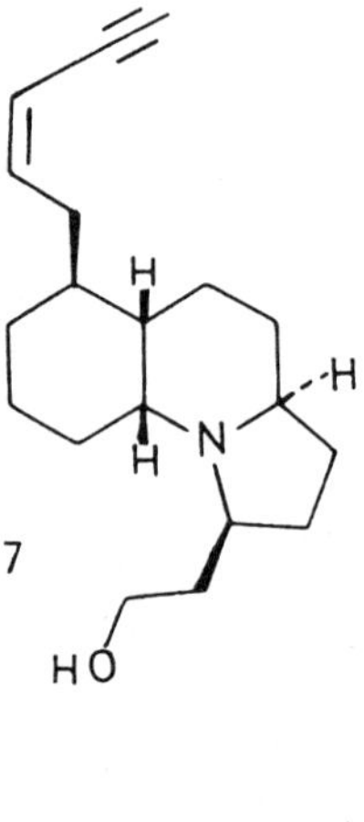

97

Chapter 32

THE ACRIDINE ALKALOIDS

MALCOLM SAINSBURY

The acridine, or acridone, alkaloids were discussed in both chapters 31 and 32 of the main work (C.C.C. 2nd Edn., Vol IVG, 1978). Here, however, these alkaloids are considered only in this section.

In the search for new alkaloids from novel sources it is inevitable that compounds of established structure are also isolated. Table 1 summarizes this work showing the alkaloid and its botanical origin; reference should be made to the second edition for details of the constitution of these alkaloids.

TABLE 1

Acridone alkaloids of known structure

Alkaloid name	Botanical Source (Genus)
Acridone	*Esenbeckia*[25], *Thamnosma*[4] *Toddalia*[15]
1-Hydroxyacridone	*Boenninghausia*[3]
1-Hydroxy-*N*-methylacridone	*Boenninghausia*[3]
1-Hydroxy-3-methoxy-*N*-methylacridone	*Boenninghausia*[23], *Esenbeckia*[16,25], *Ruta*[19]
1,3-Dimethoxyacridone	*Oricia*[13], *Vespris*[9]
1,3-Dimethoxy-*N*-methylacridone	*Diphasia*[1], *Oricia*[5], *Teclea*[2,5,7]
1,2,3-Trimethoxy-*N*-methylacridone	*Acronychia*[26], *Baurella*[22], *Sarcomelicope*[28] *Melicope*[8], *Vespris*[9]
1,2,4-Trimethoxy-*N*-methylacridone	*Melicope*[8], *Sarcomelicope*[28], *Teclea*[7]
Acronycine	*Acronychia*[26]
Noracronycine	*Acronychia*[26], *Boenninghausia*[11], *Murraya*[18]
Des-*N*-methylacronycine	*Murraya*[18]
Des-*N*-methylnoracronycine	*Murraya*[18]
Arborininine	*Oricia*[20], *Glycomis*[12,17], *Monnieria*[27], *Ruta*[6], *Teclea*[7], *Vespris*[9]

Evoxanthine	*Oricia* [13], *Teclea*[7]
Gravacridindondiol	*Ruta*[24]
Melicopidine	*Sarcomelicope*[28]
Melicopicine	*Baurella*[22], *Sarcomelicope*[28]
Melicopine	*Sarcomelicope*[28]
N-Methylatataphylline	*Atalantia*[21]
Rutacridone	*Boenninghausia*[11], *Ruta*[6]
Tecleanthine	*Teclea*[5,7]
Xanthevodine	*Melicope*[8], *Sarcomelicope*[28]

References

1. P.G. Waterman, Phytochem., 1975, *14*, 2092.
2. J. Vaquette *et al.*, Planta Med., 1975, *9*, 304.
3. Z. Rozsa *et al.*, Pharmazie, 1975, *30*, 753.
4. P.T.O. Chang *et al.*, Lloydia, 1976, *39*, 134.
5. F. Fish, I.A. Meshal and P.G. Waterman, J.pharm. Pharmacol., 1976, *28*,Suppl., p.72P.
6. G.A. Conzalez *et al.*, Anales de Quim, 1976, *72*, 94.
7. J. Vaquette *et al.*, Planta Med., 1978, *33*, 78.
8. A. Ahond *et al.*, Phytochem., 1978, *17*,166.
9. R. Hansel and E.-M. Cybulski, Arch.Pharm., 1978, *311*, 135.
10. G.A. Gonzalez *et al.*, Anales de Quim, 1977, *73*, 430.
11. Z. Rozsa *et al.*, Phytochem., 1978, *17*, 169.
12. I.H. Bowen, K.P.W.C. Perera and J.R. Lewis, Phytochem., 1978, *17*, 2125.
13. F. Fish, I.A. Meshal and P.G. Waterman, Planta Med., 1978, *33*, 228.
14. F. Tillequin *et al.*, Lloydia, 1980, *43*, 498.
15. J. Reisch and H. Strobel, Pharmazie, 1982, *37*, 862.
16. D.L. Dreyer, Phytochem., 1980, *19*, 941.
17. K. Rastogi, R.S. Kapil and S.P. Popli, Phytochem., 1980, *19*, 945.
18. M.Th. Flauvel *et al.*, Planta med.Phythother., 1978,*12*, 207.
19. J. Reisch *et al.*, Phytochem., 1977, *16*, 151.
20. S.A. Khalid and P.G. Waterman, Phytochem., 1981, *20*, 2761.
21. J.S. Shah and B.K. Sabata, Ind.J.Chem., 1982, *21B*, 16.
22. B. Counge *et al.*, Planta med.Phythother., 1980, *14*, 208.
23. Z. Rozsa *et al.*, Fitoterapia, 1981, *52*, 37.
24. Z. Rozsa *et al.*, Fitoterapia, 1981, *52*, 93.
25. F. Bevalot *et al.*, Planta Med., 1984, *50*, 523.
26. S. Funayama and G.A. Cordell, J.nat.Prod., 1984, *47*, 285.
27. J. Bhattacharyya and L.M. Serur, J.nat.Prod., 1984, *47*, 285.
28. G. Baudousri *et al.*, J.nat.Prod., 1985, *48*, 260.

1. New alkaloids

Gravacridone triol (1, R=H) represents a new alkaloid from
Ruta graveolens, where it co-occurs with its monoglucosyl
derivative (1, R=Glu) and the glucoside (2, R=Glu) of a known
alkaloid gravacridone diol (2, R=H) (J. Reisch *et al.*,
Phytochem., 1976, *15*, 240). A new chlorine containing acridone
(3) known as isogravacridonchlorine is present in this plant,
together with the unnamed alkaloid (4) (*idem.*, *ibid.*, 1977,
16, 151). This last alkaloid is also found in *Glycosmis
citrifolia* (T.-S. Wu, H. Furukawa and K.-S. Hsu, Heterocycles,
1982, *19*, 1227).

It is likely that hydroxyrutacridone (5) has gravacridone diol (2, R=H) as its immediate biological precursor and the counterpart to gravacridone triol - the hydroxy epoxide (6) - has been isolated from the roots of R. graveolens and also from cultures of the roots, where it is accompanied by gravacridonal (7) (Z. Rozsa et al., Fitoterapia, 1981, 52, 93; V. Eilert et al., Z. Naturforsch.,1982, 37 C., 132).

The original structure (8) for rutacridone from R. graveolens has been revised, following a reinterpretation of the [1]H n.m.r. spectrum of a sample isolated from a sister species, R. chalepensis. In particular the resonance of the C-5 hydroxyl group proton occurs at very low field (δ 15.2 ppm) and the hydroxyl group must therefore be hydrogen bonded to the acridone carbonyl function. At one time the linear representation (9) (G.A. Gonzalez et al., Anales de Quim., 1976, 72, 94) was preferred over the alternative (10), again on the basis of chemical shift data, but this suggestion was subsequently refuted (J. Reisch, Z. Rozsa and I. Mester, Z.Naturforsch., 1978, *33* B, 957). Structure (9) is now accepted for the alkaloid since it has been synthesised by several groups (Reisch *et al.*, Ann., 1981, 85; Mester *et al.*, Heterocycles, 1981, *16*, 77; J.H. Adams *et al.*, Tetrahedron, 1981, 7, 209) and shown to be the same as the natural product. Yet another alkaloid is rutacridone epoxide (11), which is isolated from callus tissue cultures, of *R. graveolens* (A. Nahrstedt *et al.*, Z.Naturforsch., 1981, *36* C, 200).

The genus *Teclea* is another rich source of alkaloids; from *T. bovincana* come the new acridones, 6-methoxytecleanthine (12), 1,3,5-trimethoxy-10-methylacridone (13) (J. Vaquette *et al.*, Plant.med.Phytother., 1974, *8*, 57) and 1,3,4-trimethoxy-10-methylacridone (14) (*idem.*, Planta Med., 1978, *33*, 78).

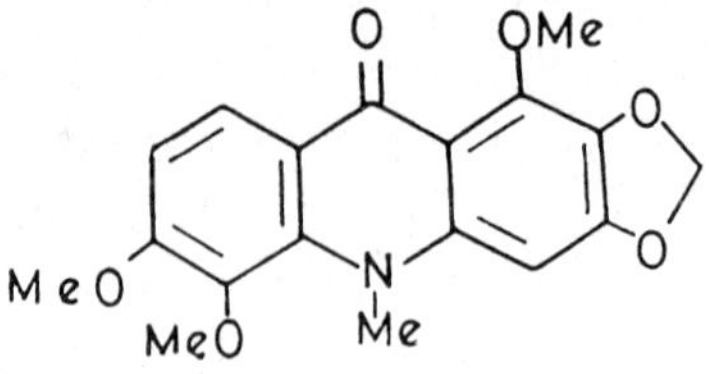

11-Hydroxynoracronycine (16) occurs in the plant *Atalantia coylonica*, it is also a metabolite of acronycine in mammals. A part synthesis from 1,3-dihydroxy-5-methoxy-9-acridone (15) (outlined in Scheme 1) confirms this structural assignment (J.H. Adams, P.T. Bruce and J.R. Lewis, Lloydia, 1976, *39*, 399).

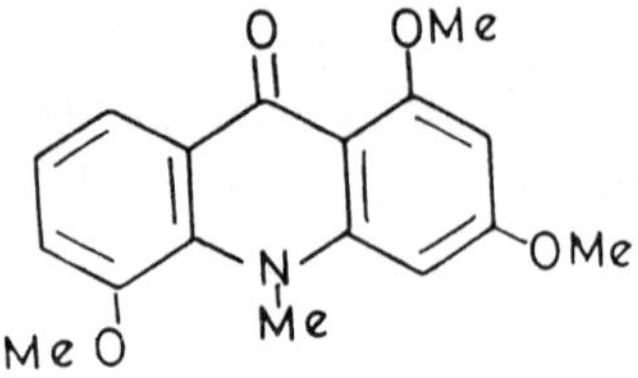

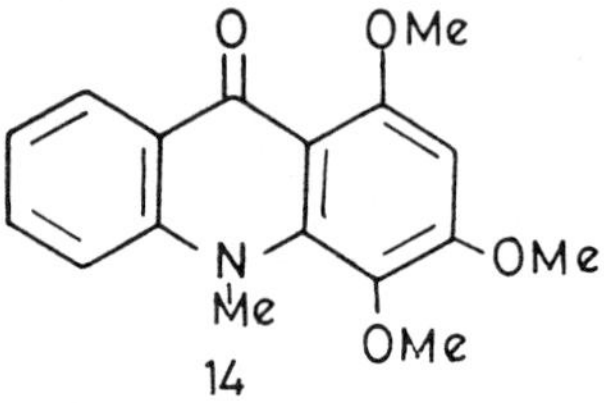

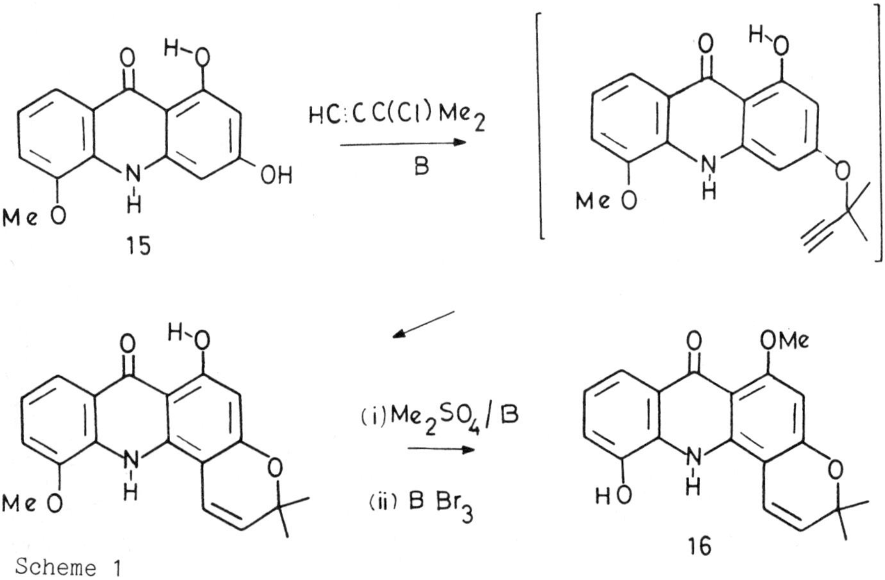

Scheme 1

The alkaloid (17; R=H) is an extractive from *A. monophylla*;
it was given the name atalaphyllidine (A. Chatterjee and D.
Ganguly, Phytochem., 1976, *15*, 1303), but this has already
been reserved for the different acridone (18) isolated from
the same plant (S.C. Basa, Experientia, 1975, *31*, 1387).
The 3, 5-dimethyl ether (17, R=Me) is also present in *A.
monophylla* (G. H. Kulkarni and B. K. Sabata, Phytochem., *1981,
20,867*)

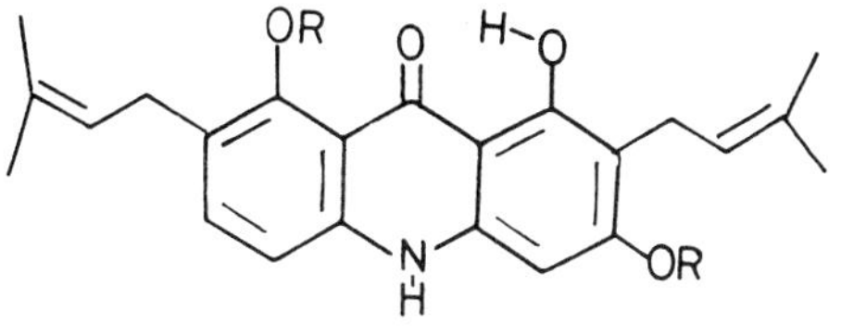

17

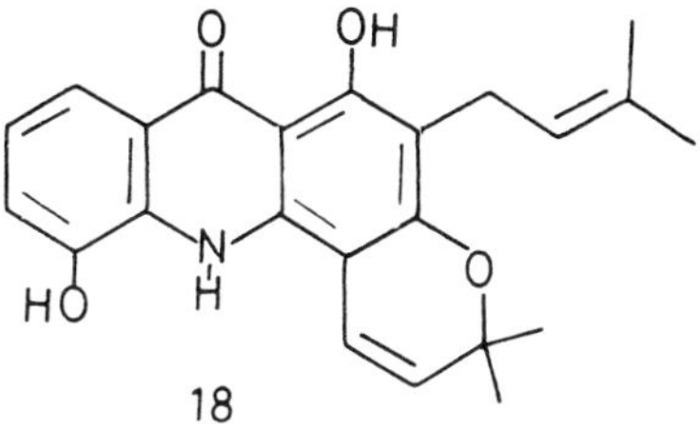

18

Atalaphyllidine also occurs in *Severinia buxifolia*
together with its *N*-methyl derivative (19, R^1=OH, R^2=Me),
severifoline (19, R^1=R^2=H) and *N*-methyl severifoline (19, R^1=H,
R^2=Me) T.-S. Wu, C.-S. Kuoh and H. Furukawa, Phytochem., 1982,
21, 1771).

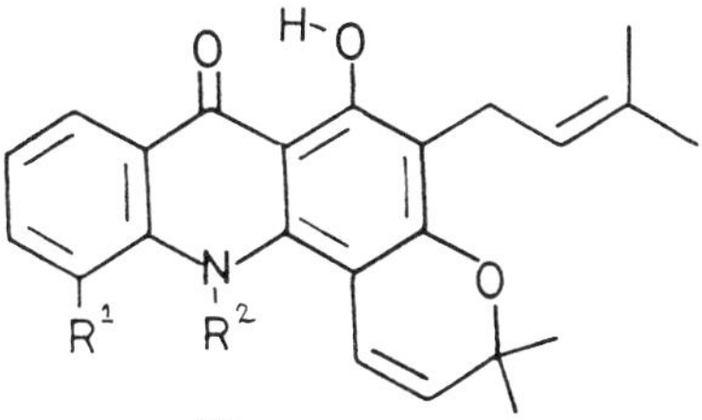

19

Normelicopine (20) is a new alkaloid from *Acronychia baueri*
(S. Funayama and G.A. Cordell, J.nat.Prod., 1984, *47*, 285).

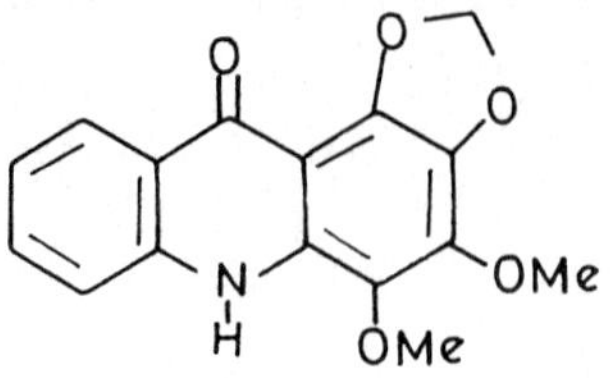

20

The 5-hydroxy derivative (21, R=OH) of the known alkaloid
arborinine (21, R=H) has been obtained for the first time
from the leaves of *Glycosmis bilocularis* (I.H. Bowen, K.P.W.C.
Perera and J.R. Lewis, Phytochem., 1978, *17*, 2125). Two
other acridones differing only in the presence or absence of
methoxyl groups are the alkaloids (22; R=H) and (22; R=OMe)
isolated from the leaves of *Bauerella simplicifolia* (F.
Tillequin *et al.*, J.nat.Prod., 1980, *43*, 498).

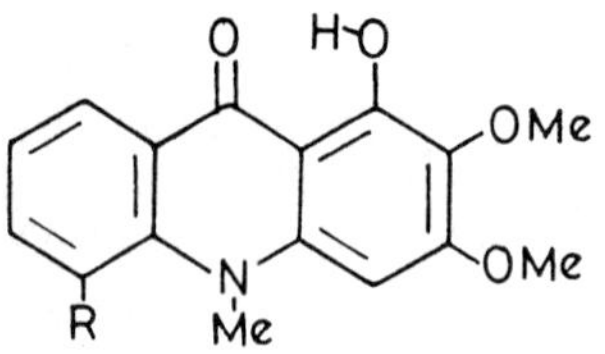

21

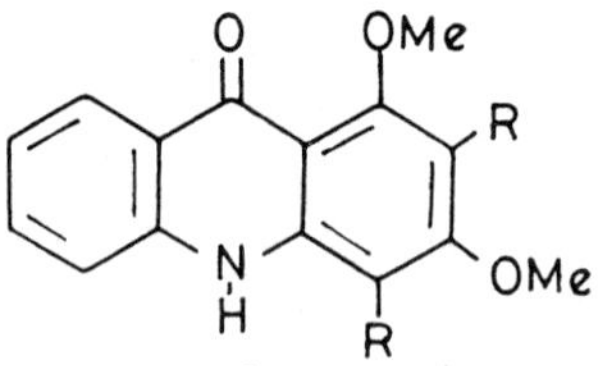

22

Boenninghausenia albiflora produces 1-hydroxyacridone (23) and
possibly also 1,7-dihydroxyacridone (24) (Z. Rozsa *et al.*,
Pharmazie, 1975, *30*, 753). Two new acridones, 1-hydroxy-3,4-
dimethoxy-10-methylacridone (25) and 1-hydroxy-3-geranyloxy-4-
methoxy-10-methylacridone (26) occur in extracts of the plant
Sarcomelicope leiocarpa indigenous to New Caledonia together
with eight other known structures (G. Baudouin *et al.*, J.nat.
Prod., 1985, *48*, 260).

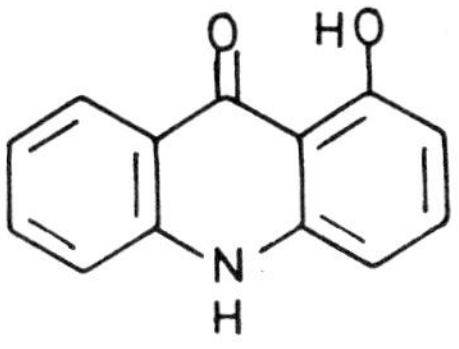

23

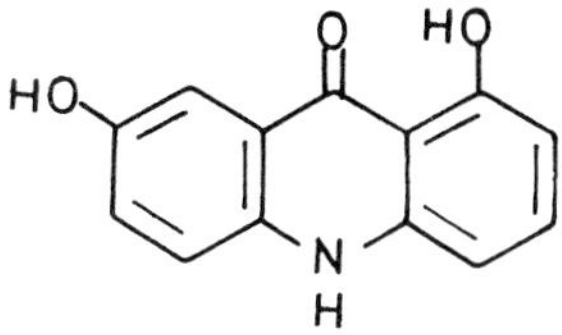

24

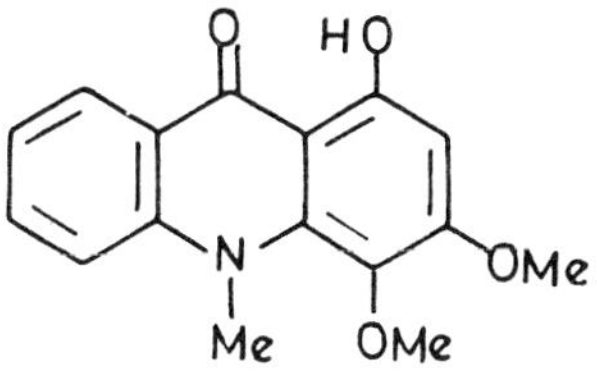

25

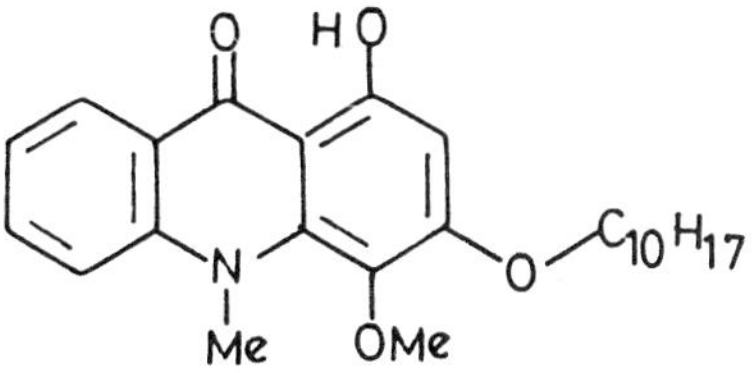

26

Another series of related alkaloids are produced by the plant *Glycosmis citrifolia*; these include the acridones glycocitrine-I (27), glycocitrine-II (28), its *O*-methyl ether (29), glyfoline (30), furofoline-II (31) (T.-S. Wu, H. Furukawa and C.-S. Kuoh, Heterocycles, 1982, *19*, 1047), pyranofoline (32) (*idem., ibid.*, p.1227) and the unnamed structure (33) (Wu and Furukawa *ibid.*, p.825).

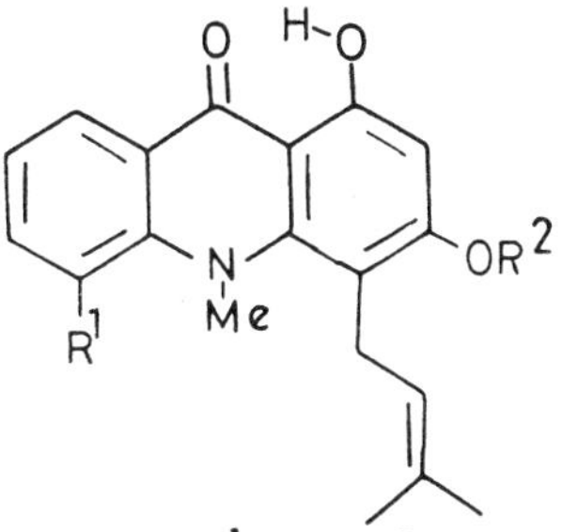

27, R¹=OH, R²=Me
28, R¹= R²= H
29, R¹=H, R²= Me

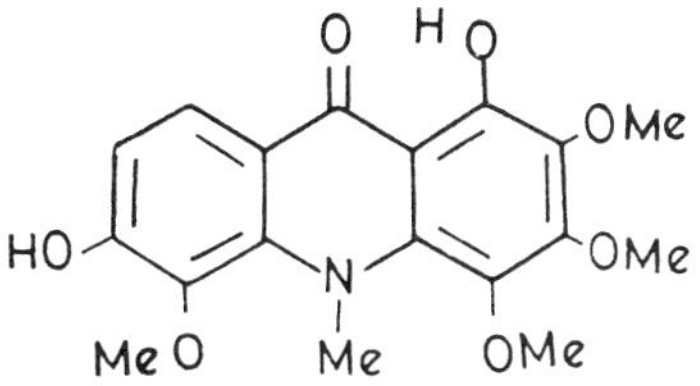

30

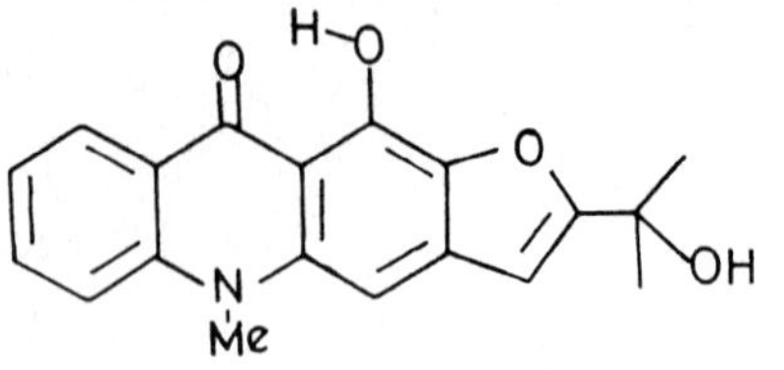

31

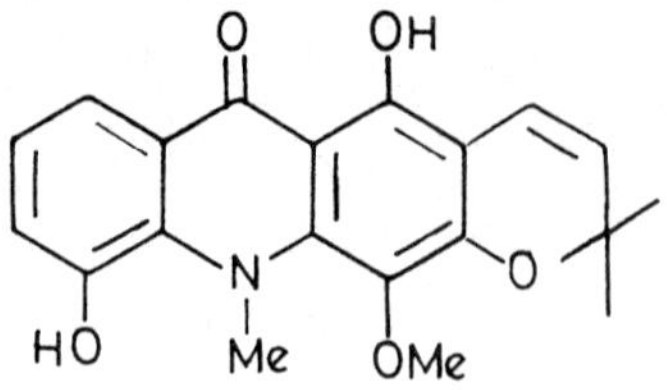

32

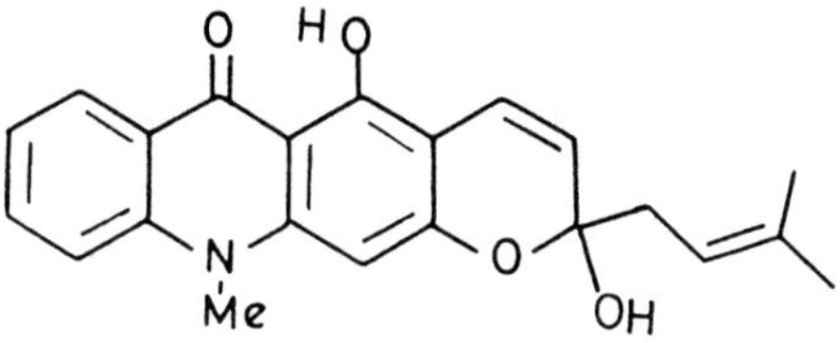

33

The root bark of *Citrus depressa* is the source of six
alkaloids, 5-hydroxy-noracryonycine (34), citpressine-I
(35) and citpressine-II (36), citracridone-I (38),
citracridone-II (39) and prenylcitpressine (37) (T.-S. Wu,
H. Furukawa and C.-S. Kuoh, Heterocycles, 1982, *19*, 273;
Chem.pharm.Bull., 1983, *31*, 895). These compounds also
occur in *C.grandis* accompanied by five others - grandisine-
I (41) and grandisine-II (42), grandisinine (45),
glycocitrine-1 (46), and citrusinine-I (43) (*idem.*,
Phytochem., 1983, *22*, 1493). Citrusinine-II (44) and
citbrasine (47) have been isolated from *C. sinensis* (Wu
and Furukawa, Chem.pharm.Bull., 1983, *31*, 901), and 2',2'-
dimethyl-(pyrano 5',6':3:4)-1,5-dihydroxy-6-methoxyacridone
(40) is a new alkaloid from *C. decumana* (S.C. Basa and R.N.
Tripathy, J.nat.Prod., 1984, *47*, 379).

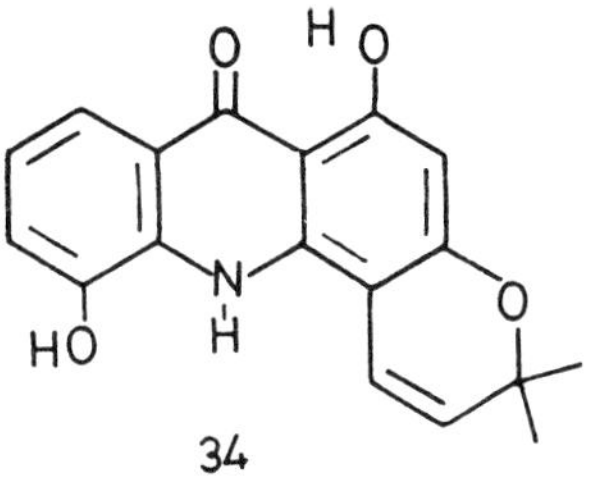

34

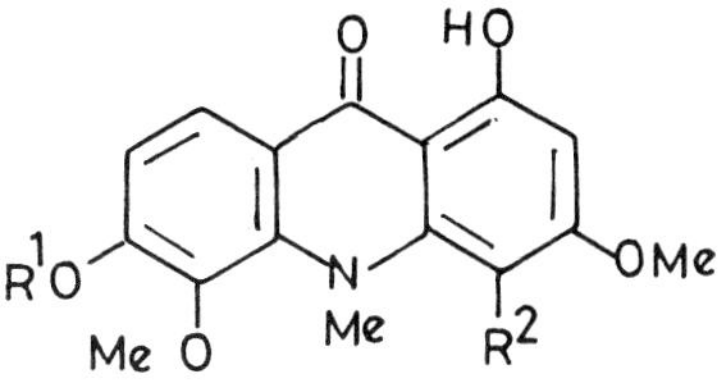

35, $R^1 = R^2 = H$
36, $R^1 = Me$; $R^2 = H$
37, $R^1 = H$; $R^2 = $ ⟶

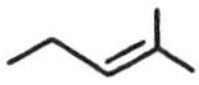

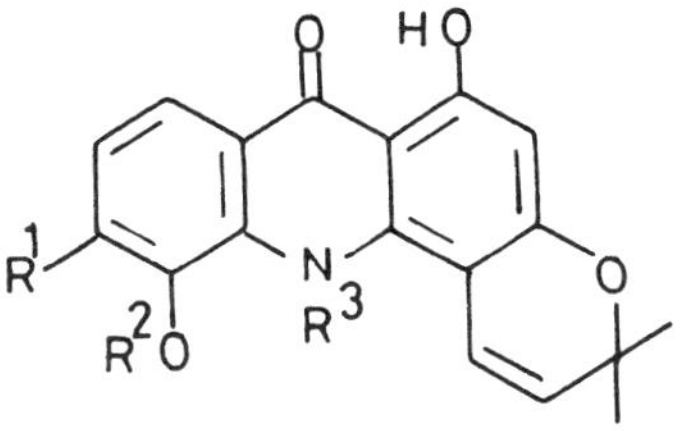

38, $R^1 = OH$; $R^2 = R^3 = H$
39, $R^1 = OMe$; $R^2 = R^3 = Me$
40, $R^1 = OMe$; $R^2 = H$; $R^3 = Me$

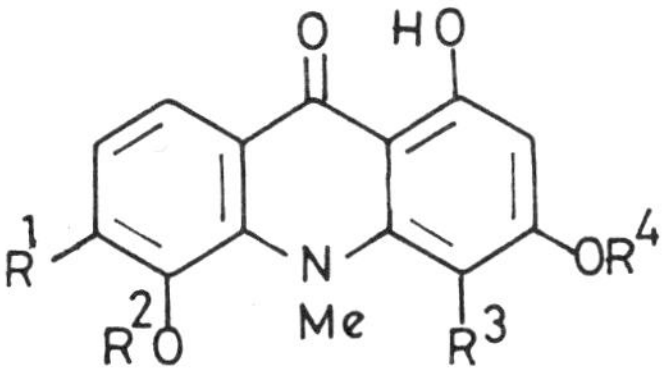

41, $R^1 = OMe$; $R^2 = R^3 = H$; $R^4 = OMe$
42, $R^1 = OMe$; $R^2 = Me$; $R^3 = R^4 = H$
43, $R^1 = R^2 = H$; $R^3 = OMe$; $R^4 = Me$
44, $R^1 = R^2 = R^4 = H$; $R^3 = OMe$

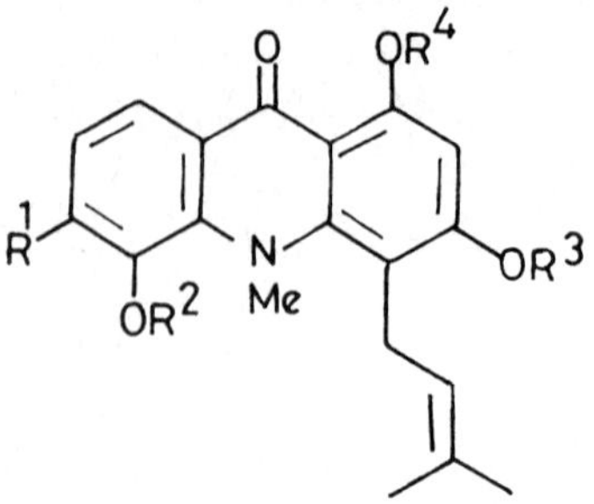

45, R^1=OH ; R^2= R^3=Me ; R^4= H
46, R^1= R^2= R^4= H; R^3= Me

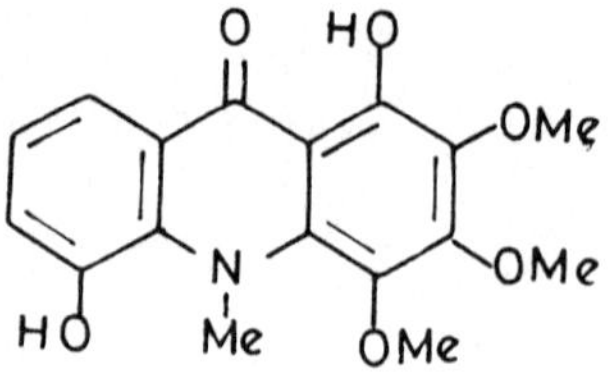

47

Natsucitrines-I (48, R=H) and -II (48, R=Me) occur in
C. natsudaidai (M. Ju-ichi, M. Inoue and Y. Fujitani,
Heterocycles, 1985, *23*, 1131), while the first tropone
derived homoacridones, citropones-A (49) and -B (50),
have been isolated from *C. grandis* (A.T. McPhail *et al.*,
Tetrahedron Letters, 1985, *26*, 3271).

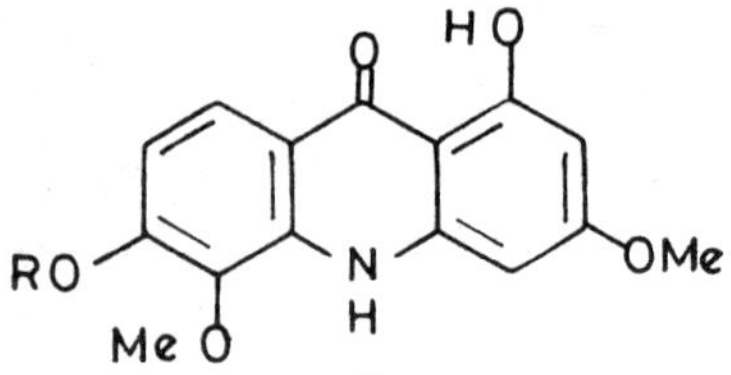

48

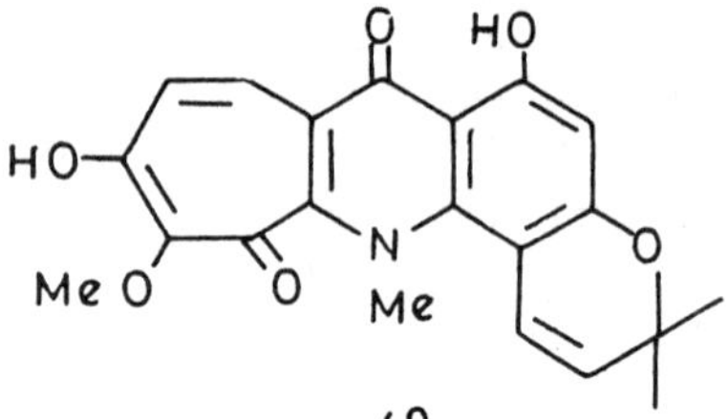

49

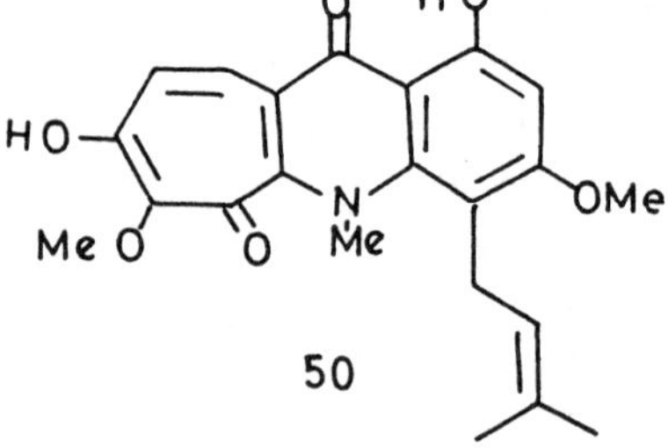

50

2. *Acronycine dimers and trimers*

Acronycine was first reported from the bark of the Australian scrub ash *Acronychia baueri* Schott. (see C.C.C. 2nd Edn.,Vol. IVG, p.191) and has subsequently generated a good deal of interest because of its activity against a broad spectrum of tumours. However, there are few reports of its actual use in human medicine.

An interesting development in the chemistry of this alkaloid is the discovery that when acronycine is demethylated with methanolic hydrochloric acid the product noracronycine reacts further to give dimers (51) and (52), a trimer (53) and oligomers (S. Funayama and G.A. Cordell, Planta Med., 1983, *48*, 263; J.org.Chem., 1985, *50*, 1737).

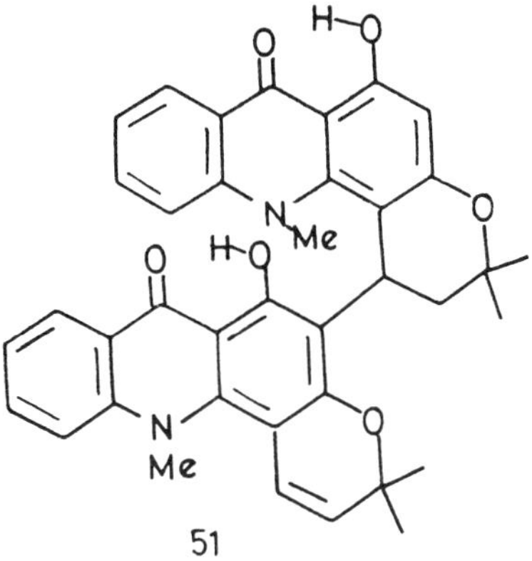

51

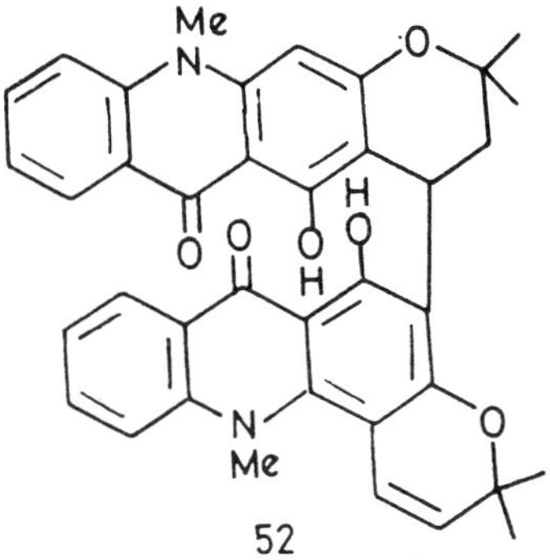

52

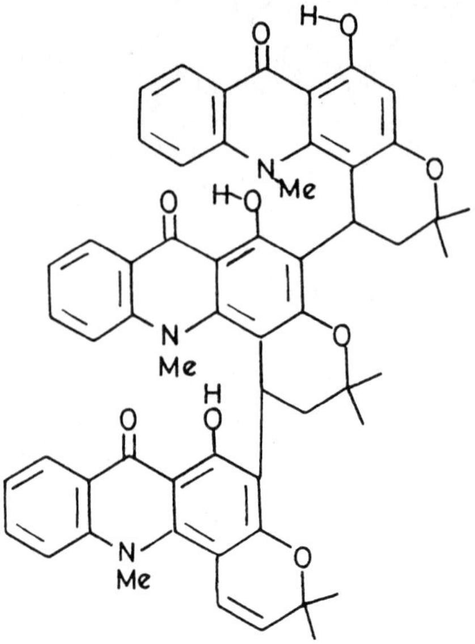

53

Related to these structures is the natural "dimer"
glycobismine A (54) first isolated from *Glycosmis citrifolia*
by H. Furukawa *et al.*, (Chem.pharm.Bull., 1984, *32*, 1647).

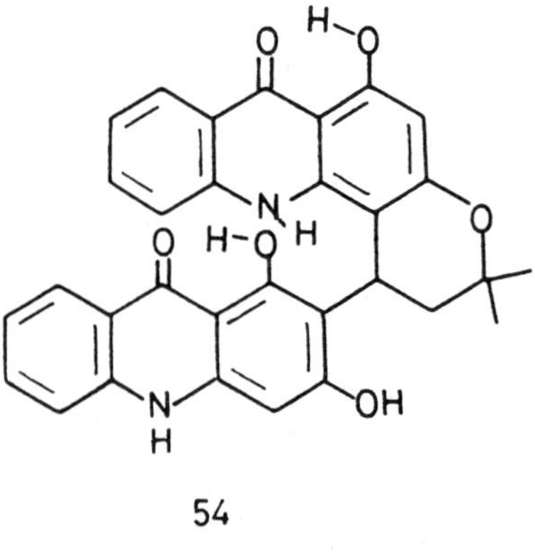

54

Chapter 33

THE ISOQUINOLINE ALKALOIDS

K.W. BENTLEY

Developments in the chemistry of the alkaloids of this group
have been annually reviewed in Specialist Periodical Reports
of the Royal Society of Chemistry "The Alkaloids" Vols. 7
(1975-6), 8 (1976-7), 9 (1977-8), 10 (1978-9), 11 (1979-80),
12 (1980-1) and 13 (1981-2) and subsequently in Natural
Product Reports 1984, 1, 355 (1982-3), 1985, 2, 81 (1983-4)
and 1986, 3, 153, (1984-5).

1. Simple isoquinolines, dihydro- and tetrahydroisoquinolines

The range of simple isoquinoline alkaloids, principally obtained from cacti, has recently been greatly expanded by the
isolation of isoquinolines, 3,4-dihydroisoquinolines and
1,2,3,4-tetrahydroisoquinolines bearing one, two, three and

TABLE 1

Isoquinoline Alkaloids

	R^1	R^2	R^3	R^4	R^5
Backebergine	H	OMe	OMe	H	H
Isobackebergine	H	H	OMe	OMe	H
Isosalsolidine	H	OMe	OMe	H	Me
Isonortehuanine	OMe	OMe	OMe	H	H
Isonorweberine	OMe	OMe	OMe	OMe	H
Isopachycereine	OMe	OMe	OMe	OMe	Me

TABLE 2

Dihydroisoquinoline Alkaloids

	R^1	R^2	R^3	R^4	R^5
Dehydroheliamine	H	OMe	OMe	H	H
Dehydrolemaireocereine	H	H	OMe	OMe	H
Dehydrosalsolidine	H	OMe	OMe	H	Me
Dehydronortehuanine	OMe	OMe	OMe	H	H
Dehydronorioeberine	OMe	OMe	OMe	OMe	H
Dehydropachycereine	OMe	OMe	OMe	OMe	Me

TABLE 3

Tetrahydroisoquinoline Alkaloids

	R^1	R^2	R^3	R^4	R^5	R^6
Weberidine	H	H	OMe	H	H	H
Heliamine	H	OMe	OMe	H	H	H
N-Methylheliamine	H	OMe	OMe	H	H	Me
Lemaireoereine	H	H	OMe	OMe	H	H
Carnegine	H	OMe	OMe	H	Me	Me
Nortehuanine	OMe	OMe	OMe	H	H	H
Tehuanine	OMe	OMe	OMe	H	H	Me
Anhalidine	H	OMe	OMe	OH	H	Me
Anhalonidine	H	OMe	OMe	OH	Me	H
Pellotine	H	OMe	OMe	OH	Me	Me
O-Methylpellotine	H	OMe	OMe	OMe	Me	Me
Norioeberine	OMe	OMe	OMe	OMe	H	H
Weberine	OMe	OMe	OMe	OMe	H	Me
Pachycereine	OMe	OMe	OMe	OMe	Me	H
N-Methylpachycereine	OMe	OMe	OMe	OMe	Me	Me

four oxygen substituents in positions 5,6,7 and 8, with and
without a methyl group at position 1 and as secondary bases,
tertiary bases, quaternary salts and N-oxides. Using tandem
mass spectrometry for both separation and identification of

alkaloids, very small quantities of plant material can be examined and in this way the twenty seven alkaloids listed in tables 1, 2 and 3 were identified in *Pachycereus weberi* (R.A. Roush *et al.* Anal.Chem., 1985, 57, 109). Quaternary salts are exemplified by pycnarrhine (1) and *N*-oxides by tehuanine-*N*-oxide and deglucopterocereine-*N*-oxide (2).

MeO
HO
Me
+N—Me

(1)

OH
MeO
MeO
Me
+N—O-
CH$_2$OH

(2)

MeO
MeO
O
O

(3)

R^{1}O
R^{2}O
NMe

(4)

O
MeO
AcO
NMe

(5)

OAc
MeO
O
NMe

(6)

MeO
MeO
+N—BH$_3$

(7)

AlBu$^i_3$
MeO
MeO
N—BH$_3$
Me

(8)

R
MeO
MeO
N—CO$_2$Et
Me

(9)

N-Methylheliamine has been synthesised from the isochromanone (3) by ring opening with hydrogen bromide, followed by treatment with methylamine and reduction of the resulting lactam (G.D. Pandey and K.P. Tiwari, Pol.J.Chem., 1979, 53, 2159). Isocorypalline (4, R^1=H, R^2=Me) has been oxidised by lead (IV) -acetate to the quinone acetal (5) which suffers Thiele acetylation to a product that may be hydrolysed and *O*-methylated to tehuanine (H. Hara *et al.*, Heterocycles, 1982, 17 (Special Issue), 293. Corypalline (4, R^1=Me, R^2=H) with lead tetra-

acetate affords the dienone (6), which reacts with veratrole
and with corypalline in trifluoroacetic acid to give 8-vera-
trycorypalline and 8,8'-dicorypalline respectively (Hara, O.
Hoshino and B. Umezawa, *ibid*, 1981, 15, 911). 6,7-Dimethoxy-
isoquinoline and diborane afford the complex (7) which with
lithium methyl yields the 1-methyl-1,2-dihydro compound and
this reacts with di-isobutylaluminium hydride to give (8).
 Treatment of (8) with ethyl chloroformate gives the urethane
(9, R=AlBu$_2^2$), converted by water into (9, R=H), which is reduced by lithium aluminium hydride to carnegine (D.J. Brooks
et al., J.org.Chem., 1984, 49, 130).

(10)

(11)

(12)

TABLE 4

5-Naphthylisoquinoline Alkaloids (11)

	R^1	R^2	R^3	R^4	Variants
Ancistrocladine	OMe	H	Me	H	} rotational
Hamatine	OMe	H	Me	H	isomers
O-Methylancistrocladine	OMe	Me	Me	H	
Ancistroealaensine	OMe	Me	Me	H	C-1 α-Me
Ancistrocline	OMe	H	Me	Me	
Ancistrocladonine	OMe	Me	Me	Me	C-1 α-Me
Ancistrocladinine	OMe	H	Me	-	1,2-dehydro
Ancistrocladeine	OMe	H	Me	-	1,2,3,4-dehydro
Ancistrocongolensine	OMe	H	Me	H	} 3,4-dehydro
Ancistrocongine	H	H	H	H	no C-1 Me

TABLE 5

7-Naphthylisoquinoline Alkaloids (12)

	R^1	R^2	R^3	R^4	Variants
Ancistine	Me	OMe	H	H	
Ancistrine	Me	OH	Me	H	
Ancistrocladisine	Me	OMe	Me	-	1,2-dehydro
Triphyophylline	Me	H	H	H	
Isotriphyophylline	Me	H	H	H	C-1 α-Me
N-Methyltriphyophylline	Me	H	H	Me	
O-Methyltriphyophylline	Me	H	Me	H	
Dehydro-O-methyltriphyo-phylline	Me	H	Me	-	1,2-dehydro
Bisdehydro-O-methyltri-phyophylline	Me	H	Me	-	1,2,3,4-dehydro
Triphyopeltine	H	H	H	H	C-3' CH_2OH
O-Methyltriphyopeltine	Me	H	H	H	C-3'· CH_2OH

A group of naphthylisoquinoline alkaloids has been identified of which twenty one have so far been isolated from *Ancistrocladus* and *Triphyophyllum* species, these being the 7-β-naphthylisoquinoline ancistrocladidine (10) and the 5 and 7 α-naphthyl-compounds listed in tables 4 and 5 respectively. The alkaloids are chiral as a result of restricted rotation of the biaryl system as well as of the possession of asymmetric carbon atoms; the methyl at C-3 is on the α-face in all alkaloids and that at C-1 on the β-face in most, but not all cases. The structure of ancistrocladine has been determined by oxidation to the acid (13) the methyl ester of which is prepared by two routes, and by Hofmann degradation of its O,N-dimethyl-derivative successively to the methine base (14) and the nitrogen-free products (15) and (16), the second of which on ozonolysis gives an aldehyde which yields the lactone (17) on oxidation. The position of the phenolic hydroxyl group in the alkaloid has been confirmed by Claisen rearrangement of the allyl ether (T.R. Govindachari and P.C. Parthasarathy, Tetrahedron, 1971, 27, 1013).

The structures of other alkaloids of the group have been determined by similar methods and X-ray crystallographic studies have confirmed both gross structures and absolute configurations. (Govindachari *et al.*, J.chem.Soc., Perkin I, 1974,

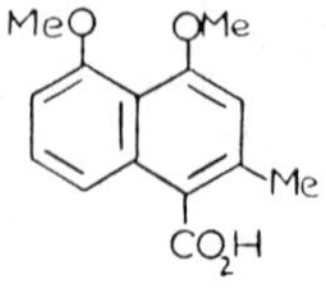

(13) 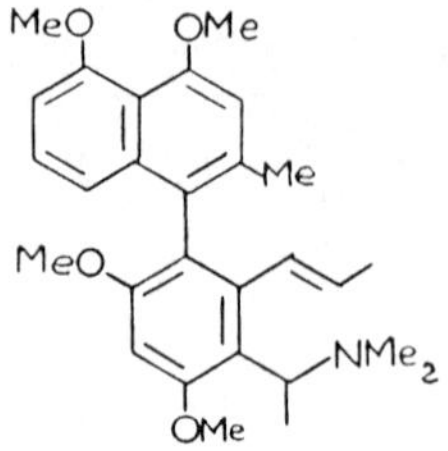 (14) 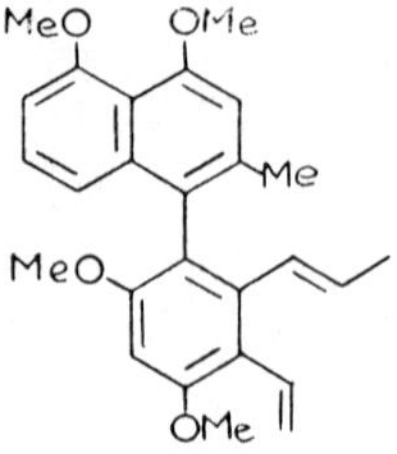 (15)

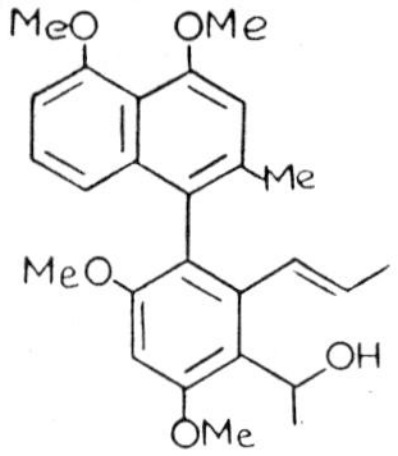

(16) 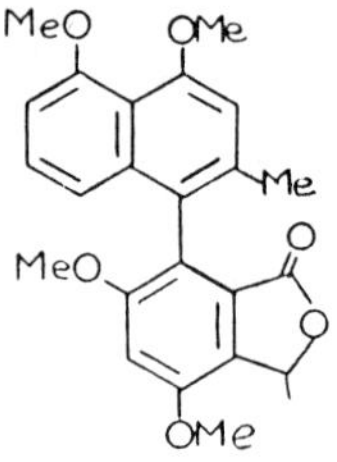 (17) 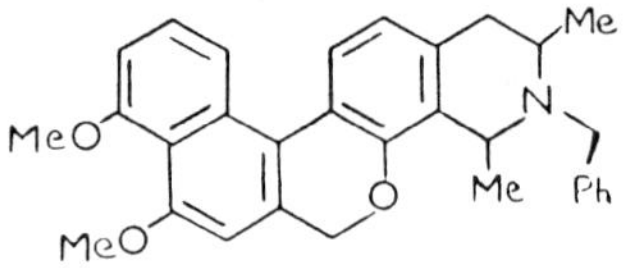 (18)

1413; Pasathasarathy and G. Karthy, Indian J.Chem., 1983, 22B, 590). Bisdehydro-*O*-methyltriphiophylline has been synthesised from the base (18) (prepared by the Bischler-Napieralsky procedure) by hydrogenolytic cleavage, *O*-methylation and dehydrogenation (G. Bringman and J.R. Jansen, Tetrahedron Letters, 1984, 25, 2537.

A new 4-arylisoquinoline alkaloid, latifine, isolated from *Crinum latifolium*, has been identified as an isomer (21, R^1= OH, R^2=H) of cherylline (21, R^1=H, R^2=OH) (S. Kobayashi, T. Tokomuto and Z. Taira, Chem.Comm., 1984, 1043). A novel synthesis of cherylline has been achieved from the iminium salt (19) by treatment with diazomethane, cyclisation of the resulting aziridinium salt (20) and hydrogenolytic removal of the benzyl groups (T. Kametani, *et al*., J.chem.Soc., Perkin I, 1982, 2935.

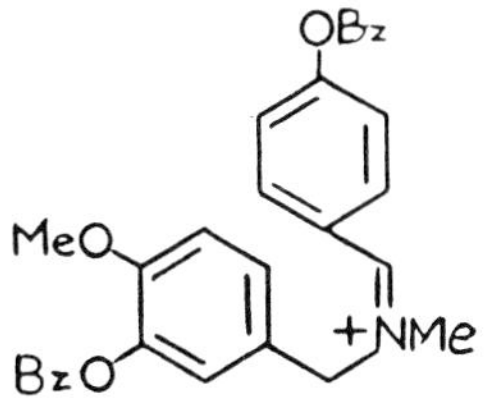

(19)

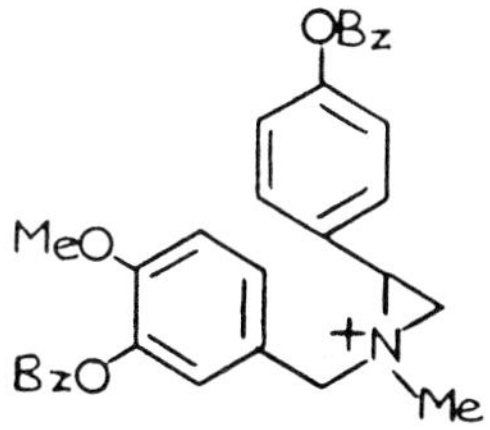

(20)

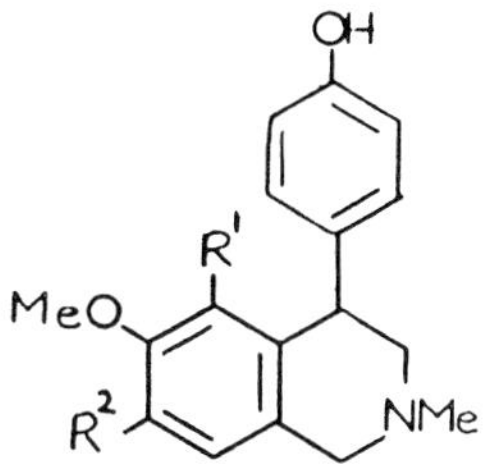

(21)

2. *Benzylisoquinolines*

Of the many new benzylisoquinoline alkaloids discovered, unusual types are juziphine (22, R=OH), juziphine *N*-oxide and
gorchacoine (22, R=Me), from *Corydalis gortschakovii* (T.
Irgashev *et al.*, Khim.Prir.Soedin., 1977, 127), quettamine
chloride (23), secoquettamine (24) and the related benzofuran
dehydrosecoquettamine, from *Berberis baluchistanica* (M.H.A.
Zarga, G. Miana and M. Shamma, Heterocycles, 1982, 18 (Special
Issue), 63). All of these are related to petaline, which is
N-methylgorchacoine chloride, previously the only known benzylisoquinoline alkaloid with oxygen substituents only at positions 7 and 8 of the isoquinoline system.

(22) (23) (24)

Other unusual alkaloids are the 4-pyrrolidyl-compounds mac-
rostomine (25, R=Me), two of its *N*-oxides, arenine (25, R=H)
and dehydronormacrostomine (26), obtained from *Papaver macros-
toma* and *Papaver arenarium* (V.A. Muatsakanyan *et al.*, Coll.
Czech.chem.Commun., 1977, 42, 1421; I. Israilov, M.A.Manush-
akayan and M.S. Yunusov, Khim.Prior.Soedin., 1978, 417). Mac-
rostomine has been synthesised from the ketone (27) by treatment with 1-nitroso-2-pyrrolidyllithium to give the alcohol
(28), which was subjected to hydrogenolysis, reduction with
lithium aluminium hydride and *N*-formylation to give the base
(29), followed by dehydrogenation and reduction of the *N*-for-
myl group to *N*-methyl (W. Wykypiel and D. Seebach, Tetrahedron
Letters, 1980, 21, 1927). It has also been synthesised by a
conventional Bischler-Napieralski procedure from 2-(3,4-dim-
ethoxyphenyl)-2-(1-methyl-2-pyrrolidyl)-ethylamine (W.Wiegrebe
et al., Chimia 1981, 35, 288; R.B. Sharma and R.S. Kapil,
Indian J.Chem., 1982, 21B, 141).

(25) (26)

(27)   (28)   (29)

The benzylic carbanion from papaverine has been converted into a variety of substituted bases of general structure (30) by treatment with alkyl halides, aminoalkyl halides and ethyl acrylate (A. Buzas and G. Lavielle, Patent, Chem.Abs., 1981, 95, 25378, 150978) and the ion from dihydropapaverine (31) has been found to react with cyclohexan-1,2-dione in the presence of excess base to give the spiro ketone (35) presumably via the intermediates (32), (33) and (34).

(30)   (31)   (32)

(35)   (34)   (33)

An analogue of (35) has been prepared from the anion (31) and benzil. (S. Ruchirawat and V. Somchitman, Tetrahedron Letters, 1976, 4159).

N-Methylpapaverinium salts on treatment with sodium hydroxide are deprotonated to the enamine (36), and this can be oxidised by singlet oxygen in the presence of copper (I) chloride to give *N*-methylisoquinolone and veratric aldehyde. The 3,4-dihydro-analogue of the enamine (36) under similar conditions is, however, oxidised to the hydroxyketone (37) which can be reduced to hydroxylaudanosine (38) though fission to the iso-quinoline and veratric aldehyde can be accomplished by photo-sensitised oxidation (Ruchirawat, Heterocycles, 1977, 6, 1724; Ruchirawat *et al.*, ibid. 1119).

(36)   (37)   (38)

Both *cis* and *trans* laudanosine *N*-oxides (39) have been prepared. On pyrolysis the *trans* oxide gives only the product of Cope degradation (40), whereas the *cis* oxide gives in addition the rearranged bases (41), (42) and (43). (J.B. Bremner and Le Van Thuc, Austral.J.Chem., 1980, 33, 379).

(39)   (40)   (41)

(42)   (43)

Bisquaternary salts of laudanosine with itself and with other tetrahydroisoquinolines and long chain dihalides have been prepared as potential neuromuscular blocking agents of the same type as tubocurarine.   The most widely studied of such salts is atracurium (44) (R. Hughes and D.J. Chapple, Brit.J.Anaesth., 1981, 53, 31;   J.B. Stenlake *et al*., Eur.J. med.Chem.-Chim.Ther., 1981, 16, 503 and 515;   Brit.J.Anaesth., 1983, 55 (Suppl.1), 3).

(44)

The *N*-benzyltetrahydroisoquinoline alkaloids sendaverine (45, R=Me) and corgoine (45, R=H) have been synthesised by several different methods.   The isochromone (46) with 4-meth - oxybenzylamine gives the amide (47), which is reduced to the amine debenzylated and converted into the chloromethyl-compound (48), which readily cyclises to sendaverine (Kametani *et al*., J.chem.Soc., Perkin I, 1979, 1836).   A basically similar syn - thesis starts from ethyl 4-benzyloxy-2-bromomethyl-phenylacet-

ate instead of the isochromone (Pandey and Tiwari, Indian J.
Chem., 1980, 193, 160). Cyclodehydration of the alcohol (49)
in polyphosphoric acid followed by reductive debenzylation
also affords sendaverine (M. Masood and Tiwari, Acta Chim.
Hung. 1983, 113, 177) as does the insertion of carbon monoxide
between the nitrogen atom and the benzene ring of the bromo-
amide (50) in the presence of tributylamine, triphenylphosphine and palladium (II) acetate, followed by reduction of the
resulting imide to the base and removal of the benzyl group
(Masood and Tiwari, Indian J.Chem., 1983, 22B, 825). Repetit-
ion of each of these syntheses using 4-benzyloxybenzylamines
in place of their 4-methoxy analogues results in the production of corgoine (Masood and Tiwari, *loc.cit.*, Pandey and
Tiwari, Pol.J.Chem., 1980, 54, 763.

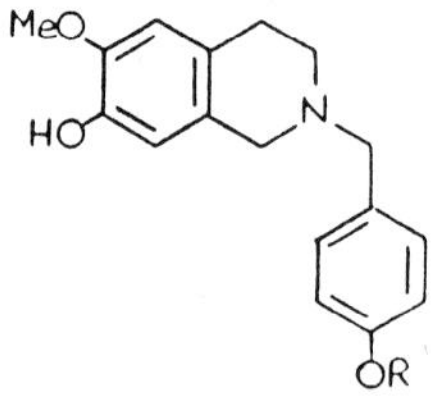

(45)

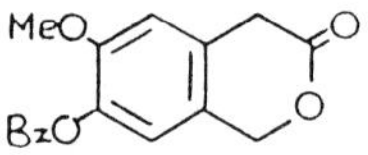

(46)

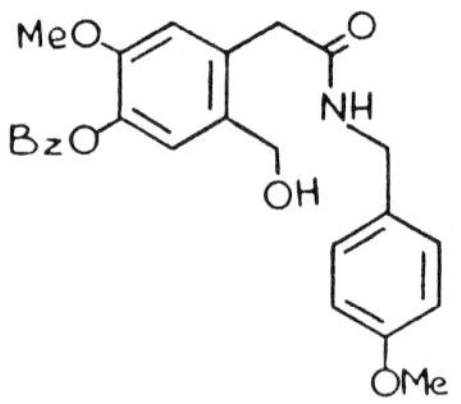

(47)

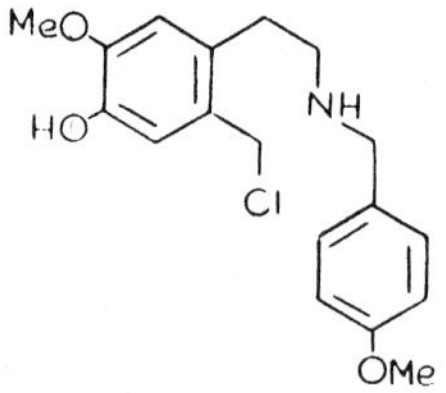

(48)

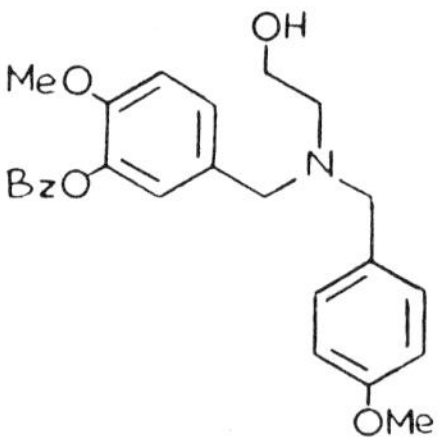

(49)

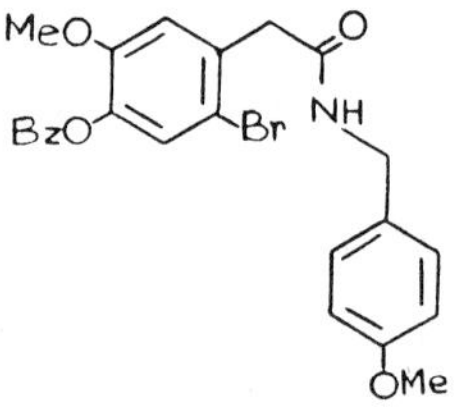

(50)

An attempt to prepare a diastereoisomer of the 13-methyl-tetrahydroberberine alkaloids cavidine and thalictrifoline by Maunich cyclisation of the secondary base (51) with formalde-hyde, using the bromine atom to direct cyclisation into the desired position was frustrated by rearrangement to the benz-oxazepine (56), which is in equilibrium with the iminium salt (55), in which form it may be reduced to the 6'-hydroxyethyl-benzyltetrahydroisoquinoline by sodium borohydride. The re-arrangement has been rationalised through the intermediates (52), (53) and (54). (S. Natarajan *et al.*, J.chem.Soc., Per-kin I, 1979), 283).

(51)   (52)   (53)

(56)   (55)   (54)

3. *Bisbenzylisoquinolines*

Several alkaloids of this group have been discovered in which one of the isoquinoline systems rather than being in the di-hydro- or tetrahydro-state is fully aromatic as in thalphine

(57), thalictrinine (58, $R^1R^2=O$), dihydrothalictrinine (58, $R^1=H$, $R^2=OH$), trigilletimine (60) and sciadoline (59). Two alkaloids, daphnine (61) and phaeantharine (77), are fully dehydrogenated in both halves of the molecule. Several mono and di-*N*-oxides of the series have also been found to be natural products.

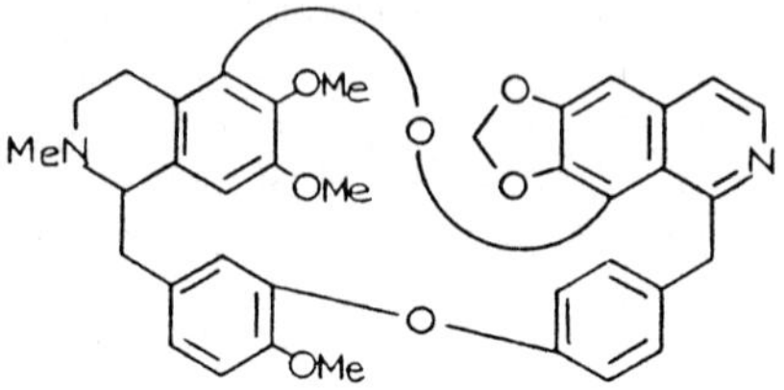

(57)

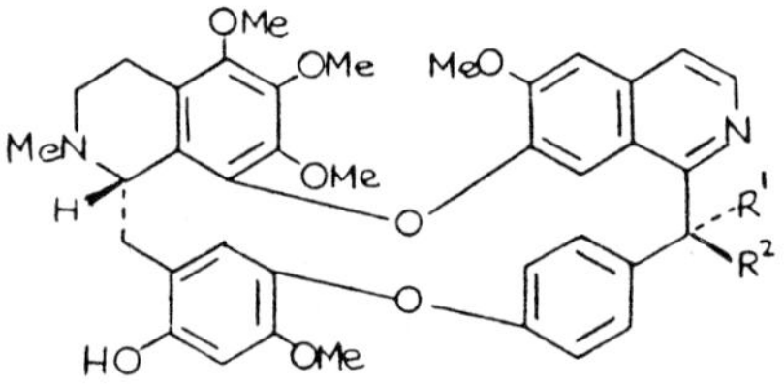

(58)

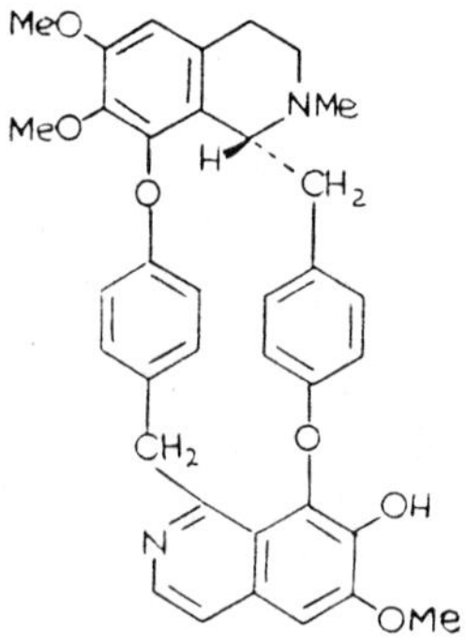

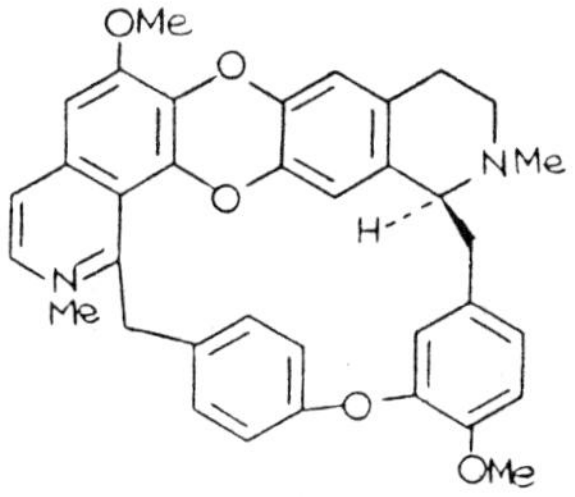

(59)

(60)

Thalictrinine and dihydrothalictrinine are two bases of a closely related series found in *Thalictrum rochebrunianum*, the other members of which are thalibrunine, N'-northalibrunine, thalibrunimine and oxothalibrunimine. Their interrelationships have been established as follows. Thalibrunimine, identified largely on spectroscopic grounds as the dihydroisoquinoline (62, $R^1=R^2=H$), on aerial oxidation affords the ketone oxothalibrunimine (62, $R^1R^2=O$), which is 3',4'-dihydrothalictrinine and can be dehydrogenated to thalictrinine (58, $R^1R^2=O$). The reduction of thalictrinine with sodium borohydride yields dihydrothalictrinine (58, $R^1=H$, $R^2=OH$). Reduction of thalibrunimine and thalibrunimine methiodide affords the corresponding tetrahydroisoquinolines *N'*-northalibrunine and thalibrunine respectively (J. Wu, J.L. Beal and R.W. Doskotch, J.org.Chem., 1980, 45, 208, 213). All of these alkaloids have the unusual 2", 4", 5" pattern of oxygen substitution, presumably arising from the oxidative coupling of norco-

(61)

(62)

(63)

claurine with a 2'-hydroxycoclaurine and an alternative coupling of two such bases is represented by the alkaloids calfati-
mine (63) and calfatine, which is *N*-methyldihydrocalfatimine
(V. Fajardo, M. Garrido and B.K. Cassels, Heterocycles, 1981,
15, 1137; Fajardo, A. Urzua and Cassels, *ibid.*, 1979, 12,
1559).

A key reaction in the elucidation of these structures is
the recently developed process for the fission of bisbenzylisoquinolines by oxidation with ceric nitrate. This reagent
oxidises laudanosine to veratric aldehyde and the *N*-methyl-6,
7-dimethoxy-3,4-dihydroisoquinolinium ion, conveniently isolated as veratryl alcohol and *N*-methyl-6,7-dimethoxytetrahydro-
isoquinoline after reduction of the mixture with sodium borohydride. Using this procedure *O*-acetylthalibrunine, hernande-
zine (desoxythalibrunine), tetrandrine (desmethoxyhernandez-
ine) and *O*-methylmicranthine have been degraded to the isoquinolines (64, R=OMe), (64, R=OMe), (64, R=H) and (65) respectively, the non-basic fragments being (66, R=OAc), (66, R=H),
(66, R=H) and (66, R=H) in the four cases. (Wu, Beal and
Doskotch, *loc. cit.*; I.R.C. Bick, Bremner and M.P. Cava, Aus-
tral.J.Chem., 1978, 31, 321).

Oxidation of bisbenzylisoquinoline alkaloids with potassium
permanganate has been found to be controllable enough to per-
mit fission of only one of the two benzylisoquinoline systems. In this way the *O*-acetyl derivatives of the isomeric alkaloids tiliacorine and dinklacorine have been oxidised to the
aldehydo-isoquinolones (67, R¹=Me, R²=Ac) and (67, R¹=Ac,R²=Me)
respectively (Shamma and J.E. Foy, J.org.Chem., 1976, 41, 1293;
D. Dwuma-Badu *et al.*, J.nat.Prod., 1979, 42, 116). Similar
fissions of other bases have been accomplished.

(64) (65)

(66) (67)

Products of such oxidative fission of bisbenzylisoquinolines with one, two and three diphenyl ether linkages have
been found to be natural products forming a small sub-group
of secobisbenzylisoquinoline alkaloids. Those so far identified are karakoramine (68), jhelumine (69, R=H), chenabine (69,
R=Me), sindamine (70), punjabine (73, R^1=H, R^2=CHO) and gilgi-
tine (73, R^1=H, R^2=COOMe), isolated from *Berberis lycium* (J.E.
Leet *et al*., Heterocycles, 1982, 19, 2355; 1983, 20, 425),
baluchistanamine (71), from *Berberis baluchistanica* (Shamma,
Foy and G.A. Miana, J.Amer.chem.Soc., 1974, 96, 7809), O-methylpunjabine (73, R^1=Me, R^2=CHO), O-methyldeoxopunjabine (73,
R^1=Me, R^2=CH$_3$), secocepharanthine (72, R=CHO and dihydroseco-
cepharanthine (72, R=CH$_2$OH) from *Stephania sasakii* (H. Ishii
et al., Tennen Yuki Kagobutsu Toronkai Koeu Yoshishu 26th.,
1983, 102) and secantioquine (74) from *Pseudoxandra lucida* (D.
Cortes *et al*., Compt.Rend., 1984, 298, 591).
    These structures have been deduced from spectroscopic data
and confirmed by the preparation of key alkaloids by the oxidation of known bisbenzylisoquinolines and by interconversions.
Baluchistanamine and sindamine have been obtained by oxidation
of the isomeric alkaloids oxyacanthine and berbamine, secocepharanthine from cepharanthine and O-methylpunjabine from

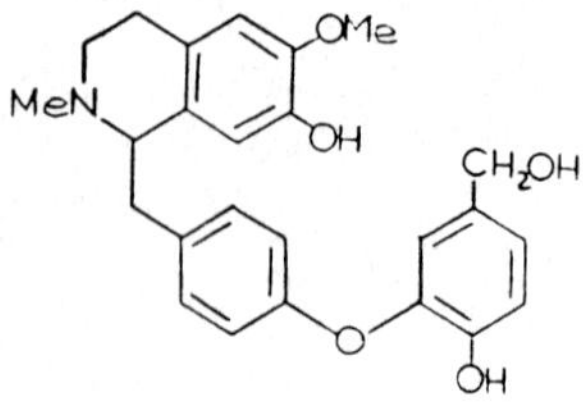

(68)

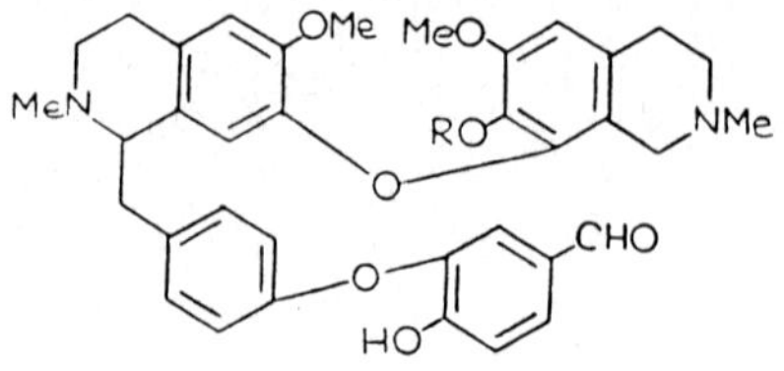

(69)

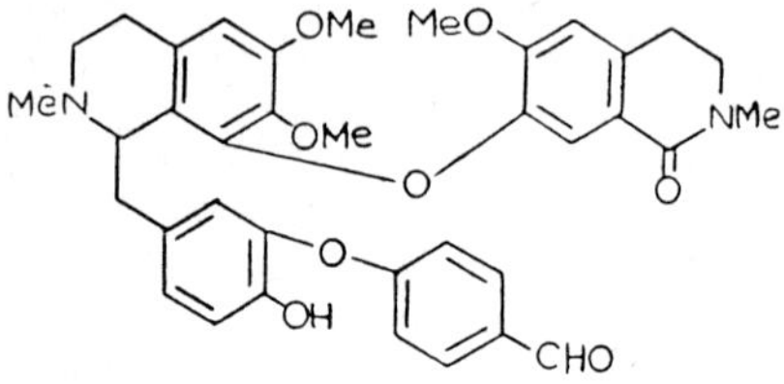

(70)

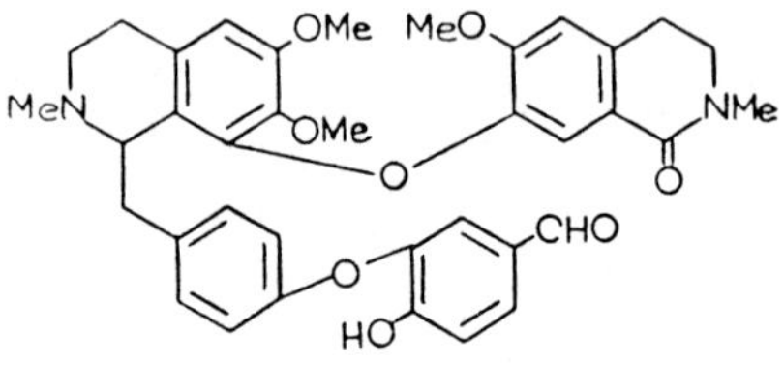

(71)

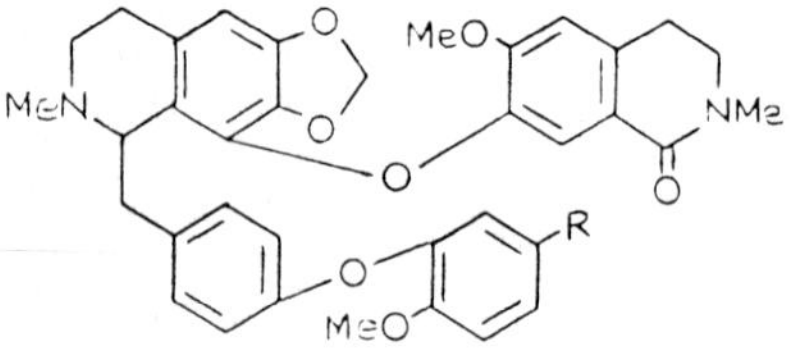

(72)

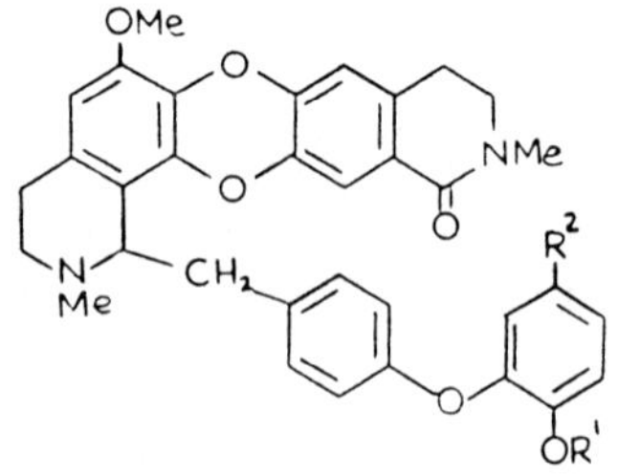

(73)

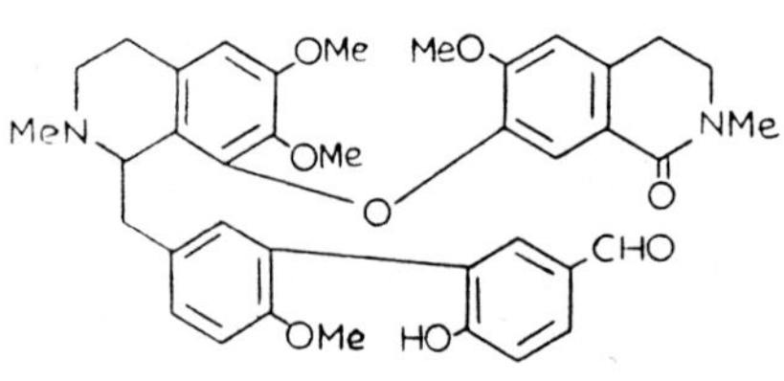

(74)

isotrilobine; reduction of secocepharanthine with sodium borohydride gives the primary alcohol (72, R=CH$_2$OH) and Wolff-Kishner reduction of O-methylpunjabine gives (73, R^1=Me, R^2 = CH$_3$). It may be noted that sindamine and chenbabine represent products of alternative fissions of berbamine.

Nuclear magnetic resonance spectroscopy plays a major part in the elucidation of the structures of these bases, and continues to be the most valuable technique in structural work in the series. Pulsed nmr difference spectroscopy of small nuclear Overhauser effects between aromatic protons and O-methyl and N-methyl groups has permitted differentiation between two possible isomeric structures for dihydrodaphnine (D. Neuhaus *et al.*, Tetrahedron Letters, 1981, 22, 2933.

Total syntheses of bisbenzylisoquinoline alkaloids include the confirmation of a revised structure for the fully aromatic phaearanthine (77). Condensation of the Reissert compound (75) with the dibromide (76) followed by elimination to give the bisbenzylisoquinoline and N-methylation leads to the bis-quaternary salt (77) (J. Knabe and W. Wigand, Arch.Pharm. (Weinheim), 1983, 316, 445; Knabe and Hanke *ibid.*, 1983, 317, 92). A synthesis of O-methyldauricine (which is octahydro-phaearanthine) has been achieved from the diketone (78) by a Willgerodt reaction with homoveratrylamine and sulphur to give the bisthioamide (79), followed by Bischler-Napieralski ring closure, N-methylation and reduction (E.P. Nakova, O.N. Tol'-kachev and R.P. Evstigneeva, Khim.Prir.Soedin., 1981, 457).

(75)   (76)

(77)

(78)   (79)

Synthesis of the alkaloids trilobine and obaberine, containing the diphenyl ether linkages, have been accomplished
from (S)-(+)-(0)-benzyl-8-bromo-N-benzoylnorarmepavine (80)
by Ullmann condensation with the phenolic amide (81) followed
by reductive debenzylation to give the phenolic ether (82).
The latter when subjected to a second Ullmann condensation
with bromomethoxyphenylacetic acid produces the acid (83).
Removal of the N-butyryl group and internal amide formation,
followed by Bischler-Napieralski cyclisation, N-methylation
and reduction gives a mixture of diastereoisomeric bases of
structure (84), one of which has been reduced to the N-benzyl
compound and completely 0-demethylated to the tetrahydric phenol (85). This is cyclised to the dioxin by hydrobromic acid
at 140°, methylated with diazomethane and reductively debenzylated to trilobine (86). The other isomer of the base (84)
has been similarly converted into obaberine, the C-1' epimer
of trilobine (Y. Inubushi *et al.*, Tetrahedron Letters, 1976,
2857; Chem.Pharm.Bull., 1977, <u>25</u>, 1636).
6'Bromolaudanosine undergoes a reatively easy Ullmann coupling with phenols in solution in pyridine in the presence of
tetra(pentafluorophenyl)copper. Using this reaction between
the bromo-alkaloid and phenolic benzylisoquinolines and aporphines unnatural bisbenzylisoquinolines, such as (87), have
been prepared from armepavine and (88) from N-methylcassyfi-
line (Cava and A. Afzali, J.org.Chem., 1975, <u>40</u>, 1553).
The biogenesis of bisbenzylisoquinoline alkaloids from coclaurine and its close relatives has been confirmed by the incorporation of labelled coclaurine. N-methylcoclaurine and
norcoclaurine into a number of different bases containing one,
two and three dipehnyl ether linkages. (R) and (S) forms of
these basses are incorporated only into halves of the dimers
of the same absolute configuration, there being no interconver-
ion of isomers at C-1 and bases labelled with tritium at this

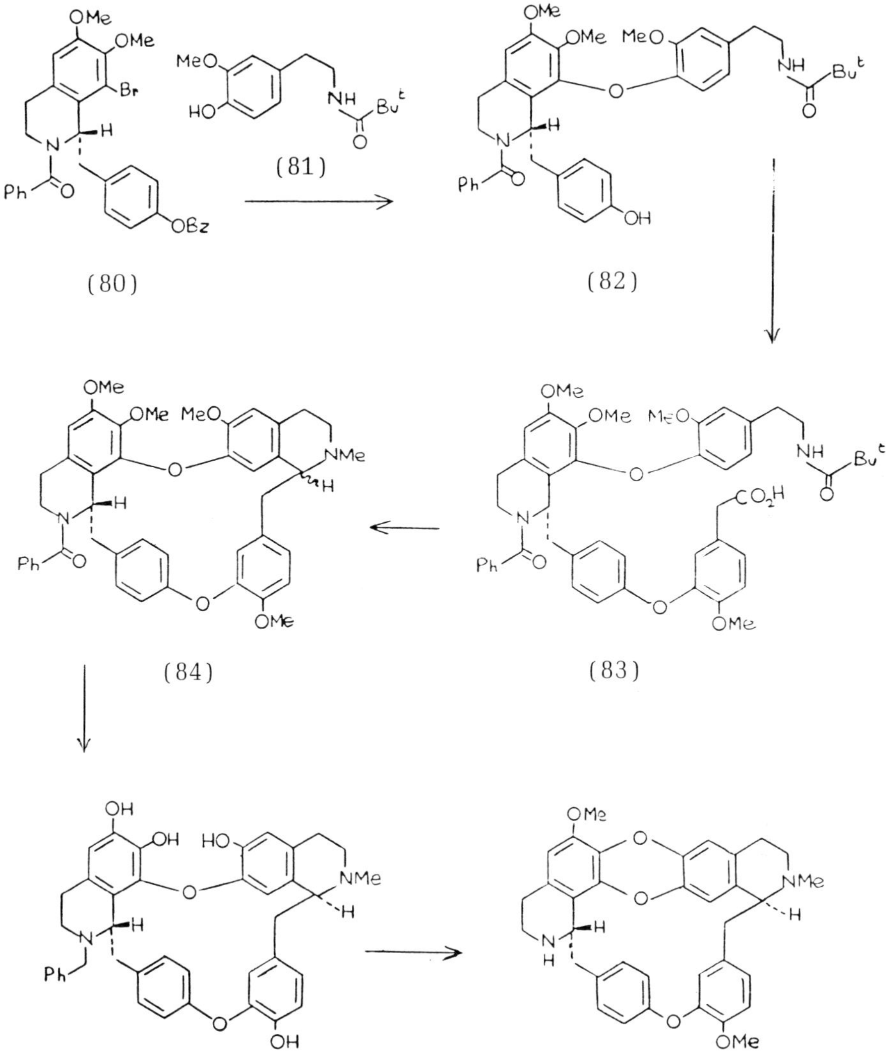

(80)

(81)

(82)

(84)

(83)

(85)

(86)

(87)　　　　　　　　　　　　　(88)

position undergo no loss of this isotope. The ready incorporation of 1,2-dehydro-coclaurine into these alkaloids is assumed to proceed through the (±)-reduced forms. (D.S. Bhakuni
et al., Chem.Comm., 1978, 226; J.chem.Soc., Perkin I, 1978,
121, 380, 1318; Tetrahedron, 1978, 34, 1409, 1980, 36, 2149,
1981, 37, 2651; Phytochemistry 1980, 19, 2347).

4. *Cularines*

Alkaloids of the cularine group have always been assumed to
be derived by internal oxidative coupling of a diphenolic base
or triphenolic base of general structure (89, R=Me or H) though
no such base was known until the isolation of crassifoline (89,
R=Me) from *Corydalis claviculata* (G. Blaschke and G. Scriba,
Z.Naturforsch., 1983, 38C, 370) and from *Sarcocapnos crassifo-
lia* (J.M. Boente *et al.*, Tetrahedron Letters, 1983, 24, 2303),
in both of which it is accompanied by several alkaloids of the
cularine group. Clearly internal oxidative coupling of (89)
could occur in two ways, either *ortho* or *para* to the 4'-hydro-
xy group and until recently only the products of *para* coupling
were known except in the morphine-cularine dimers cancentrine
and dehydrocancentrine. The range of bases now known however
is much larger and comprises the alkaloids listed in table 5
and secondary bases corresponding to cularine and cularicine
(Blaschke and Scriba, *loc.cit.*; Boente *et al.*, *loc.cit.*; D.P.
Allais and H.G. Guinaudeau, Heterocycles, 1983, 20, 2055).
    In addition a range of oxidised cularines matching that of
the oxidised aporphines is also known. The bases limousamine
(92) and 4-hydroxysarcocapnine (93) represent hydroxy substitution unknown in the benzylisoquinoline and bisbenzylisoquinoline series (probably because of the ease with which such
compounds can be converted into pavines and isopavines) but

(89)　　　　　　　(90)　　　　　　　(91)

TABLE 6

Cularine Alkaloids (90) and (91)

	Type	R^1	R^2	R^3
Cularine	90	Me	Me	Me
Cularidine	90	H	Me	Me
Cularicine	90	H	——CH$_2$——	
O-Methylcularicine	90	Me	——CH$_2$——	
Culacorine	90	H	Me	H
Celtine	90	Me	H	Me
Celtisine	90	H	H	Me
Sarcocapnine	91	Me	Me	Me
Sarcocapnidine	91	Me	H	Me
Claviculine	91	H	H	Me

observed in the aporphine and tetrahydroberberine groups (Boente *et al.*, Tetrahedron Letters, 1983, 24, 2295. More highly
oxidised alkaloids are represented by oxocularine (94, $R^1=R^2=$
Me) oxocompostelline (94, $R^1R^2=CH_2$), oxosarcocapnine (95, $R^1=$
$R^2=$Me) and oxosarcocapnidine (95, $R^1=$Me, $R^2=$H), yagonine (96)
and aristoyagonine (97), which can be prepared from yagonine
by the benzil-benzilic acid rearrangement (M.J. Campello *et
al.*, Tetrahedron Letters, 1984, 25, 5933).
    In this group of alkaloids claviculine and sarcocapnidine
are both methylated by diazomethane to sarcocapnine. Oxocularine and oxocompostelline have been synthesised by internal
Ullmann coupling of the bromophenols (98, $R^1=R^2=$Me) and (98,
$R^1R^2=CH_2$) followed by oxidation (Boente *et al.*, *loc.cit.*).

OH

HO NMe

O

MeO OH

(92)

OH

MeO NMe

O

MeO

MeO

(93)

MeO N

O

O

R¹O OR²

(94)

R¹O N

O

O

R²O

MeO

(95)

O O

MeO NMe

O

MeO

MeO

(96)

O

MeO NMe

O

MeO

MeO

(97)

Similarities to the aporphine group extend to the occurrence of secocularine (99, R=Me) and secocularidine (99, R=H) in *C. claviculata* and *S.crassifolia*; the latter can be methylated to the former, which is identical with the product of Hofmann degradation of cularine methiodide (Boente *et al.*, Tetrahedron Letters, 1984, 25, 889, 1829; Campello *et al.*, *loc cit.*).

Three alkaloids of unusual type in this group have also been identified. Gouregine (100), found in *Guatteria ouregou* is C-dimethylated in the same way as the aporphine alkaloids melosmine and melosmidine (122) (isolated from *Guatteria melosma*) and presumably do not arise from crassifoline (M. Leboeuf *et al.*, Tetrahedron, 1982, 38, 2889). Two alkaloids linaresine (101) and dehydrolinaresine (102), isolated from *Berberis valdiviana* , differ from all of the other cularine alkaloids in having a 6,7,8-oxygen substitution pattern in the isoquinoline system. It seems probable that they are derived from a diphenolic analogue of rugosinone (103), or dihydrorugosinone, both of which have been identified in *B.valdiviana* (S. Firdous *et al.*, J.Amer.chem.Soc., 1984, 106, 6099).

(98)

(99)

(100)

(101)

(102)

(103)

5. Aporphines

Interest in the aporphine alkaloids has principally centred on the range and number of oxidised bases that have been dis- covered. A considerable number of 7-hydroxy and 7-methoxy aporphines have been isolated and both 7α and 7β forms have been encountered. These probably represent intermediates in the formation of dehydroaporphines, into which they are easily converted in the laboratory, by dehydration. Alkaloids of this type isolated include oliveroline (104, R^1=Me, R^2=H) and its N-oxide; ushinsunine which is the 7α-hydroxy isomer of oliveroline; pachypodanthine (104, R^1=, R^2=Me), N-methylpach- ypodanthine (104, R^1=R^2=Me) and its N-oxide; oliverine (105, R^1=R^2=Me) and oliveridine (105, R^1=Me, R^2=H) and their N-oxi- des; the related secondary bases noroliverine (105, R^1=H, R^2 =Me) and noroliveridine (105, R^1=R^2=H); polysuavine (106); guatterine (107, R=H) and its N-oxide polyalthine (107, R=

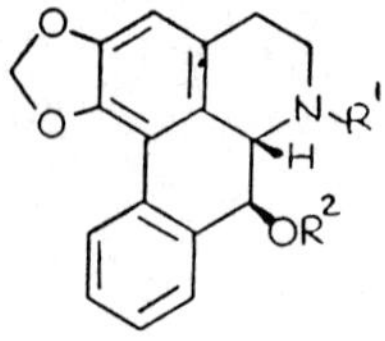

(104) 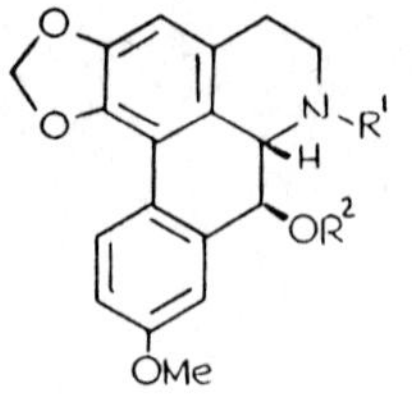   (105) 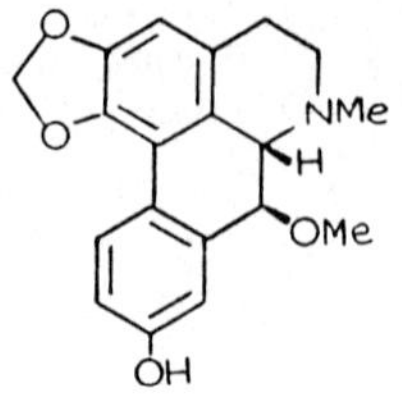   (106)

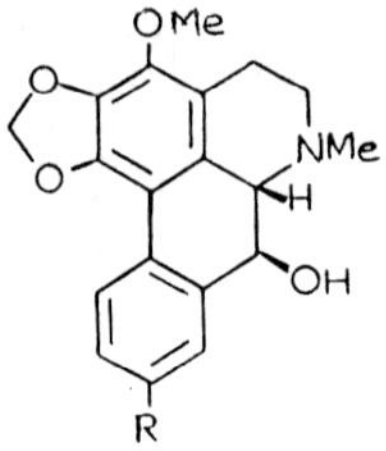

(107) 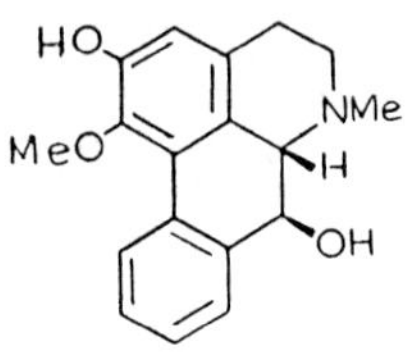   (108) 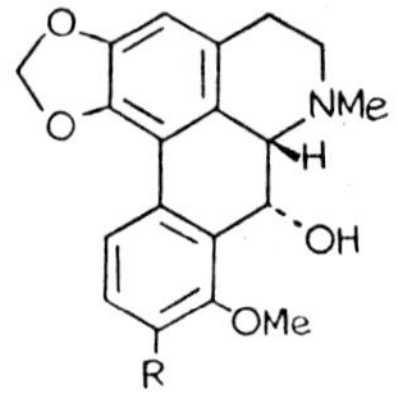   (109)

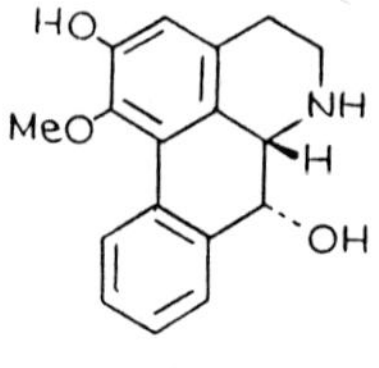

(110) 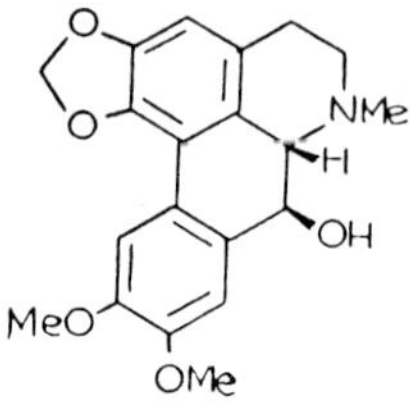   (111) 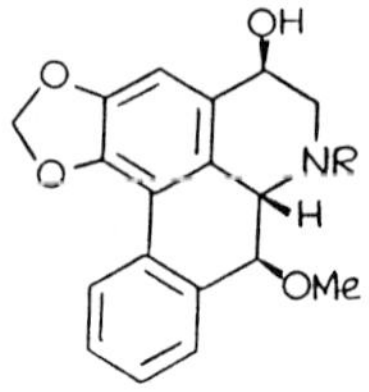   (112)

OMe); pachyconfine (108); ayuthiamine (109, R=H), sukhodiam-
ine (109, R=OMe); anaxagorine (110); duguetine (111), pachy-
staudine (112, R=Me) and norpachystaudine. (F. Bevalot, Lebe-
ouf and A. Cavé, Compt.Rend., 1976, 282, 865; Cavé *et al.*,
Planta Med., 1978, 33, 243; R. Hocquemiller *et al.*, *ibid.*,
1981, 41, 48; Guinaudeau *et al.*, J.nat.Prod., 1982, 45, 355).

Derived from these bases, no doubt, by dehydration, several
dehydroaporphines have been identified as natural products.
These incluyde dehydroisolaureline, dehydrostephanine and de-

hydrodicentrine which are the products of dehydration of oliveridine (105, R^1=Me, R^2=H), ayuthiamine (109, R=H) and duguetine (111) respectively; dehydroglaucine (113, R^1=R^2=Me) and
the secondary base dehydronorglaucine; dehydronancentrine
(113, R^1R^2=CH_2); dehydrophanostenine (114); dehydrocorydine
(115); dehydrostesakine (116, R^1=H, R^2=OH), dehydrocrebanine
(116, R^1=H, R^2=OMe); dehydrocopodine (116, R^1=R^2=OMe); dehy-
droroemerine (117, R^1=R^2=R^3=H), dehydro-ochoteine (117, R^1=R^2
=OMe, R^3=H) and didehydro-ochoteine (118) (Cava and A. Venkat-
eswarlu, Tetrahedron, 1971, 27, 2639; R. Ziyaev et al., Khim.
Prir.Soedin., 1977, 715; J. Kunimoto et al., Phytochemistry,
1980, 19, 2735; J.pharm.Soc. Japan, 1981, 101, 431).
Other alkaloids of the dehydro series are cabudine (119)
(M. Kurbanov et al., Dokl.Akad.Nauk Tadzh.S.S.R., 1975, 18,
20), duguenaine (120, R=H) and duguecalyne (120, R=OMe) (F.
Roblot, Hocquemiller and Cavé, Compt.Rend., 1981, 293, 373.
Cabudine is the only aporphine alkaloid currently known to

(113)

(114)

(115)

(116)

(117)

(118)

(119)   (120)   (121)

bear a hydroxymethyl group, though duguenaine and duguecalyne
as well as thalfenine (121) and the corresponding tertiary
base probably represent products of further cyclisations of
compounds containing such a group. Both hydroxymethyl comp -
ounds and their further cyclisation products are known natural
products in the berberine group and some secoberberines are
hydroxymethylbenzylisoquinolines.

Three dehydroaporphines of unusual C-methylated structures
are belmine (117, R^1=H, R^2=OH, R^3=Me) (D. Cortes *et al.*, Compt.
Rend., 1984, 299, 311) and melosmine (122, R=H) and melosmidine (122, R=Me), the last two being methylated in the same
position as the unusual cularine alkaloid gouregine (V. Zabel
et al., J.nat.Prod., 1982, 45, 94). Two other C-methylated
and oxidised aporphines are the 6α-methy(+)-guattescine (123)
and its 6β-methyl-O-desmethyl analogue (-)-guattescidine (124)
(Hocquemiller, S. Razamizafy and Cavé, Tetrahedron (1982, 38,
911).

(122)   (123)   (124)

The most highly oxidised alkaloids of the aporphine group
are the 4,5-dioxoaporphines, a small sub-group comprising pontevedrine (125, $R^1=R^2=OMe$), the structure of which has been
revised from that advanced in the second edition, dioxodehy-
dronantenine (125, $R^1R^2=OCH_2O$), cepharadione-B (125, $R^1=R^2=H$),
cepharadione-B (126, $R^1=R^2=H$), oxodehydrocrebanine (126, $R^1=R^2$
$=OMe$), and tuberosinone (127) and its N-β-D-glucoside (L. Castedo, R. Suau and A. Maurino, Tetrahedron Letters, 1976, 501;
M. Akasu, H. Itokawa and M. Fujita, *ibid.*, 1974, 3609; Kunimoto, Y. Murakami and Akaso, Yakugaku Zasshi, 1980, 100, 337;
D. Zhu *et al.*, Heterocycles, 1982, 17, 345. The structure (125,
$R^1=R^2=OMe$) for pontevedrine has been confirmed by synthesis
of the alkaloid by photo-catalysed cyclisation and oxidation
of the lactam (128) in alkaline solution, followed by N-methy-
lation (Castedo *et al.*, Tetrahedron Letters, 1978, 2179) and
norcepharadione-B has been synthesised by a Diels-Alder addition of benzyne to the masked diene 1,6,7-trimethoxy-1-methyl-
tetrahydroisoquinoline-3,4-dione (Castedo *et al.*, Tetrahedron
Letters, 1982, 23, 451).

(125)   (126)   (127)

(128)   (129)   (130)

Ring contraction of the α-ketolactam system of the 4,5-di-
oxoaporphines, essentially by the benzil-benzilic acid rearrangement, leads to the formation of a series of aristolactams of general structure (129) the known representatives of
which are listed in table 7. The precisely analogous transformation of the oxocularine alkaloid yagonine (96) into aristoyagonine (97) in the plant and in the laboratory has been
reported. Further oxidation of the aristolactams takes place
at the nitrogen atom to give the aristolochic acids, which
have the general structure (130) and represent most of the
very few known naturally-occurring nitro-compounds; individual acids are listed in table 8. (C.R. Chen *et al.*, Lloydia,
1974, 37, 493; R. Crohare *et al.*, Phytochemistry, 1974, 13,
1957; S.C. Pakrashi *et al.*, *ibid.*, 1977, 16, 1103; S.F. Dyke
and E. Gellert, *ibid.*, 1978, 17, 599; F. Zhou *et al.*, Yaoxue
Xuebao, 1981, 16, 638; D. Zhu *et al.*, Heterocycles, 1982, 17,
345; H.A. Priestap, Phytochemistry, 1985, 24, 849).

TABLE 7

Aristolactams (129)

	R^1	R^2	R^3	R^4
Aristolactam	— CH_2 —		H	OMe
Aristolactam - Ia	— CH_2 —		H	OH
Aristolactam - III	— CH_2 —		OMc	H
Aristolactam - C	— CH_2 —		OH	H
Aristolactam - D	— CH_2 —		OMe	OMe
Cepharanone - A	— CH_2 —		H	H
Aristolactam - AIa	H	Me	H	OH
Aristolactam - AII	H	Me	H	H
Aristolactam - AIII	H	Me	OMe	H
Aristolactam - AIIIa	H	Me	OH	H
Aristolactam - BII	Me	Me	H	H
Aristolactam - BIII	Me	Me	OMe	H
Doryflavine	Me	H	OH	H
Taliscanine	Me	Me	H	OMe

TABLE 8

Aristolochic Acids (130)

	R^1	R^2	R^3
Aristolochic Acid	H	H	OMe
Aristolochic Acid B	OH	OMe	H
Aristolochic Acid C	OH	H	H
Aristolochic Acid D	OH	H	OMe
Aristolochic Acid II	H	H	H
Aristolochic Acid IV	OMe	H	OMe
7-Hydroxyaristolochic Acid	H	OH	OMe
7-Methoxyaristolochic Acid	H	OMe	OMe

A few β-aminoethylphenanthrenes have been isolated from natural sources. Typical examples are atherospermidine (131, $R^1=R^2=H$), thalicthuberine (131, $R^1R^2=OCH_2O$), thaliglucine (132, X=2H), thaliglucinone (132, X=O) and thalflavidine, which is 2-methoxythalglucinone. These bases are clearly products of Hofmann degradation of N-methylaporphinium salts and many have been prepared in the laboratory in this way. For example degradation of thalfenine (121) yields thalglucine and of N-methylnuciferine iodide yields atherospermidine. The anti-inflammatory alkaloid taspine (133) clearly arises from the oxidation of such a phenanthrene (Shamma, The Isoquinoline Alkaloids, Academic Press, New York, 1975, p.260).

(131)   (132)   (133)

Synthesis of aporphines

Substantial improvements have been made in methods of synthesis of aporphines, particularly in the oxidative coupling
of the two aromatic nuclei, for which vanadium oxytrifluoride
in trifluoroacetic acid is one of the most effective reagents.
It gives the best yields with monophenolic non-basic benzylisoquinolines. For example N-trifluoroacetylnorcodamine (134,
R=Me) can be converted in this way into N-trifluoroacetylnor-
thaliporphine (135, R=COCF$_3$) in 70% yield. Easily accessible
non-basic starting materials for these oxidations are complexes of tertiary bases with diborane and the codamine-borane
complex (136) gives an 80% yield of thaliporphine (135, R=Me)
when oxidised in this way (S.M. Kupchan and C.-K. Kim, J.org.
Chem., 1976, 41, 3210). Bracteoline which is 10-O-desmethyl-
thaliporphine, has been synthesised in the same way from the
monobenzyl ether of the benzylisoquinoline alkaloid orientaline (Kupchan, O.P. Dhingra and Kim, *ibid.*, 1978, 43, 4076).
The cyclisation can be effected with nonphenolic and with basic materials, for example the same reagent converts tetrahyd-
ropapaverine into a mixture of norglaucine (137, R=H) (30%)
and 4-hydroxynorglaucine (137, R=OH) (38%). The same 4-hydro-
xylation was observed when glaucine was treated with the same
reagent (J. Hartenstein and G. Satzinger, Angew.Chem., intern.
Edn., 1977, 16, 730).

Diphenylselenoxide has also been found to be a valuable oxidant for the synthesis of aporphines from non-basic diphenols
and (134, R-H) is oxidised by this reagent in methanol at room
temperature and after methylation of the crude product with
diazomethane, N-trifluoroacetylnorthaliporphine is obtained
in 80% yield. The process presumably proceeds through the
orthoquinone and a non-basic nitrogen is essential if the pro-

(134)   (135)   (136)

(137)   (138)   (139)

duction of benzopyrrocolines is to be avoided (J.P. Marino and A. Schwartz, Tetrahedron Letters, 1979, 3253). Thallium (III) trifluoroacetate is another effective reagent for the oxidative coupling of rings *para* rather than *ortho* to oxygen, oxidising the nonphenolic base (138) to ocoteine (139, R=H) in 46% yield in methanol and 3',4'-methylenedioxybenzyl-6,7-methylenedioxytetrahydro- *N*—methylisoquinoline to neolitsine in 68% yield in trifluoroacetic acid. Thallium (III) acetate, however, effects acetoxylation as well as aromatic coupling, converting (138) into (139, R=OAc) (E.C. Taylor *et al*., J.Amer. chem.Soc., 1980, 102, 6513).

Tetraethylammonium diacyliodates have been found to oxidise the non basic norreticuline derivatives (140, R=CHO) and (140, R=COOEt) to the corresponding derivatives (141, R=CHO) and (141, R=COOEt) of norisoboldine in good yield (C. Szantay *et al*., Tetrahedron Letters, 1980, 21, 3509). Oxidation by copper (I) chloride and oxygen, by contrast favours *ortho*-coupling and converts reticuline (140, R=Me) into corytuberine (142) (28%) and isoboldine (141, R=Me) (8%). Reticuline *N*-oxide can provide the oxygen for the coupling process, being converted by copper (I) chloride alone into corytuberine in 61% yield. The *N*-oxide of orientaline, which is the 4'-hydroxy-3'-methoxy isomer of reticuline cannot couple *ortho* to the 4'-hydroxyl group for obvious stereochemical reasons and is coupled by copper (I) chloride *para* to the hydroxyl group in very poor yield giving the pro-aporphine alkaloid orientalinone (143) in 5% yield. (Kametani *et al*., Heterocycles, 1976, 5, 175; Kametani and M. Ihara, *ibid*., 1979, 12, 893; J.chem.Soc. Perkin I, 1980, 629.)

The quinol acetates (144) and (146), which are prepared by the oxidation of 6-hydroxy and 7-hydroxytetrahydroisoquinolin-

(140)   (141)   (142)

es with lead (IV) acetate can be converted into aporphines in good yield in strong acids. The structure (144) represents two diastereo-isomeric mixed ketals and the mixture can be cyclised to a mixture in which the principal product is *O*-acetyl-predicentrine (145, R=Me) or the *O,O*-diacetyl compound (145, R=Ac) according to the conditions employed. (O. Hoshino, M. Ohtani and B. Umezawa, Chem.Pharm.Bull., 1978, 26, 3920; Heterocycles, 1981, 16, 793). The acetoxydienone (146, R=H) can be cyclised to a mixture of isothebaine (147, R¹=OMe, R²=H) and the isomeric base (147, R¹=H, R²=OMe), which is not identical with the alkaloid lirinine (148) for which the structure had been regarded as possible (H. Hara *et al*., Chem.Pharm. Bull., 1981, 29, 1083).

A novel synthesis of the aporphine system has been achieved from the diphenylethylamine (149) by condensation with diethyl oxalate followed by Bischler-Napieralski ring closure to the dihydroisoquinoline (150) and further cyclisation with polyphosphoric acid to the oxoaporphine (151). *N*-methylation and reduction of this gave the trimethoxyaporphine *O,N*-dimethyl-tuduranine (M. Grecke and A. Brossi, Helv., 1979, 62, 1549).

An unusual reaction of 3'-nitro-*O*-methylarmepavine (152) when reduced with zinc in trifluoroacetic and trifluromethane-sulphonic acids in coupling to the aporphine (153, R=NH₂), which has been converted into *N*-methyllaurotetanine (153, R= OH) (T. Ohta *et al*., J. Amer.chem.Soc., 1980, 102, 6385). Another potentially useful synthesis involves the Diels-Alder condensation of benzynes with 1-methylene-isoquinolines; *N*-acetyldehydronornuciferine (154) has been prepared in this way from 2-acetyl-6,7-dimethoxy-1-methylenetetrahydroisoquinoline (Castedo *et al*., Tetrahedron Letters, 1982, 23, 457).

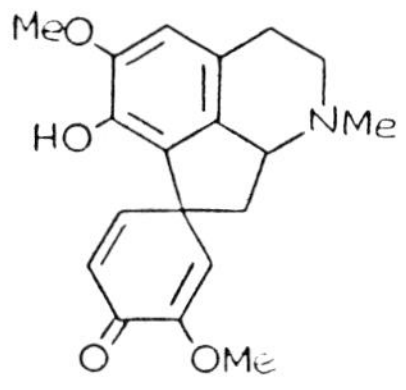

(143)

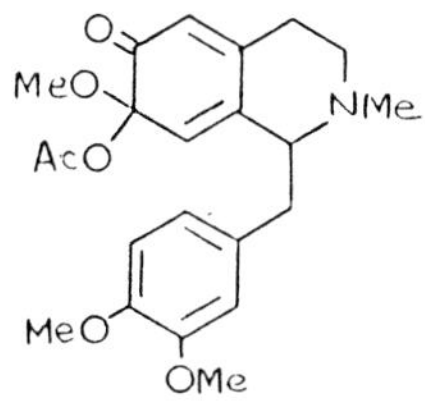

(144)

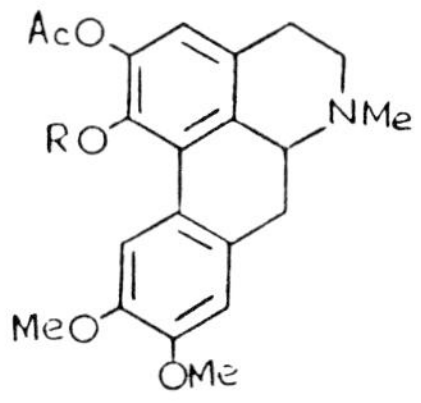

(145)

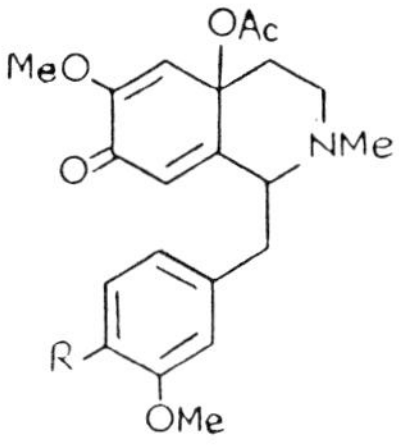

(146)

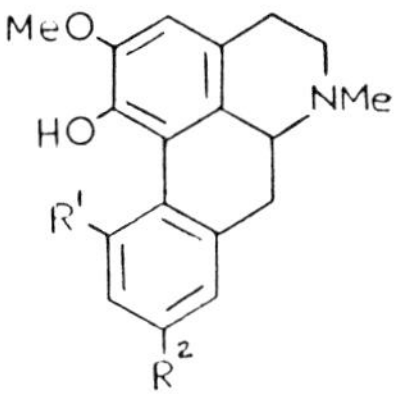

(147)

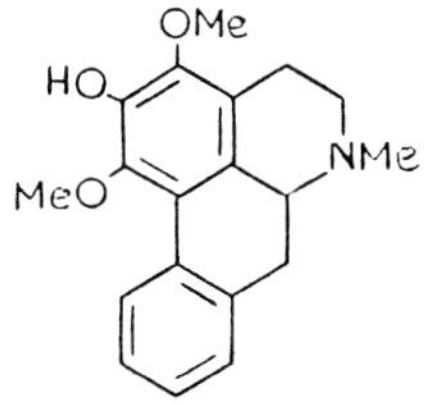

(148)

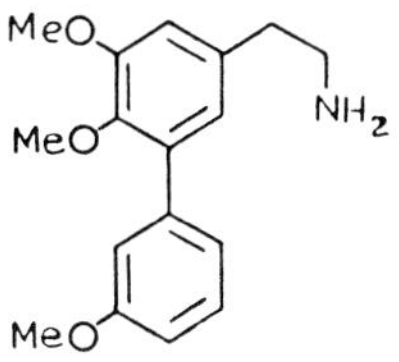

(149)

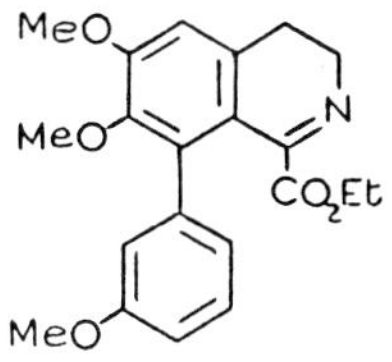

(150)

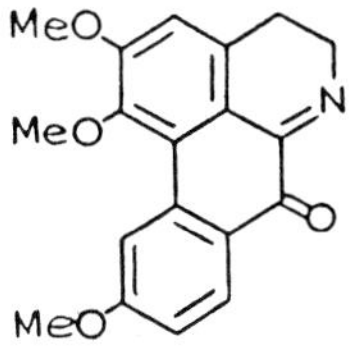

(151)

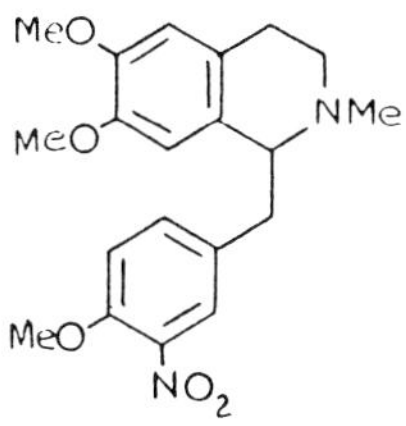

(152)

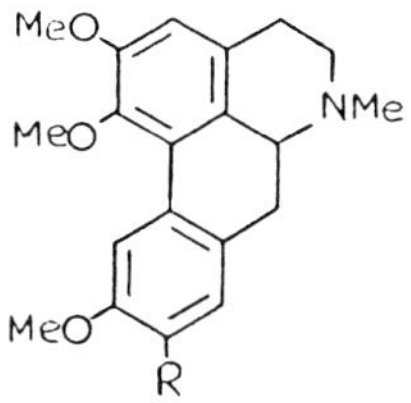

(153)

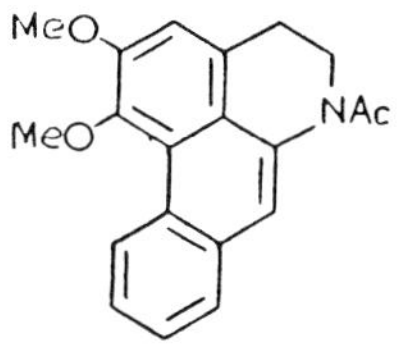

(154)

Dehydroaporphines have been prepared from aporphines by photo-oxidation (Castedo *et al.*, Heterocycles, 1981, 15, 915; 1982, 19, 245) and by microbiological oxidation (P.J. Davis and J.P. Rozza, Bioinorg. Chem., 1981, 10, 97). Photo-irradiation of the hydroxybenzylisoquinolines (155, R¹=H, R²=OH), (155, R¹=OH, R²=H) and (156) affords oliveroline (104, R¹=Me, R²=H), its epimer ushinsunine and oliveridine (105, R¹=Me) R²=H) respectively and similarly the 4-hydroxy compound (157) has been cyclised to 4-hydroxyroemerine (S.V. Kessar *et al.*, Tetrahedron Letters, 1980, 21, 3307; Indian J. Chem., 1981, 20B, 984. Both 7α and 7β hydroxyaporphines have been obtained by the complete reduction of *N*-methyl-7-oxoaporphinium salts (S. Chakalamamul and D.R. Dalton, Tetrahedron Letters, 1980, 21, 2029).

(155)   (156)   (157)

Some aporphines are believed to arise in plants by the rearrangement of proaporphines. This process, known to occur in acid, has now been found to be induced by light in the abscence of acid (S.F. Hussain *et al.*, Tetrahedron Letters, 1980, 21, 723).

Azafluoranthenes

A few azafluoranthenes occur as natural products. They bear the same relationship to 1-phenylisoquinolines as aporphines do to 1-benzylisoquinolines and may have a similar origin to that of aporphines, though an alternative biogenesis by the decarbonylation of 7-oxoaporphines is also plausible. The alkaloids of the group are imenine (158, R=H), telitoxine (158, R=OH), norrufescine (159, R¹=R²=H), rufescine (159, R¹=Me, R²=H) and imelutine (159, R¹=Me, R²=OMe). Rufescine and imelutine have been synthesised by routes involving conventional

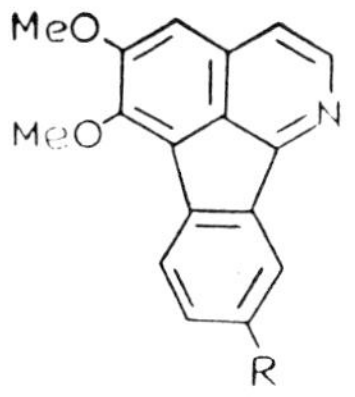

(158) 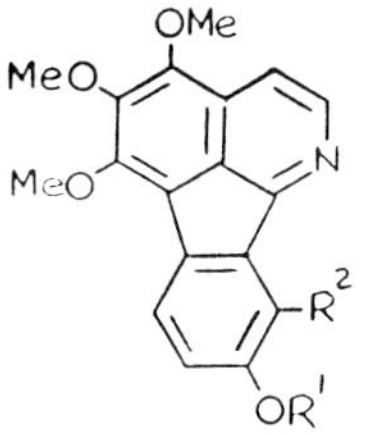   (159) 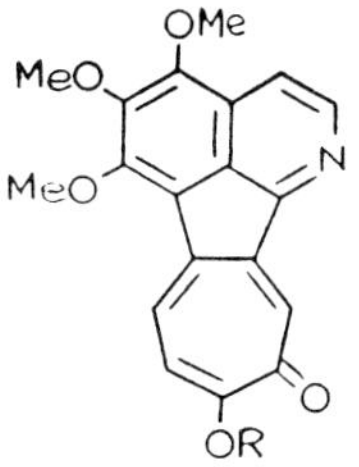   (160)

Bischler-Napieralski and Pschorr ring-closures. The structure of norrufescine has been confirmed by an X-ray crystallographic study (Cava, K.T. Buck and A.I. da Rocha, J.Amer.chem.Soc., 1972, 94, 5931; M.D. Klein *et al.*, *ibid.*, 1978, 100, 662; M.D. Menachery and Cava, J.nat.Prod., 1981, 44, 320). Two related tropolones, the orange-red imerubine (160, R=Me) and the red-brown grandirubine have also been obtained from the same plants as the azafluoranthenes (*Abuta imene* and *A.rufescens*). Their structures were determined crystallographically (J.V. Silverton *et al.*, J.Amer.chem.Soc., 1977, 99, 6708; Menachery and Cava, Heterocycles, 1980, 14, 943).

6. *Pavines and Isopavines*

Few new alkaloids of these relatively small sub-groups have been reported, but there have been significant developments in the synthesis of both structurers.

Quinol acetates, easily obtained by the oxidation of phenolic benzylisoquinolines with lead (IV) acetate can be cyclised to isopavines in good yield in trifluoro-acetic acid. For example the quinol acetate (146, R=OMe) is converted into thalisopavine (162, R=H), presumably through the quinone methide (161). *O*-Methylthalisopavine (162, R=Me) has been obtained by Stevens' rearrangement of the quaternary tetrahydro-6, 12-methanobenz[c,f]azocine (163). This same salt has also been converted into the pavine alkaloid argemonine by the following reaction sequence. Hofmann degradation of the quaternary salt (163) affords the olefin (164), which yields the ketone (166) when subjected to an oxidative rearrangement. Reduction of this ketone to the secondary alcohol and heating of the alcohol with acetic anhydride and acetic acid results in cyclisation by a reversed Hofmann reaction to give

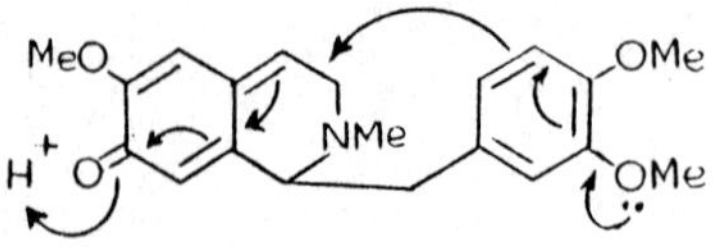

(161)

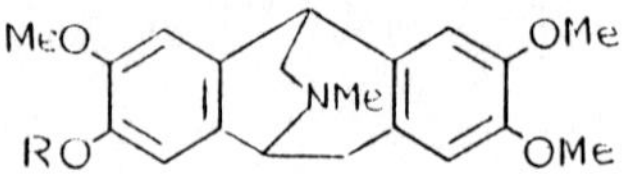

(162)

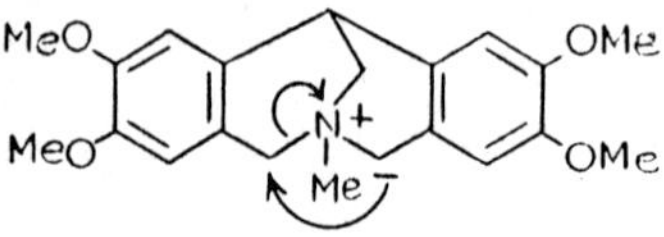

(163)

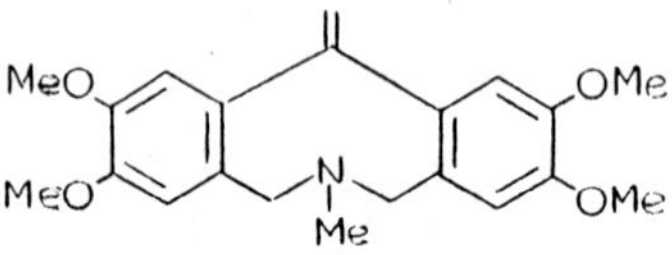

(164)

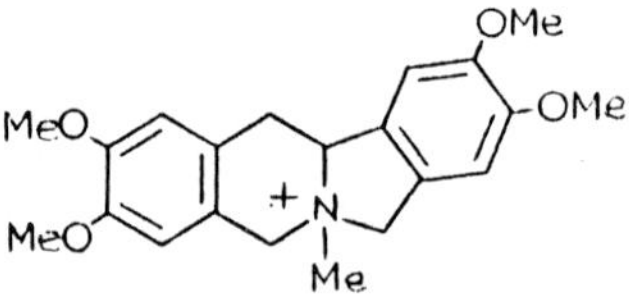

(165)

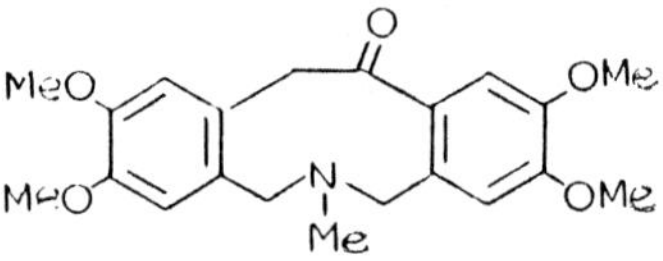

(166)

(167)

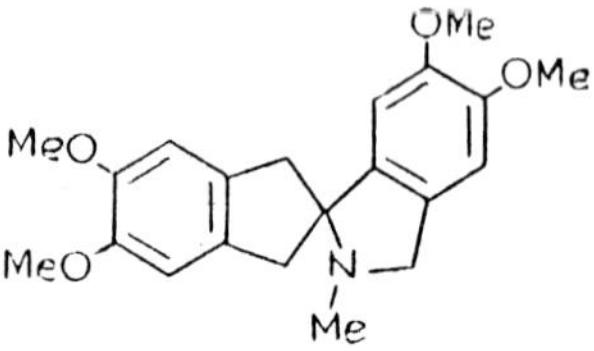

(168)

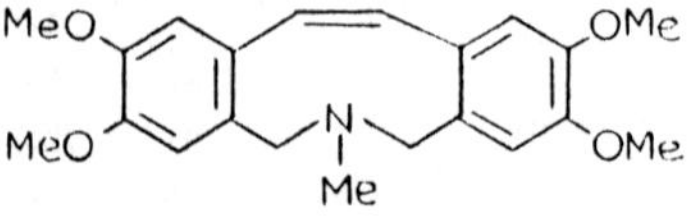

(169)

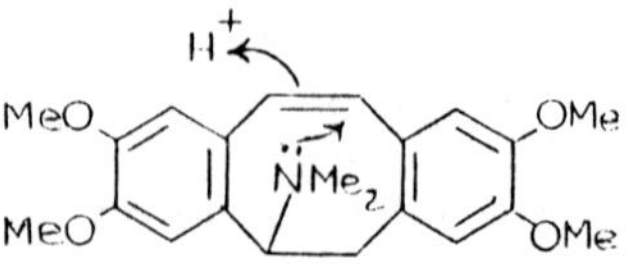

(170)

the quaternary salt (165), isomeric with the original salt
(163).   When Stevens' rearrangement is applied to this salt
(165) the products are argemonine (167) and the spiro-base
(168).   A better   yield of argemonine is obtained from the
salt (165) by Hofmann degradation to the olefinic base (169),
quaternisation of this and Stevens' rearrangement of the resulting salt to the olefinic base (170), which is then subjected to transannular cyclisation in acid and N-demethylation of
the resulting quaternary salt (H. Takayama, M. Takamoto and T.
Okamoto, Tetrahedron Letters, 1978, 1307;   T. Nomoto and Takayama, Chem.Comm., 1981, 1113).

N-Formylnorreticuline has been oxidised by dichlorodicyanobenzoquinone in methanol to 4-methoxy-N-formylnorreticuline,
which may be cyclised in mineral acid to the pavine and in
methanesulphonic acid to the N-formylisopavine (K.C. Rice et
al., J.org.Chem., 1980, $\underline{45}$, 801).

Two aporphine-pavine dimers, pennsylpavoline (171, R=H) and
pennsylpavine (171, R=Me), together with their aporphine-benz-
ylisoquinoline dimer precursors, pennsylvamine and pennsylvan-
amine, have been isolated from Thalictrum polyganum (Shamma
and Moniot, J.Amer.chem.Soc., 1974, $\underline{96}$, 3338).

(171)

7.   *Benzopyrrocolines*

O-Methylcryptaustoline iodide (175, R=Me) has been synthesised
from the keto-acid (172, R=H), via the nitro-compound (172, R=
NO_2) and the amine (173), which on heting is converted into
the lactam (174) which is reduced and quaternised to (175, R=
Me) (P.W. Elliott, J.org.Chem., 1982, $\underline{47}$, 5398).   Cryptaustol-
ine iodide (175, R=H) has been obtained by photolysis of the
enamine (176) followed by reductive debenzylation and N-methy-

(172)   (173)   (174)

(175)   (176)   (177)

lation (I. Ninomiya, J. Yasui and T. Kiguchi, Heterocycles, 1977, 6, 1855) and the same alkaloid and its methylenedioxy-analogue cryptowoline have been synthesised by the cyclisation of the appropriate 2'-bromobenzyltetrahydroisoquinoline in liq-uid ammonia containing potassamide. In this last process the products are only stable if they are phenolic; non-phenolic salts suffer Hofmann degradation to such bases as (177) (I. Ahmad and M.S. Gibson, Canad.J.Chem., 1975, 33, 3660; Kessar *et al.*, Indian J.Chem., 1975, 13, 1109, 1116).

8.   Berberines and Tetrahydroberberines

Four unusual alkaloids of this group, with bridged ring syst-ems, have been discovered. These are solidaline (178), from *Corydalis solida* (R.H.F. Manske *et al.*, Canad.J.Chem., 1978, 56, 383), staudine (179), from *Pachypodanthium staudii* (Cavé *et al.*, J.nat.Prod., 1980, 43, 103), karachine (180, R=Me), from *Berberis aristata* (G. Blasko *et al.*, J.Amer.chem.Soc., 1982, 104, 2039) and valachine (180, R=H) from *B.valdiviana* (Firdous *et al.*, Chem.Comm., 1984, 1371). The structures of these bases were determined by spectroscopic and crystallogra-phic methods. Karachine is the first alkaloid of the group to

incorporate two acetone residues and valachine incorporates
one acetone and one acetaldehyde residue.

(178)   (179)   (180)

The oxidation of alkaloids of this group with a variety of
reagents has been studied in detail. With lead (IV) acetate
phenolic tetrahydroberberines are oxidised to quinol acetates
in the same way as benzylisoquinolines. Govanine (181, R^1=Me,
R^2=H) is converted in this way into the dienone (182) and this
is transformed by acetic anhydride and sulphuric acid into ace-
toxyacetylgovanine (183, R^1=Me, R^2=R^3=Ac), which on hydrolysis
in methanol gives (183, R^1=R^3=Me, R^2=H) and its C-5 epimer.
The isomeric phenol (181, R^1=H, R^2=Me) gives the acetoxy compound (183, R^1=R^3=Ac, R^2=Me) and its C-5 epimer directly on
treatment with lead (IV) acetate. The phenol (184, R^1=H, R^2=
Me) behaves in the same way as govanine, giving first the acetoxydienone (185), but this is converted by acetic anhydride
and sulphuric acid into a mixture of the diastereoisomers (186,
R^1=Ac, R^2=OAc, R^3=H) and (186, R^1=Ac, R^2=H, R^3=OAc), which are
both hydrolysed in methanol to the single compound (186, R^1=R^3
=H, R^2=OMe). In contrast with all of these reactions the phe-

(181)   (182)   (183)

nol (184, R^1=Me, R^2=H) is converted by lead (IV) acetate into the acetoxyphenol (189), presumably by way of ring opening of the acetoxydienone (187) and Mannich-type cyclisation of the resulting iminium salt (188); only a very small amount of an isomer of (186, R^1=Ac, R^2=H, R^3=OAc) is formed (Hara *et al.*, J.chem.Soc., Perkin I, 1980, 1169).

(184)   (185)   (186)

(187)   (188)   (189)

Betaines of 13-hydroxyberberine and its analogues of general type (190) are easily prepared. Berberine betaine (190, R=H) is obtained by the oxidation of berberine-acetone (191) with potassium permanganate, in a process that invoves a retro-Mannich loss of acetone, and methoxyberberine betaine (190, R= OMe) results from the controlled photo-oxidation of berberine. Other analogues can be prepared by the oxidation of dihydro- and 8-alkyldihydro-berberines. Both (190, R=H) and (190, R= OMe) can be reduced to a mixture of ophiocarpine (192) and the C-13 epimeric epiophiocarpine (M. Hanaoka. C. Mukai and Y. Arata, Heterocycles, 1977, <u>6</u>, 895). Photo-oxidation of the betaine (190, R=H) in the presence of Rose Bengal gives the peroxide (193), which can be photolysed further at elevated temperatures to the lactone (194) and the aldehyde (195) (Y. Kondo, J. Imai and S. Nozoe, J.chem.Soc., Perkin I, 1980, 911 and 919). Similar products, though in a higher state of oxidation can be prepared from the betaine (190, R=OMe). Acid-catalysed hydrolysis of this betaine, followed by autoxidation in pyridine, affords the carbinolamine (196), which can also be obtained by rearrangement of the peroxide (193) by pyridine

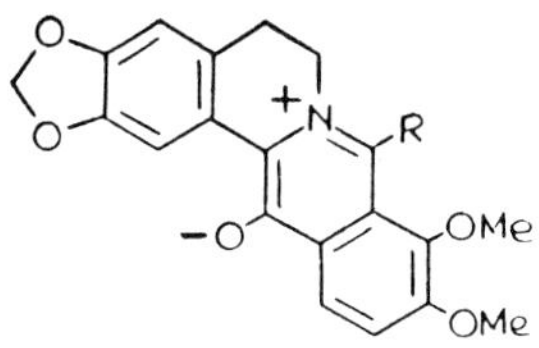

(190) 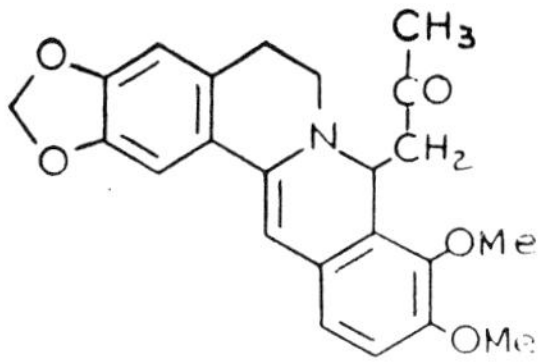　　　　　　　(191) 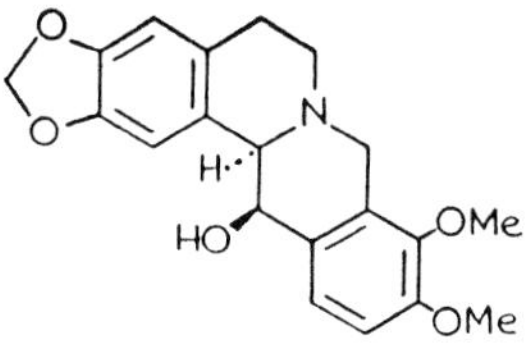　　　　　　　(192)

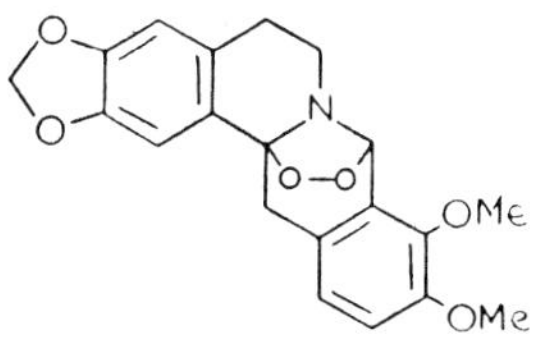

(193) 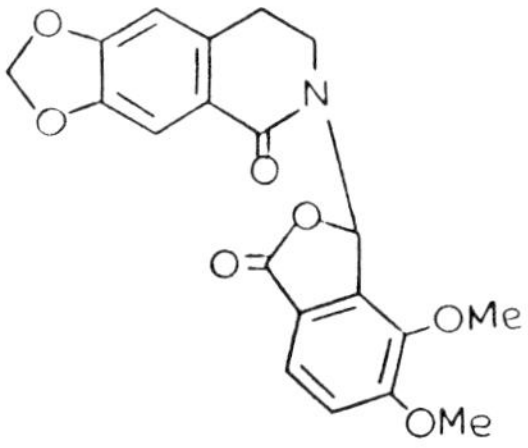　　　　　　　(194) 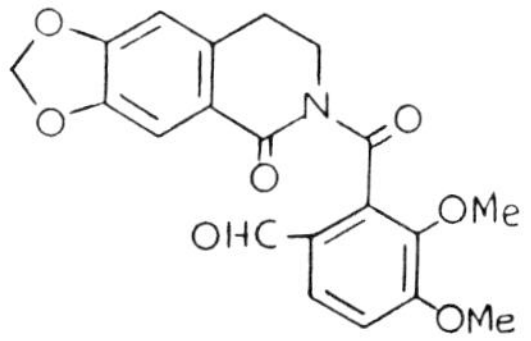　　　　　　　(195)

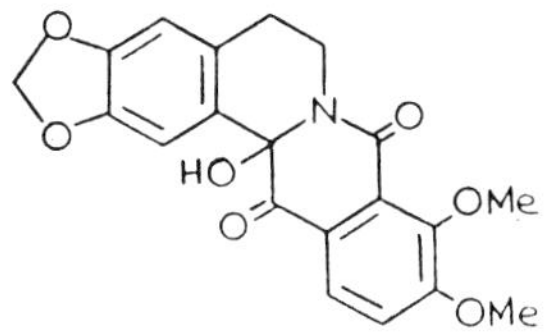

(196) 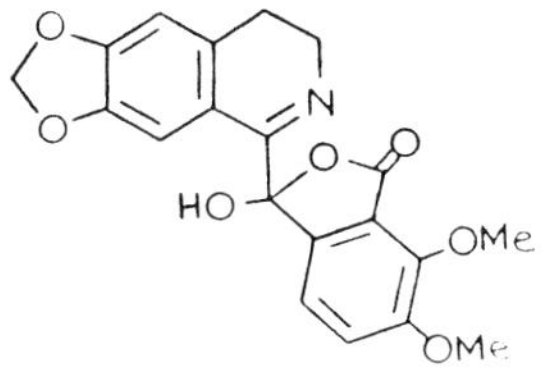　　　　　　　(197) 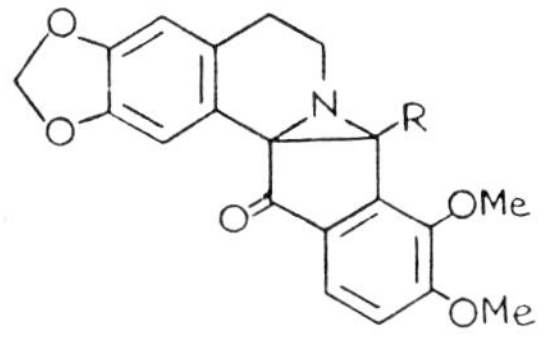　　　　　　　(198)

hydrochloride. This carbinolamine (196) can be rearranged by
ammonium chloride to the isoindoloisoquinoline alkaloid chilenine (see section 16) and converted into the ring-opened base
(197), related to hydrastine, by 25% sulphuric acid (Moniot,
D.M. Hindenlang and Shamma, J.org.Chem., 1979, 44, 4343 and
4347).

Non-oxidative photolysis of the betaines (190, R=H) and
(190, R=OMe) affords the aziridines (198, R=H) and (198, R=
OMe), which can be ring-opened in both possible ways to spiro-
benzylisoquinolines (see section 13) and indanobenzazepines
(see section 14) (Hanaoka *et al*., Tetrahedron Letters, 1979,
3749; Heterocycles, 1982, 18, 31). The methoxybetaine (190,
R=OMe) undergoes Diels-Alder addition to acetylenic dienophils
to give (199, $R^1=R^2=$COOMe), (199, $R^1=R^2=$COPh), (199, $R^1=$COOMe,
$R^2=$H), (199, $R^1=$H, $R^2=$COOMe), (199, $R^1=$COMe, $R^2=$H) and (199,
$R^1=$H, $R^2=$COMe) and these adducts have been rearranged by heat
to the related lactams (200). Similar additions of the betaine (190, R=H) to acetylenes have been observed (Hanaoka *et
al*., Heterocycles 1979, 12, 511).

Berberine (201) is degraded by prolonged boiling with acetic anhydride and sodium acetate to the naphthalene derivatives (205) and (206). The process doubtless involves the enone (204) or its equivalent and this may be formed through
the intermediates (202) and (203) (Shamma *et al*., Tetrahedron
Letters, 1975, 3803; Tetrahedron, 1977, 23, 2907). The reaction appears to be a general one of *N*-alkylisoquinolinium
salts.

Tetrahydroberberine is cleaved by ethyl chloroformate to
give (207) and (208) as minor products and (209) as the major
product, in marked contrast to cleavage by cyanogen bromide,
which gives an *N*-cyano-analogue of (209) only as a minor product. Hydrolysis and *N*-methylation of (209), followed by ox-

(199)   (200)

(201)   (202)   (203)

(205)   (204)   (206)

idation of the CH$_2$OH group gives the secoberberine alkaloid
canadaline (210) and hydrolysis and condensation with formaldehyde gives the hydroxymethyltetrahydroberberine demethoxy-
mecambridine (211) (Hanaoka *et al.*, Chem.Pharm.Bull., 1983,
31, 2685). The aziridine (198, R=H) is cleaved by ethyl chloroformate to the chloro-compound (212, R^1=H, R^2=Cl), whereas
the cleavage product from the aziridine (198, R=Me) loses hydrogen chloride to give (212, R^{1}R^2= CH$_2$) (Hanaoka *et al.*,
Tetrahedron Letters, 1979, 3749).

(207)   (208)   (209)

(210) (211) (212)

Several new approaches to the synthesis of tetrahydroberberines have been developed. The photolysis of N-benzoyl-1-
methylenetetrahydroisoquinolines has been found to give oxy-
berberines, reducible to tetrahydroberberines. Initially the
starting materials are brominated in the benzoyl group and the
reaction generally proceeds with the loss of bromine as in the
sequence (213)→(214)→(215), where R=Br, R^1=H, R^2=OMe, though
in some cases the bromine is retained and methoxyl lost as in
the conversion of (213, R=Br, R^1=OMe, R^2=H) into the lactam
(216). The bromine, however, is not essential for the cyclisation to occur and neither is the amide system. For example
the photolysis of N-benzyl-7-benzyloxy-6-methoxy-1-methylene-
tetrahydroisoquinoline, followed by reductive debenzylation,
affords the unusually substituted tetrahydroberberine alkaloid
bharatamine (218) which is believed to be formed from tyrosine
and a monoterpene rather than from two moles of tyrosine. In
one case an unbrominated compound failed to aromatise (213, R^1
=R^2=R^3=H) giving the diene (217) (Kametani *et al.*, J.Chem.Soc.,
Perkin I, 1977, 1151; Pakrashi *et al.*, Tetrahedron Letters,
1983, 24, 291; T. Naito, Y. Tada and Ninomiya, Heterocycles,
1983, 20, 853).

(213) (214) (215)

(216)   (217)   (218)

The cyclisation of appropriately substituted N-benzyl quaternary salts of 3,4-dihydroisoquinolines has afforded tetrahydroberberines. For example the electrolytic reduction of the salt (219, R=CH$_2$Br) and the treatment of the salt (219, R=CH$_2$SiMe$_2$) with sodium fluoride yields xylopinine (220, R^1=H, R^2=OMe) (T. Shono et al., Tetrahedron Letters, 1978, 4819; S. Takano, H. Numata and K. Ogasawa, Heterocycles, 1983, 20, 417). The condensation of 6,7-dimethoxy-3,4-dihydroisoquinoline with methyl 2,3-dimethoxy-6-bromomethylbenzoate in acetonitrile in the presence of zinc as a reducing agent affords tetrahydropalmatine (220, R^1=OMe, R^2=H) (H. Hamaguchi et al., J.org. Chem., 1983, 48, 1621) and the condensation of the same dihydroisoquinoline with the anion of m-meconine yields the hydroxy amide (221), which has been reduced to 13-hydroxytetrahydropalmatine (R. Marsden and D.B. MacLean, Tetrahedron Letters, 1982, 24, 2063).

Homophthalic anhydrides can be condensed with 3,4-dihydroisoquinolines or the equivalent 1-hydroxytetrahydroisoquinolines to give lactam acids from which tetrahydroberberines can be prepared by decarboxylation and reduction. 13-Methyltetrahydroberberines are accessible by reduction to 13-hydroxymethyl compounds, tosylation and further reduction; for example (222), prepared from norhydrastinine, can be converted into canadine and thalictricavine. 1-Chloroisoquinolines in a similar reaction gave 6,7-dehydro analogues of compounds such as (222) (M. Cushman, J. Gentry and F.W. Dekow, J.org. Chem., 1977, 42, 1111; M. Attaimova et al., Tetrahedron, 1977, 33, 331; B.R. Pai and S. Natarajam, Indian J.Chem., 1982, 21B, 607; V. Ongayanov and M. Haimova, Heterocycles, 1982, 19, 1069).

The berberine ring system has been completed by carbonyl insertion into 6'-bromotetrahydropapaverine (223, R^1=R^2=Me) and (223, R^{1}R^2=CH$_2$) by treatment with carbon monoxide, lead

(219)

(220)

(221)

(222)

(223)

(224)

(IV) acetate and triphenylphosphine in tributylamine to give 8-oxoxylopinine (224, $R^1=R^2=$Me) and its analogue (224, $R^1R^2=$ CH_2) (Pandey and Tiwari, Synth.Comm., 1979, 10, 895; Tetrahedron, 1981, 37, 1213).

9. Azaberberines

Eight 10-azaberberine alkaloids have been isolated from *Alangum lamarckii*, namely alamarine (225, R^1=Me, R^2=H) and iso-alamarine (225, R^1=H, R^2=Me), alangimarine (226, R^1=Me, R^2=H) and isoalangimarine (226, R^1=H, R^2=Me), alangimarinone (227), dihydroalamarine (228, R^1=Me, R^2=H) and dihydroisoalamarine (225, R^1=H, R^2=Me) and alangimaridine (229). Their structures follow from their spectra and from the dehydration of alamarine and the oxidation of alangimaridine to alangimarine (226, R^1=Me, R^2=H) (Pakrashi *et al.*, Tetrahedron Letters, 1980, 21, 2667; Indian J.Chem., 1985, 24B, 19) and have been confirmed by the synthesis of alamarine by the photolytic and thermal cyclisation of the enamide (230), followed by reduction and hydrogenolysis (Naito *et al.*, Heterocycles, 1981, 16, 725.

(225)　　　　　　　(226)　　　　　　　(227)

(228)　　　　　　　(229)　　　　　　　(230)

10. Secoberberines

A group of ring-opened berberines structurally different from
the protopine alkaloids and not so highly oxidised as the phthalide-isoquinolines has been identified. They can be formally regarded as arising from tetrahydroberberines by oxidation to carbinolamines (231) followed by ring opening to aldehydes such as (232) which can then be reduced or further oxidised to other bases in the group. Alkaloids so far identified
are canadaline (232, $R^1=R^2=Me$) aobamine (232, $R^1R^2=CH_2$), corydalisol (233), macrantaline (234, $R=CH_2OH$), macrantoridine
(234, R=COOH), peshawarine (235), hypecorimine (236, X=O) and
hypecorine (236, $X=H_2$), obtained from *Corydalis incisa*, *C.och-
otensis*, *Papaver pseudo-orientale* and *Hypecoum procumbens*.
　　The structures and absolute configurations of the alkaloids
have been determined by interconversions between them and between them and alkaloids of the tetrahydroberberine and phthalide-isoquinoline groups. Canadaline (232, $R^1=R^2=Me$) has been
prepared from tetrahydroberberine by treatment with ethyl
chloroformate to give (209) followed by hydrolysis, *N*-methyl-
ation and oxidation (Hanaoka, Nagami and Imanishi, Heterocyc-

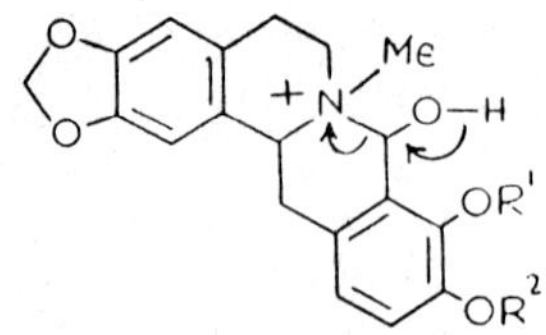

(231)

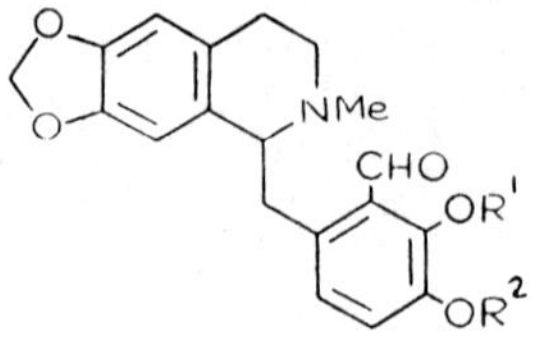

(232)

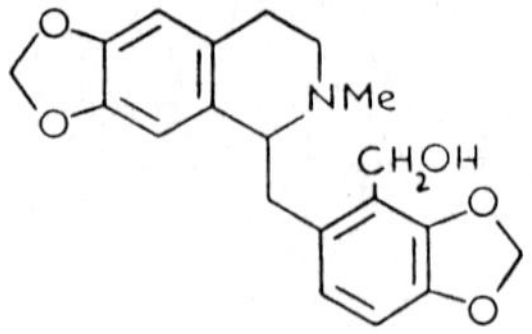

(233)

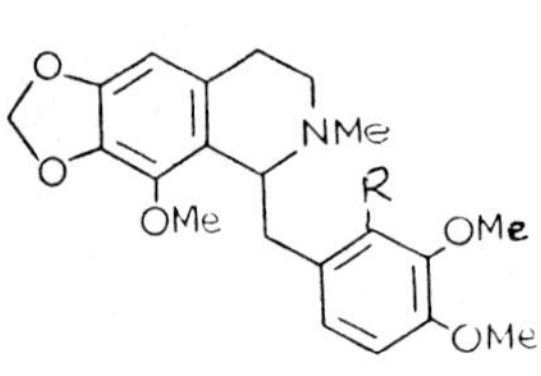

(234)

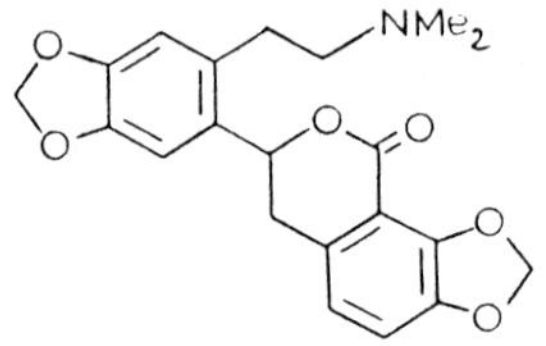

(235)

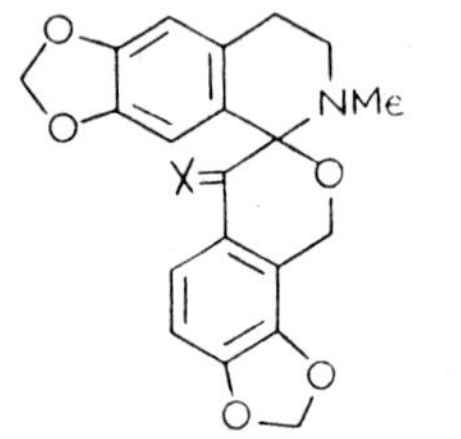

(236)

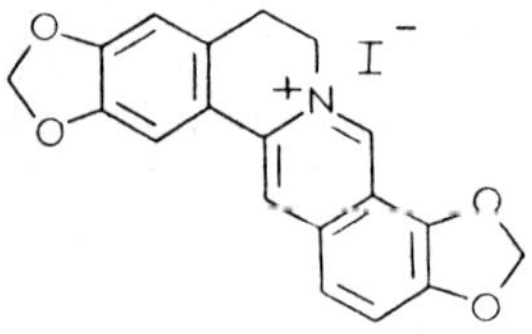

(237)

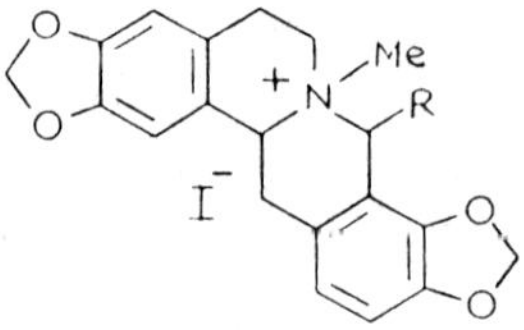

(238)

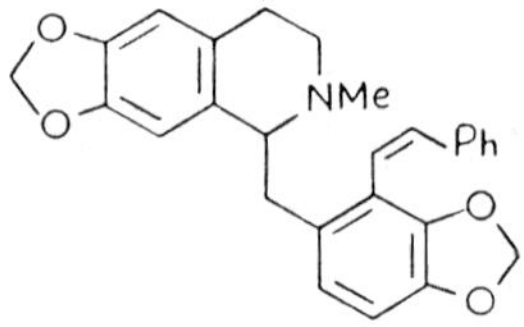

(239)

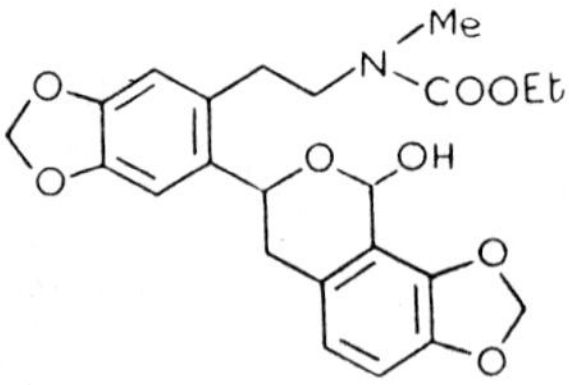

(240)

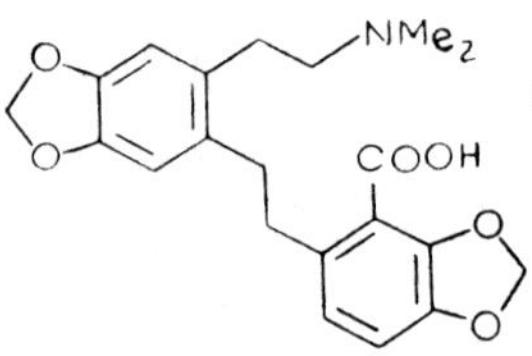

(241)

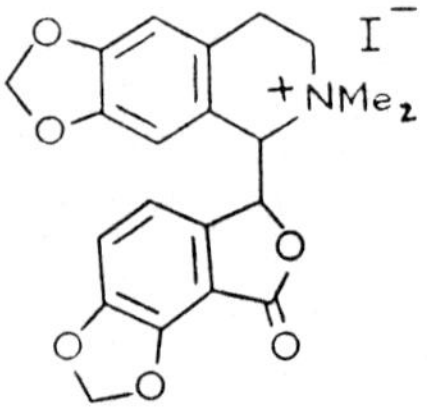

(242)

les, 1979, 12, 497; H. Roensch, Z.Chem., 1979, 19, 447).
The treatment of coptisine iodide (237) with benzylmagnesium
iodide, followed by reduction and N-methylation gives 8-benzyl-
tetrahydrocoptisine methiodide (238, R=CH$_2$Ph). Hofmann degradation of this salt gives the stilbene derivative (239), which
is oxidised by osmium tetroxide to aobamine (232, R^{1}R^2=CH$_2$)
which gives coralydisol (233) on reduction with sodium borohydride (Shamma, A.S. Rothenberg and Hussain, Heterocycles,
1977, 6, 707). Coralydisol is converted into tetrahydriocopt-
isine methiodide (238, R=H) by treatment with phosphorus (IV)
chloride and sodium iodide (G. Nonaka and I. Nishioka, Chem.
Pharm.Bull., 1975, 23, 294).

    Cleavage of aobamine (232, R^{1}R^2=CH$_2$) results in the format-
ion of the hemi-acetal (240), which is converted into peshawarine (235) by reduction with lithium aluminium hydride followed by oxidation of the resulting N-methyl primary alcohol with
chromic acid. Hydrogenolytic cleavage of the benzylic ester
system of peshawarine affords the acid(241), also obtained by
the hydrogenolysis of bicuculline methiodide (Shamma et al.,
Tetrahedron, 1978, 34, 635). The absolute configuration of
peshawarine is demonstrated by the production of the alkaloid
by the Emde reduction of rhoeadine methiodide (243, R=Me) to
the mixed acetal (244), followed by hydrolysis and oxidation
(V. Simanek, V. Preininger and F. Santavy, Heterocycles, 1977,
6, 711) and by the production of peshawarine diol (245) by the
reduction of peshawarine with lithium aluminium hydride and by
the Emde reduction of rhoeagenine methiodide (243, R=H) (Shamma et al., Tetrahedron, 1978, 34, 635).

    Macrantoridine (234, R=COOH) has been reduced to macrant-
aline (234, R=CH$_2$OH) by lithium aluminium hydride and macrantaline has been converted into the base (234, R=CH$_3$) obtainable
also from narcotine (246) via narcotine diol (G. Sariyar and

(243)   (244)   (245)

310

J.D. Phillipson, Phytochemistry, 1977, 16, 2009). Hypecorinine (236, X=O) has been reduced by lithium aluminium hydride
to a mixture of bicuculline diol (247, R^1=OH, R^2=H) and adlumine diol (247, R^1=H, R^2=OH), obtained by the reduction of the
diastereoisomeric phthalide-isoquinoline alkaloids bicuculline
and adlumine (see section 12) with the same reagent (Nonaka
and Nishioka, loc.cit.). Hypecorinine has been synthesised
from dehydrobicuculline by reduction with lithium aluminium
hydride and re-oxidation with mercury (II) acetate (B.C. Nalliah and MacLean, Canad.J.Chem., 1978, 56, 1378).

(246)   (247)   (248)

(249)   (250)   (251)

(252)   (253)   (254)

Protonation of hypecorine (236, X=2H) gives the iminium salt (249), in which form it can be reduced by sodium borohydride to corydalisol (233). Basification of the salt (249) regenerates hypecorine and analogues of hypecorine and hypecorinine have been prepared by basifying the salts (250, X=2H) and (250, X=O) (Simanek *et al.*, Heterocycles, 1978, 9, 1233).

Both hypecorine and corydalisol have been obtained by the reduction of the base (251), which is one product of the pyrolysis of protopine *N*-oxide (section 11). The base (252), structurally similar to hypecorine and hypecorinine, is a minor product of the photolytic oxidation of the alkaloid fumaricine (see section 13).

6'-Hydroxymethyllaudanosine, which is an analogue of macrantaline, gives the oxirane (253) on treatment with ethyl chloroformate, whereas macrantaline under the same conditions gives the stilbene (254) (S.Prior, W. Wiegrebe and G. Sariyar, Arch.Pharm. (Weinheim), 1982, 315, 273).

11. *Protopines*

Allocryptopine (256) has been obtained by the photo-oxidation of tetrahydroberberine methiodide (255) (Hanaoka *et al.*, Pol. J.Chem., 1979, 53, 79).

A novel synthesis of the protopine ring system has been achieved from indenobenzazepines. Friedel-Crafts cyclisation of the chloroacetamide (257) gives the lactam (258) which when treated with base and the appropriate benzyl bromides give (259, $R^1 = R^2 = Me$) and (259, $R^1R^2 = CH_2$), which are then cyclodehydrated by phosphorus oxychloride to the indenobenzazepines (260). Photo-catalysed oxidation of the indenobenzazepines gives the 8-oxoprotopines (261), which are completely reduced to the alcohols (262) and these re-oxidised by manganese (IV) oxide to fagarine-II (263, $R^1 = R^2 = Me$) and ψ-protopine (263, $R^1 R^2 = CH_2$) (K. Orito *et al.*, Heterocycles, 1980, 14, 11; Orito, Y. Yurokawa and M. Itoh, Tetrahedron, 1980, 36, 617).

A reverse transformation of the protopine system to an indanobenzazepine is represented by the formation of the base (264) from 13-oxo-allocryptopine (265) by irradiation in *t*-butanol in the presence of potassium *t*-butoxide (Blasko *et al.*, Chem.Comm., 1981, 1246).

The pyrolysis of protopine *N*-oxide has been shown to give, in addition to the oxamine (251), the normal product (266) of Cope degradation and the nitrogen-free cyclised product (267). Photolysis of the *N*-oxide gives the oxamine (251) and the benzophenanthridine alkaloid sanguinarine (268). Analogous prod-

(255)   (256)   (257)

(258)   (259)   (260)

(261)   (262)   (263)

ucts have been obtained by the pyrolysis and the photolysis of allocryptopine *N*-oxide (K. Iwasa and N. Takao, Heterocycles, 1983, 20, 1535).

Studies of the ^{13}C-nmr spectra of the salts of protopine, allocryptopine, 1-methoxyallocryptopine, hunnemarine and thal-

(264)   (265)   (266)

(267)   (268)   (269)

ictrine show that the *cis*-protopinium salt (269) predominates over the *trans*-salt in all cases except for thalictrine, where the *trans*-salt predominates (Hussain *et al.*, J.nat.Prod., 1983, 46, 251).

12. *Phthalideisoquinolines*

Six new phthalide-isoquinoline alkaloids, adlumidine (270, R^1 R^2=CH$_2$) hydrastidine (270, R^1=H, R^2=Me), corftaline (270, R^1=Me, R^2=H), adlumine (271, R^1=R^2=Me), corledine (271, R^1=Me, R^2=H) and severzine (271, R^1=H, R^2=Me) have been discovered. In addition several modified phthalide-isoquinolines have been isolated as natural products. These comprise the hemi-acetal egenine (272) (B. Gozler, T. Gozler and Shamma, Tetrahedron, 1983, 39, 577) and the ring-opened isoquinoline Z-aobamidine (273, $\overline{R^1}$=H, R^2R^3=CH$_2$), which is the enol-lactone of adlumidiceine (Kametani *et al.*, Heterocycles, 1976, 4, 723) and the related base microcarpine (273, R^1=OH, R^2=R^3=$\overline{Me}$) (Boente *et al.*, Tetrahedron Letters, 1984, 25, 889), the amides fumschleicherine (274) Kh.G. Kiryakov *et al.*, Phytochemistry, 1980, 19, 2507), Z-fumaramine (275), *E*-fumaramine (276) and narceine imide (277)

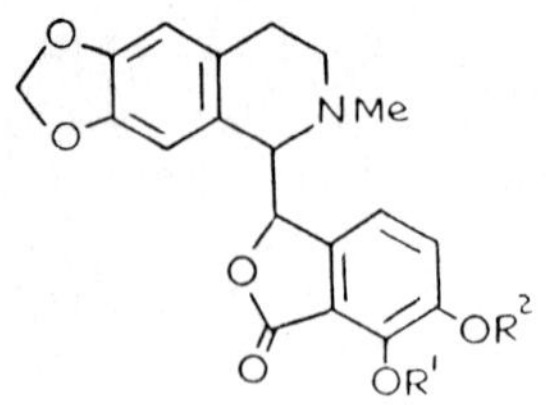

(270)

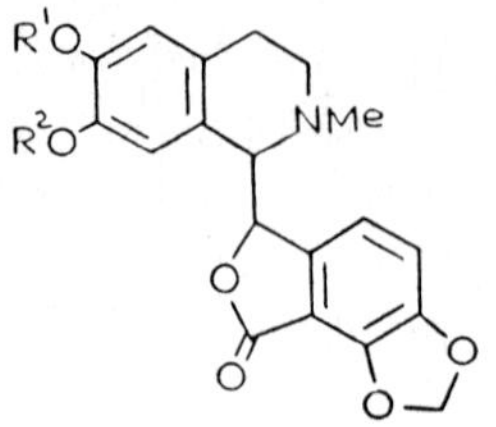

(271)

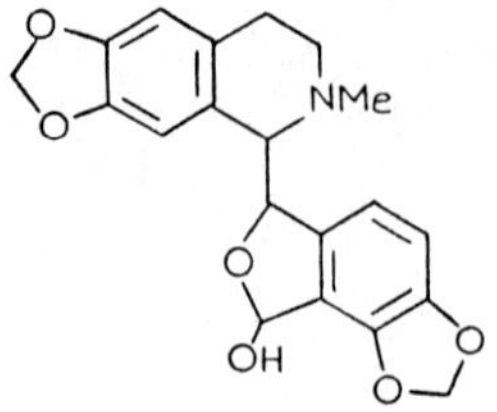

(272)

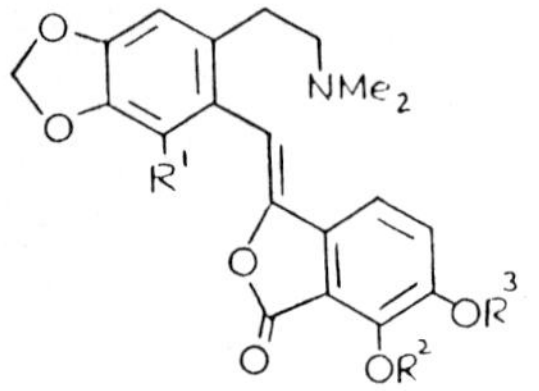

(273)

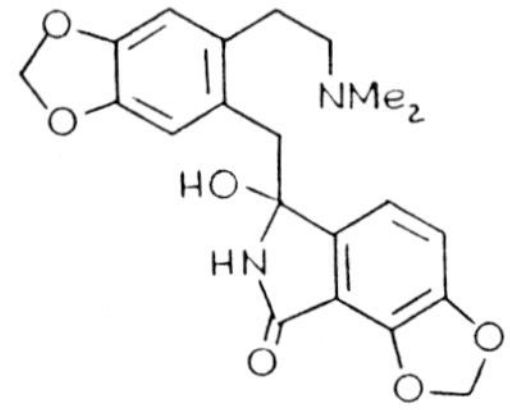

(274)

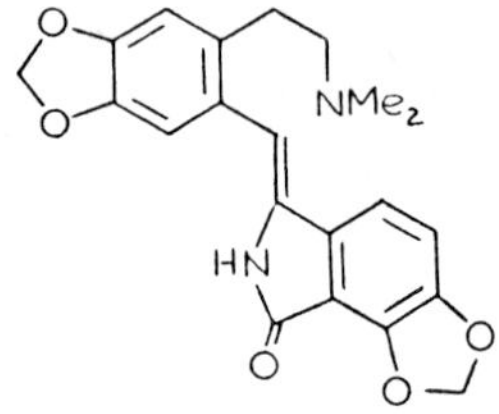

(275)

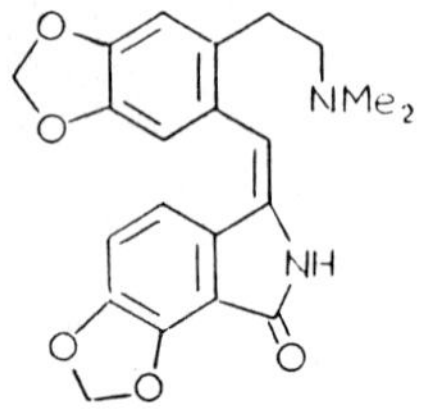

(276)

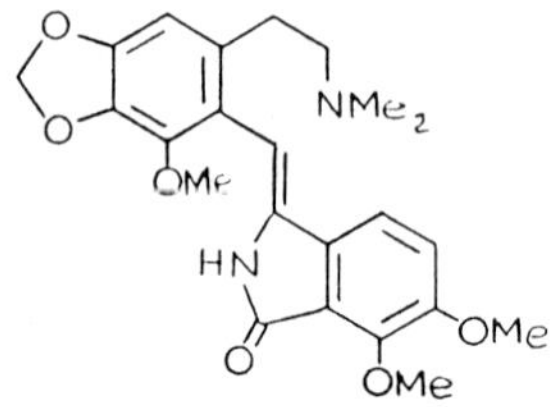

(277)

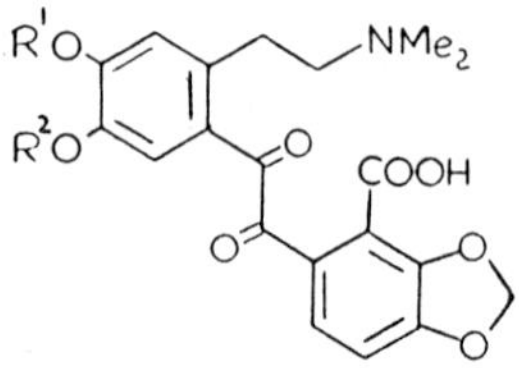

(278)

(Shamma and Moniot, Chem.Comm., 1975, 89; Gozler, Gozler and Shamma *loc.cit*.), the diketones bicucullinine (278, $R^1R^2=CH_2$) and bicucullinidine (278, $R^1=R^2=Me$) (R.G.A. Rodrigo *et al*., Canad.J.Chem., 1976, 54, 471) and the seco-phthalide-isoquinoline narlumidine (K.K. Seth *et al*., Chem.Ind.(London), 1979, 744; Dasgupta *et al*., Planta Med., 1984, 50, 481).

Several conversions of derivatives of berberine into hydra-

stine (270, $R^1=R^2=Me$) have been reported. Dye-catalysed photo-
oxidation of oxidoberberine (280, R=H) affords the lactol (281)
the methiodide of which can be reduced with sodium borohydride
to give (±)−β−hydrastine. (Shamma, *et al.*, Tetrahedron Letters, 1977, 4285; Y. Kondo, J. Imai and S. Nozoe, J.Chem.Soc.,
Perkin I, 1980, 919). Methoxyoxidoberberine (280, R=OMe) on
hydration and *N*-methylation gives the keto-ester (282) which

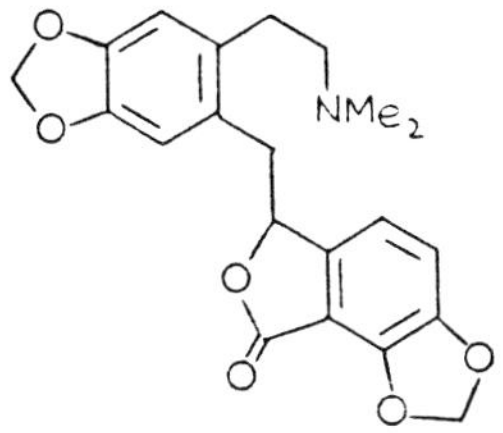

(279)

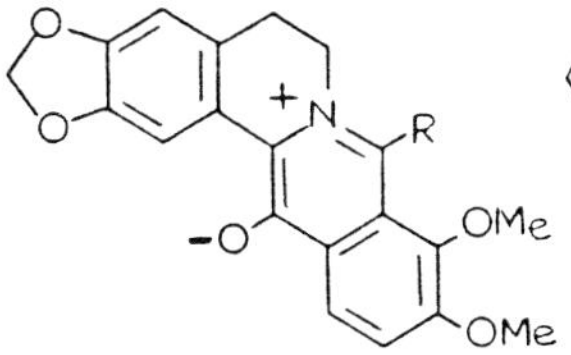

(280)

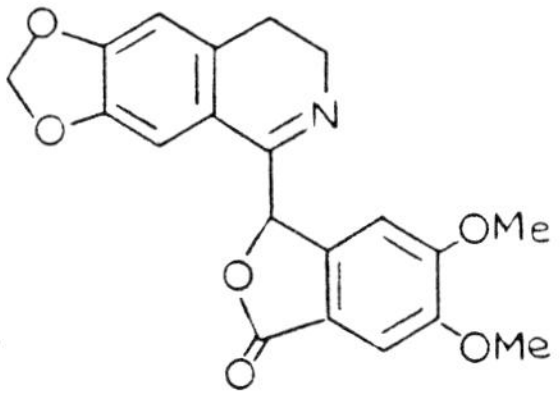

(281)

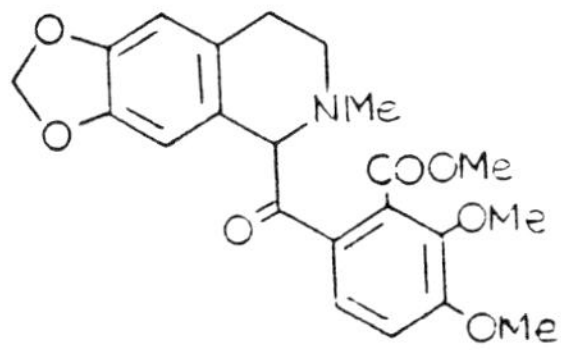

(282)

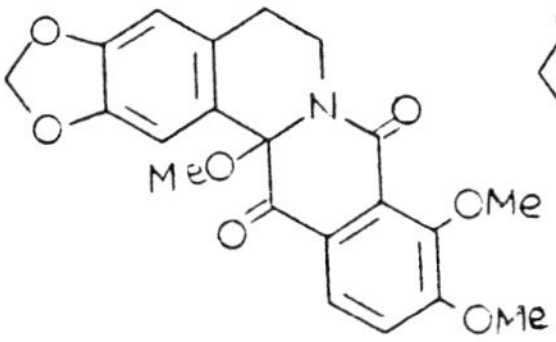

(283)

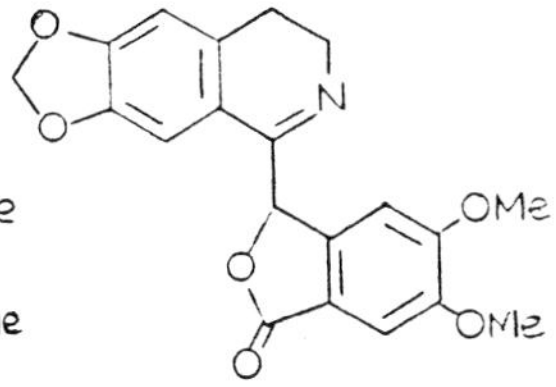

(284)

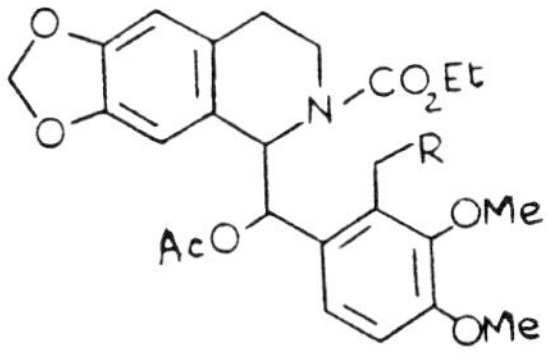

(285)

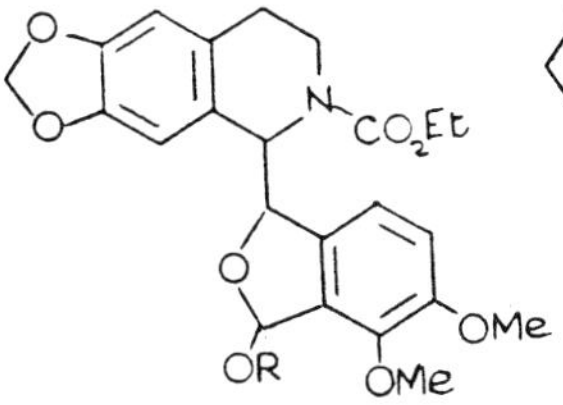

(286)

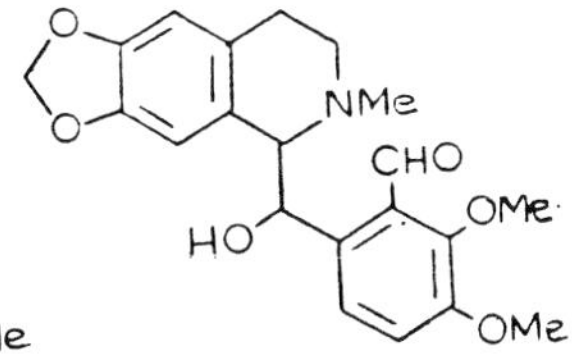

(287)

can be reduced and hydrolysed to a mixture of α and β-hydrast-
ines (J.L. Moniot and Shamma, J.Amer.chem.Soc., 1976, 98, 6714).
Methoxydioxocanadine (283) on pyrolysis affords a 15% yield
of the lactone (284), convertible by N-methylation and reduction into racemic α and β-hydrastine (Elango and Shamma, J.org.
Chem., 1983, 48, 4879). Treatment of O-acetylophiocarpine
with ethyl chloro-formate yields (285, R=Cl), hydrolysable to
(265, R=OH) and oxidation of this followed by hydrolysis of
the acetoxy group yields the hemi-acetal (286, R=H). Conversion of this into the mixed acetal (286, R=Et) allows the N-
COOEt group to be reduced to NMe and subsequent hydrolysis
gives the aldehyde (287) which can be oxidised to α-hydrastine.
A similar reaction sequence starting from epiophiocarpine
yields β-hydrastine (Hanaoka, K. Nagami and Imanishi, Chem.
Pharm.Bull., 1979, 27, 1947). The reverse conversion of a
phthalide-isoquinoline into a berberine derivative has been
reported in the dye-catalysed photolysis of β-narcotine into
the methoxyoxidoberberine (288) (V. Chervenkova and Z. Mardir-
osyan, Nauchni Tr.-Plovdivski Univ., 1983, 21, 121).
    Novel syntheses of the phthalide-isoquinoline alkaloids
have been achieved as follows. Condensation of the Reissert
compound (289, $R^1R^2=CH_2$) with methyl opianate (290) in the
presence of sodium hydride gave the isoquinoline (291), which
is hydrolysed, reduced and then N-methylated to racemic α and
β hydrastine. The isomeric cordrastines are prepared in the
same way from the compound (289, $R^1=R^2=Me$) (P. Kerekes, et al.,
Acta Acad.Sci.Hung., 1978, 97, 353; 1980, 105, 283). Hydra-

(288)

(289)

(290)

(291)

stine, cordrastine and narcotine have been prepared by the reductive condensation of the appropriately substituted 3,4-di-
hydro-*N*-methylisoquinolinium salt with bromodimethoxyphthalide
in aceto-nitrile in the presence of zinc (C.E. Slemon *et al.*,
Canad.J.Chem., 1981, 59, 3055; H. Hamaguchi, *et al.*, J.org.
Chem., 1983, 48, 1621). 3,4-Dimethoxyphenylethylamine (292)
and the dimethylaminomethylenephthalides (293, R¹=R²=Me) and
(293, R¹R²=CH₂) have been condensed to give phthalide-isoquinolines that have been *N*-methylated to cordrastine and to adlumine respectively (S.I. Clarke *et al.*, Austral.J.Chem., 1983,
36, 2493). The indeno[2,1-a]benzazepine (294) has been oxidised to a mixture of the end-lactone (295) and the spiro-iso-
quinoline (296) and of these the enol-lactone has been reduced
to a mixture of positional isomers of cordrastine (Kametani *et
al.*, Chem.Pharm.Bull., 1977, 25, 321).

(292)

(293)

(294)

(295)

(296)

(297)

(298)

In the seco-phthalide-isoquinoline series the benzil derivatives bicucullinidine (278, $R^1=R^2=Me$) and bicucullinine (278,
$R^1R^2=CH_2$) have been prepared by oxidation of the deoxybenzoin
alkaloids adlumiceine and adlumidiceine (Kh.Kiryakov and Mard-
irosyan, Dokl.Bolg.Akad.Nauk, 1981, 34, 1717), the analogue
(297) of narceinimide has been obtained by the Hofmann degradation of the methiodide of the nitrile (278), prepared by the
dehydration of alpenigenine oxime (see section 15) (H. Roensch,
Tetrahedron, 1981, 37, 371) and narceinimide N-oxide has been
converted into a mixture of the isoindolobenzazepines (299,
R=NMe$_2$), (299, R=OH), (299, R=OAc) and (300) by heating it
with acetic anhydride in chloroform (B. Proska, J. Fuska and
Z. Voticky, Pharmazie, 1982, 37, 350).

(299)   (300)

13. Spirobenzylisoquinolines

Few new bases of this group have been isolated and some of the
alkaloids previously assigned spiro-structures have been re-
classified among the new group of indanobenzazepines (see
section 14). The alkaloid hyperectine, isolated from *Hypecoum erecta* has been identified as the base (301) on the basis of an X-ray crystallographic study (M.E. Perel'son, *et al*.,
Khim.Prir.Soedin., 1984, 628) and densiflorine, from *Fumaria
densiflora* has been identified by spectral studies as the seco-
base (302) (B. Sener, Int.J.crude drug Res., 1984, 22, 79).
Fumaritine N-oxide has been identified as a natural product
by its reduction to fumaritine (H.G. Kiryakov *et al*., Canad.J.
Chem., 1979, 57, 53).

(-)-Fumaricine (303) has been converted into the secoberberine hemi-acetal (304) in low yield by photo-catalysed oxidation (Gozler, *et al*., Heterocycles 1982, 19, 2067). Fumariline

(301)   (302)   (303)

(304)   (305)   (306)

(305 , $R^1R^2=R^3R^4=CH_2$) and parfumine (305, R^1=Me, R^2=H, $R^3R^4=$ CH_2) have been converted into the related phenols in which R^3 =Me and R^4=H by potassium hydroxide in methanol and into homo- logues in which R^3=Et and R^4=H by the hydroxide in ethanol, R^1 and R^2 being unchanged. The phenols obtained in this way have been oxidised at the benzylic position by lead (IV) acetate to the acetoxy compounds, for example (306) (Blasko, Hussain and Shamma, J.Amer.chem.Soc., 1982, 104, 1599).

The conversion of berberines into spirobenzylisoquinolines has been examined in detail. As described in section 8 photo- lysis of the betaines (190, R=H) and (190, R=Me) affords the aziridines (198, R=H) and (198, R=Me). Treatment of (198, R= Me) with methyl iodide is accompanied by Hofmann degradation to give the ochotensine analogue (307, R=Me) and treatment with ethyl chloroformate results in a von Braun degredation

(307)

(308)

(309)

(310)

(311)

(312)

and loss of hydrogen chloride to give (307, R=COOEt). With
(198, R=H) and its analogues such reactions are not possible
and treatment with methyl iodide leads to indanobenzazepines
(section 14) and ethyl chloroformate gives chloro-compounds
or the related lactones, from which secondary alcohols or
their deoxy-analogues can be prepared. For example the aziridine (308) can be converted into the halide (309) which can
be reduced catalytically and with lithium aluminium hydride
to fumaricine (303). Compound (310) on reduction before treatment with ethyl chloroformate gives the lactone (311),which
has been converted into ochrobirine by hydrolysis and N-methylation. Photolysis of the betaine (190, R=OMe) and of the
aziridine (198, R=OMe) in methanol affords the ketal (312)
(Hanaoka *et al*., Heterocycles, 1977, 6, 1981; 1980, 14, 1455;
Chem.Pharm.Bull., 1984, 32, 2230).
    The photolysis of quaternary dihydroberberinium salts un-
substituted at C-13 gives 1-benzoyldihydroisoquinolines, but

(313)   (314)   (315)

such salts bearing a methyl group at C-13, for example (313)
give analogues (314, R^1R^2=CH$_2$) of ochotensine (T.-T. Wu, Moniot and Shamma, Tetrahedron Letters, 1978, 3419). Stevens'
rearrangement of canadine methochloride (315, R=H) and thalictricavine methochloride (315, R=Me) gives the bases (314,
R^1=R^2=H) and (314, R^1=H, R^2=Me) respectively, together with
products of Hofmann degradation (Roensch, Phytochem., 1977,
16, 691).

    In total syntheses of alkaloids of this group Bischler-
Napieralski ring-closure of amides of the type (316) affords
dehydrophthalide-isoquinolines (317) and these can be reduced
to spiro-bases by di-isobutylaluminium hydride at -10°C. In
this way, starting from the amides (316, R^1=R^2=OMe) and (316,
R^1R^2=CH$_2$) used for the synthesis of the phthalide-isoquinoline
alkaloids adlumine (271, R^1=R^2=Me) and adlumidine (271, R^1R^2=
CH$_2$), the diastereoisomeric spiro-bases yenhusomidine (318,
R^1=OH, R^2=H) and raddeanone (318, R^1=H, R^2=OH) and their meth-
ylenedioxy-analogues corydaine and sibiricine have been prepared (Nalliah et al., Canad.J.Chem., 1979, 57, 1545). An alternative approach involves Pictet-Spengler cyclisation of
dopamine and the bromoindandione (319) to give (320), the N-
formyl derivative of which undergoes a stereospecific replacement of bromine by hydroxyl in the presence of silver ions to
give a secondary alcohol from which both yenhusomidine and
corydaine have been prepared (D. Dime and S. McLean, Canad.J.
Chem., 1979, 57, 1569).

14. Indanobenzazepines

In 1981 two new alkaloids lahorine and lahoramine were isolated from Fumaria parviflora and were assigned the structures
(323, R^1=R^2=CH$_2$) and (323, R^1=R^2=Me) when it was shown that
they can be prepared from dihydrofumariline (321, R^1R^2=CH$_2$)

(316)   (317)   (318)

(319)   (320)

and (321, $R^1=R^2=$Me) respectively by dehydration with rearrangement to the indenobenzazepines (322), followed by oxidation with iodine (Blasko *et al.*, Tetrahedron Letters, 1981, 22, 3127). A subsequent study of the dehydration of dihydroparfumidine (321, $R^1=R^2=$Me) and dihydroparfumine (321, $R^1=$Me, $R^2=$H) in trifluoroacetic acid showed that the quenching of the reaction in methanol gave the methoxy-compounds (324, R=Me) and (324, R=H), which were found to be identical with the alkaloids fumaritrine and fumaritridine which had previously been regarded as spiro-benzylisoquinolines (Blasko *et al.*, *ibid.*, 3143). Fumarofine was also reassigned to the indanobenzazepine group with the structure (326, R=H) following the preparation of its methyl ether (326, R=Me) from an isomer of dihydroparfumidine by rearrangement to the olefin (325), oxidation of this to the *cis*-glycol and further oxidation of the secondary alcoholic group (Blasko *et al.*, *ibid.*, 3135). More recently the alkaloid bulgaramine, isolated from a medicinal

(321)   (322)   (323)

(324)   (325)   (326)

preparation of a mixture of *Fumaria* species, has been shown
to have the structure (322, $R^1=R^2=Me$) (G.I. Yakinov, *et al.*,
J.nat.Prod., 1984, 47, 1048). The alkaloid can be prepared
by the dehydration of dihydroparfumidine (321, $R^1=R^2=Me$) and
also by the pyrolysis of *O*-acetyldihydroparfumidine, which is
O-methylfumarophycine and accompanies it in the *Fumaria* preparation (N. Muguresan, Tetrahedron Letters, 1981, 22, 3131).

The reverse transformations of the indanobenzazepine gly-
cols (327, $R^1=OH$, $R^2=H$) and (327, $R^1=H$, $R^2=OH$) into the spiro-
benzylisoquinoline alkaloids raddeanine (329, $R^1=OH$, $R^2=H$)
and yenhusomine (329, $R^1=H$, $R^2=OH$) has been accomplished by
trifluoroacetic anhydride in pyridine. *O*-Methylfumarafine
(326, R=Me) has similarly been rearranged to the ketone (330),
isomeric with raddeanone (318, $R^1=H$, $R^2=OH$), which gives raddeanine (329, $R^1=OH$, $R^2=H$) on reduction (Blasko *et al.*, *ibid.*,
1981, 22, 3139).

These transformations in both directions may be assumed to

(327)　　　　　　　　(328)　　　　　　　　(329)

(330)　　　　　　　　(331)　　　　　　　　(332)

proceed through aziridinium salts (328). Protonated aziridines have been converted into indanobenzazepines as well as
spirobenzylisoquinolines. Cleavage of the aziridine (331)
by acid in methanol, followed by *N*-methylation affords hydro-
xyfumaritine, the methanesulphonate ether of which is reduced
to fumaritrine (324, R=Me) (Hanaoka *et al.*, Tetrahedron Letters, 1983, 24, 3845). A similar cleavage of the aziridine
(332) in water, followed by *N*-methylation and hydrogenolysis
yields fumarofine (326, R=H) (Hanaoka, Chem.Pharm.Bull., 1983,
31, 2172). The methiodide of the aziridine (198, R=H) is
cleaved by hydrochloric acid to the indanobenzazepine (333),
which has also been prepared by the photolysis of oxoallocryp-
topine (334) in *t*-butanol in the presence of potassium *t*-but-
oxide (Blasko *et al.*, Chem.Comm., 1981, 1246) and has also
been identified among the products of transformation of meth-
iodides of ophiocarpine (192) and epiophiocarpine by callus
cell cultures of *Corydalis ophiocarpa* (V.I.Vinogradova *et al.*,

(333)   (334)   (335)

(336)   (337)   (338)

Khim.-Farm.Zh., 1983, 1, 44). The same aziridine (198, R=H)
reacts with formaldehyde to give the cyclic carbinolamine ether (335) which is reduced by sodium borohydride to the *cis*-
glycol (336) and oxidation of this gives the keto-lactone
(337) (Murugesan *et al.*, Tetrahedron Letters, 1981, 22, 3131).
  A structurally distinct group of indanobenzazepines, based
on a 3,4-benzazepine rather than a 4,5-benzazepine, is represented by the alkaloids ribasine (also given the names limogine
and grandiflorine) from *Corydalis claviculata* and *Argenone
grandiflora* and himalayine from *Meconopsis villosa*. An X-ray
crystallographic study has shown that ribasine has the structure (338, R=H) (Boente *et al.*, Tetrahedron Letters, 1983, 24
2029) and spectroscopic studies indicate that himalayine is
hydroxyribasine (338, R=OH) (D.P. Allais *et al.*, *ibid.*, 2445).
Nothing is known about the biogenesis of these two alkaloids.

15. *Rhoeadines*

The alkaloids of the rhoeadine group are benzazepines and essentially bear the same relationship to the indanobenzazepines
as do the secoberberines (section 10) to the spirobenzylisoquinolines (section 13).

Alpenigenine (339) has been converted into *cis*-alpinigen-
ine by hydrochloric acid (A. Guggisberg, M. Hesse and H.
Schmid, Helv. 1977, 60, 2402). *O*-Methylalpinigenine has been
subjected to Hofmann degradation to give the styrene (340)
but reacts with cyanogen bromide, unusually for a benzylamine,
to give the *N*-cyanonor-compound (341) rather than by ring fission (Roensch Symp.Pap.IUPAC Int.Symp.Chem.Nat.Prod., 11th.,
1978, 2, 38). Alpenigenine oxine (342) has been dehydrated
to the nitrile (343), the methiodide of which suffers Hofmann
degradation to the amide (344), which is an analogue of the
alkaloid *Z*-fumaramine (275) (Roensch, Tetrahedron, 1981, 37,
371).

(339)   (340)   (341)

(342)   (343)   (344)

A positional isomer of xylopinine of structure (345), prepared by synthesis, has been subjected to Hofmann degradation
to the olefin (346), which on oxidation with osmium tetroxide
gives the diol (347). Cleavage of the diol gave the dialde-
hyde (348), photolysis of which yields a mixture of *cis*-alpen-
igenine (30%) and alpenigenine (339) (1%) (S.B. Prabhakar, *et
al*., J.Chem.Soc., Perkin I 1981, 1273). The 7,8,13,14-tetra-

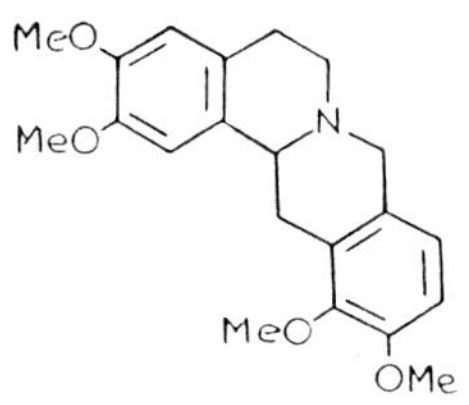

(345)

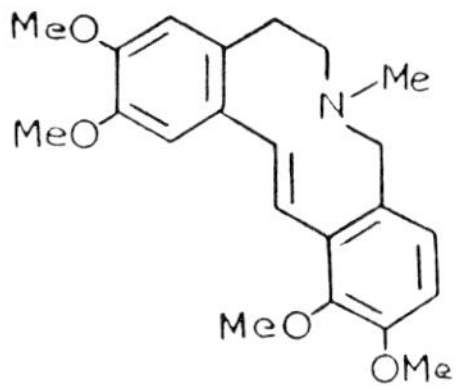

(346)

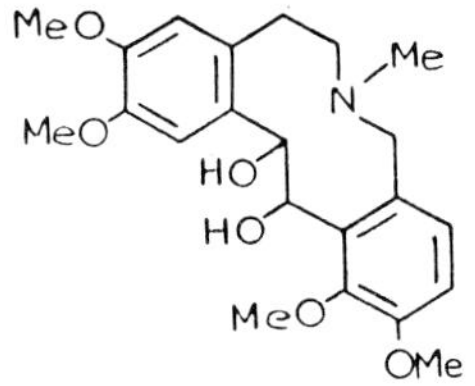

(347)

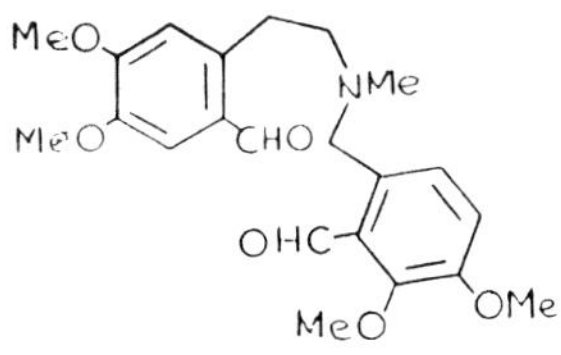

(348)

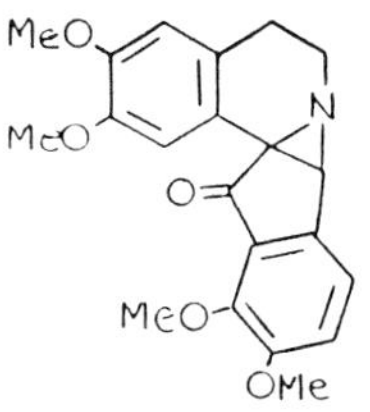

(349)

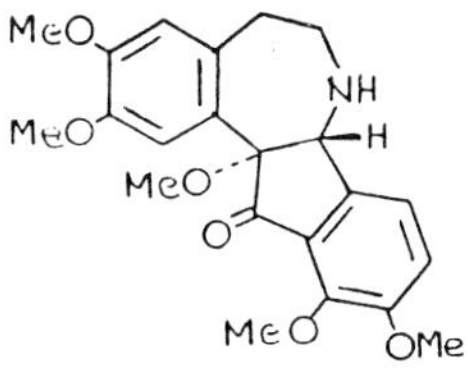

(350)

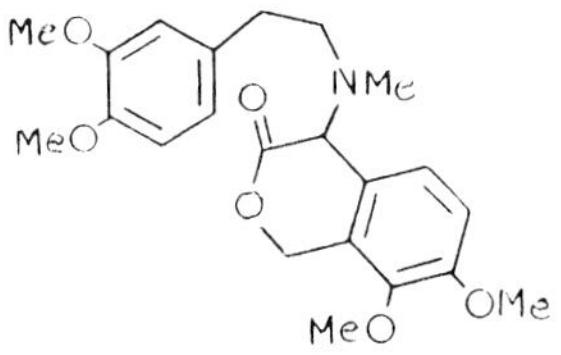

(351)

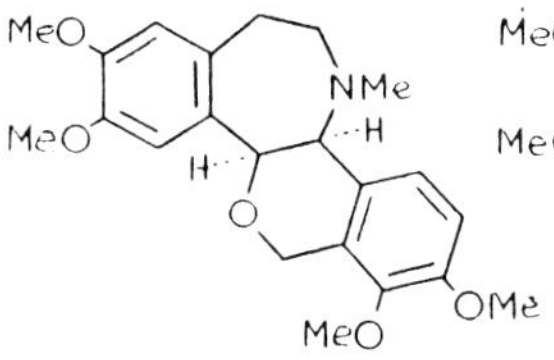

(352)

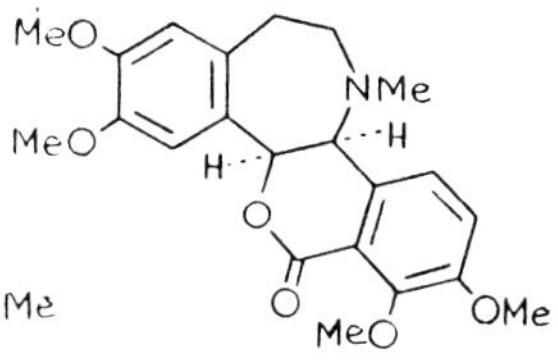

(353)

dehydro-analogue of (345) has been converted through the aziridine (349) into the indanobenzazepine (350) by methods discussed in section 14, and this has been converted into *cis*-
alpenigenine (Hanaoka *et al.*, Chem.Pharm.Bull., 1982, 30,
1110). Reduction of the lactone (351) with di-isobutyl-alum-
inium hydride gives the related hemi-acetal, cyclodehydration
of which gives *cis*-deoxyalpenigenine (352). Oxidation of
this gives the lactone (352) which may be reduced to *cis*-alp-
enigenine (I. Ahmad and V. Sniekus, Canad.J.Chem., 1982, 60,
2678).

Bicuculline (354) has been rearranged to the lactone (355,
R=Me) and conversion of this into the secondary base (355, R=
H) and the urethane (355, R=CONHTol) followed by reduction
to the hemi-acetal and *O*-methylation of this with methyl orthoformate and final hydrolysis to the secondary base yields
papaverrubine-E (356) (R. Hohlbrugger and W. Kloetzer, Ber.,
1979, 112, 849).

(354)   (355)   (356)

16. *Other Modified Berberines*

Several alkaloids based on the isoindoloisoquinoline, isoind-
olobenzazepine, isoindolobenzazocine and isoquinolinobenzazepine skeletons have been discovered. In all of these except
two the aromatic substitution pattern is the same as that in
tetrahydroberberine, the exceptions being related to tetrahydropalmatine. They are assumed to be derived from alkaloids
of the berberine series, from which some transformations have
been effected.

Chilenine, an alkaloid from *Berberis empetrifolia*, has been

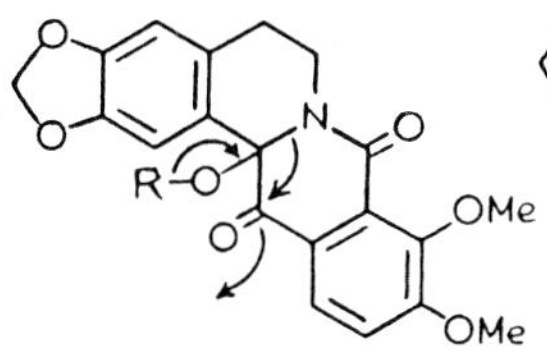

(357)

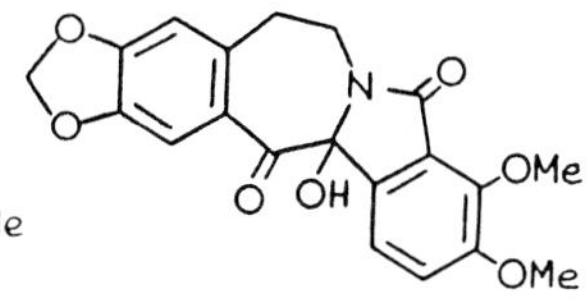

(358)

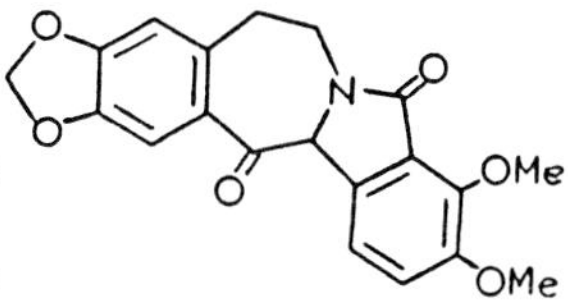

(359)

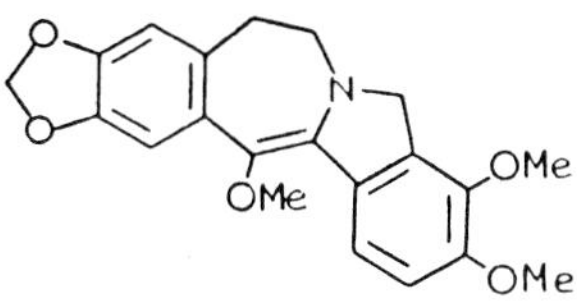

(360)

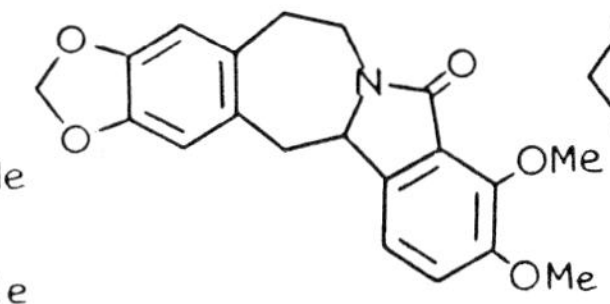

(361)

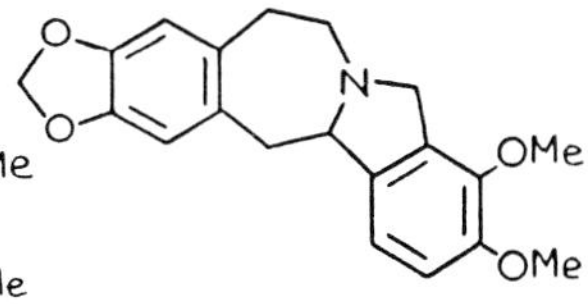

(362)

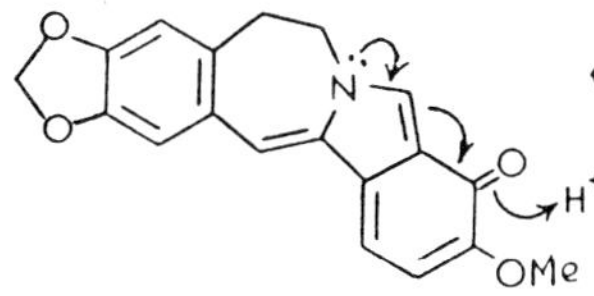

(363)

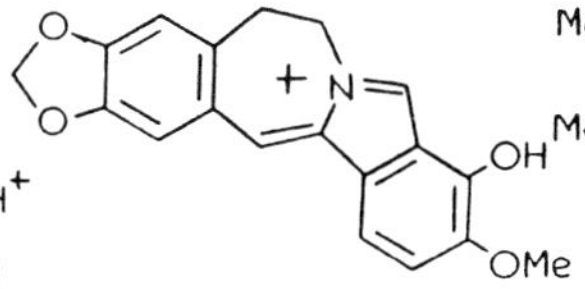

(364)

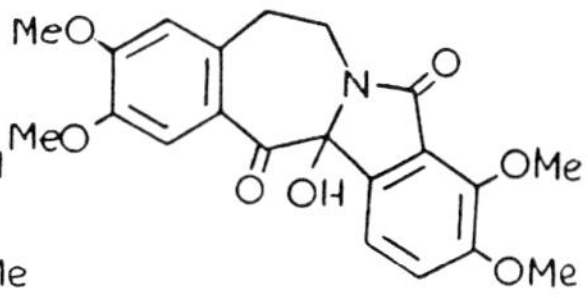

(365)

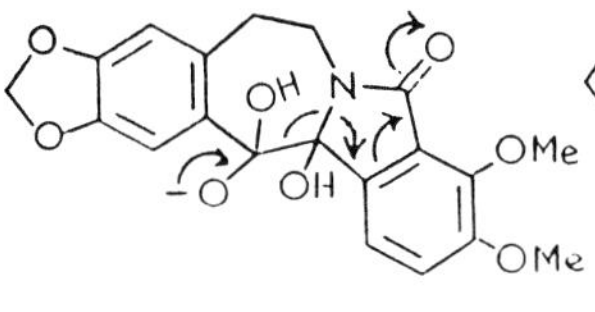

(366)

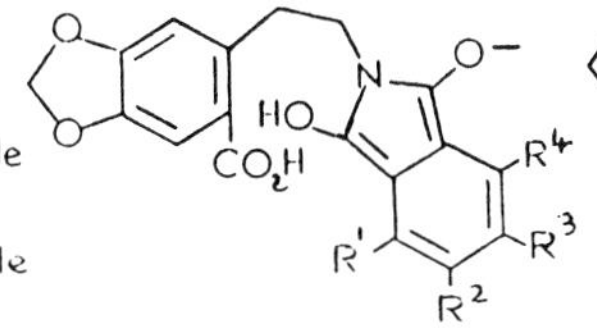

(367)

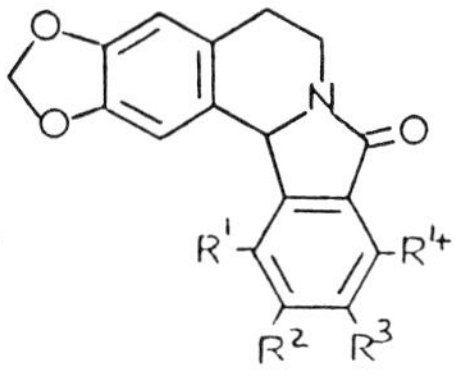

(368)

identified as the base (358) by spectroscopic studies and by
its identity with a previously known product of that structure
obtained by the rearrangement of the berberine derivative
(357, R=H) on treatment with ammonium hydroxide (Fajardo *et
al*., Tetrahedron Letters, 1982, 23, 39; Moniot, Hindenlang
and Shamma, J.org.Chem., 1979, 44, 4343; C. Manumikar and
Shamma, Heterocycles, 1980, 14, 827). It is also a minor product of the thermal rearrangement of the base (357, R=Me)
(Elanjo and Shamma, J.org.Chem., 1983, 48, 4879), which has
been found as a natural product in *Berberis actinacantha*, *B.
actinacantha* and *B.valdiviana* , where it is accompanied by
13-deoxychilenine (359), pictonamine (360), lenoxamine (361),
chilenamine (362), chilenone (363), palmanine (365) and the
isoindoloisoquinoline alkaloid nuevamine (368, $R^1=R^2=$OMe, $R^3=$
$R^4=$H). The keto-enamine system of chileninone is readily protonated to give the phenolic iminium salt (364) (E. Valencia
et al., Tetrahedron Letters, 1984, 25, 599; Tetrahedron, 1984,
40, 3957).

Nuevamine was originally assigned the structure (368, $R^1=$
$R^2=$H, $R^3=R^4=$OMe) on the basis of its preparation from chilenine by fission in alkali (366) to a keto-acid followed by cyclisation with the loss of carbon dioxide (Moniot, Hindenlang
and Shamma, J.org.Chem., 1979, 44, 4347; Valencia *et al*., Tet-
radhedron Letters, 1984, 25, 599). However total synthesis
of a compound of that structure by the cyclisation of (369,
$R^1=R^2=$H, $R^3=R^4=$OMe) gave a product that differed from nuevamine, which was finally prepared by the cyclisation of (369,
$R^1=R^2=$OMe, $R^3=R^4=$H) prepared from homopiperonylamine and 3-
bromo-ψ-meconine. The fission of chilenine must give the anion (370, $R^1=R^2=$H, $R^3=R^4=$OMe), which can add a proton to give
an intermediate capable of being cyclised to either (368, $R^1=$
$R^2=$H, $R^3=R^4=$OMe) or (365, $R^1=R^2=$OMe, $R^3=R^4=$H) (A. Ricardo,
L. Castedo and D. Dominguez, Tetrahedron Letters, 1985, 26,
2925).

Both ring-B and ring-C homologues of the isoindolobenzaze-
pine alkaloids occur naturally. The isoquinolinobenzazepine
puntarenine (370, $R^1R^2=CH_2$), the structure of which has been
determined by nmr spectrometry and X-ray diffractometry (Fajardo *et al*., Tetrahedron Letters, 1983, 24, 155; S. Sepulveda
et al., Planta Med., 1983, 39, 32), has been isolated from
Berberis empetrifolia and the tetramethoxy analogue saulatine
(370, $R^1=R^2=$Me) has been obtained from *Abuta bullata* (Hocquemiller, Cavé and A. Fournet, J.nat.Prod., 1984, 47, 539).
The isoindolobenzazocine alkaloid magallanesine (371) has
been isolated from *Berberis darwinii* (Valencia *et al*., Tetra-

(369)

(370)

(371)

hedron Letters, 1985, <u>26</u>, 993). Sodium borohydride reduces puntarenine and saulatine to the secondary alcohols and magellanesine to the dihydro-alcohol. The biogenesis of these alkaloids has not been elucidated.

The spirobenzazepine alkaloid turkiyenine (377) has been isolated from *Hypecoum procumbens* and it has been suggested that it is derived from the berberine alkaloid coptisine, *via* the base (372, $R^1R^2=CH_2$), which is an analogue of rugosinone (372, $R^1=R^2=Me$), itself known to be derived from berberine. *N*-Methylation of this base, followed by nucleophilic attack by hydroxyl ion, would give the ketocarbinolamine (373), which could undergo rearrangement similar to that of (357, R=H) to chilenine (358) to give (374). Reduction of this *via* the iminium salt (375), followed by conversion of this into the oxonium salt by reaction with formaldehyde, could then give the enol (376), the cyclisation of which to turkiyemine (377) is plausible (Gozler *et al.*, J.Amer.chem. Soc., 1984, <u>106</u>, 6101).

(372)   (373)   (374)

(377)   (376)   (375)

17. *Benzophenanthridines*

The range of structural types in the benzophenanthridine group
of alkaloids has been extended by the isolation of 6-methoxy-
sanguinarine (378) and the related tertiary base 6-methoxy-
norsanguinarine (pancorine) (V.B. Pandey, Ray and Dasgupta,
Phytochemistry, 1979, 18, 695; M. Alumova *et al*., Khim.Prir.
Soedin, 1981, 671), the carbinolamine ethers 6-methoxydihydro-
sanguinarine and 6-methoxydihydrochelerythrine (L.A. Mitscher
et al., Lloydia, 1978, 41, 143), 6-iminodihydrosanguinarine
(379) (Castedo *et al*., Heterocycles, 1981, 16, 533), luguine
(381) (Castedo *et al*., Tetrahedron Letters, 1978, 2923), 6-
acetonyldihydrosanguinarine 6-acetonyldihydrochelerythrine
(382, R=Me) and its homologue (382, R=Et) (W. Doepke, V. Hess
and V. Jimenez, Z.Chem., 1976, 16, 54; E.M. Assem, I.A. Bena-
gesand S. Albonico, Phytochem., 1979, 18, 511) and chelerydi-
merine (toddalidimerine) (380) (P.N. Sharma *et al*., Phytochem.
1982, 21, 252) and couspernine which is 6-acetonyl-*O*-acetyl-
corynoline (Q. Fang, M. Lin and Q. Weng, Planta Med., 1984,
50, 25). The ketonic condensation products (382, R=Me), (382,
R=Et) and (380) are easily prepared from chelerythrine and

(378)

(379)

(380)

(381)

(382)

the appropriate β-keto-dicarboxylic acids.

The ring opened benzophenanthridines arnottianamide (383, R^1=H, R^2=R^3=OMe), isoarnottianamide (383, R^1=R^2=Ome, R^3=H), iwamide (383, R^1=H, R^2=OH, R^3=OMe) and integriamide (383, R^1 R^2=OCH$_2$O, R^3=H) are also new alkaloids, isolated from *Xanthoxylum arnottianum*, and their structures have been confirmed by their production by Baeyer-Villiger oxidation of chelerythrine, nitidine, N-methyldecarine (as its benzyl ether) and avicine respectively with m-chloroperbenzoic acid (H. Ishii *et al.*, J.chem.Soc., Perkin I, 1984, 1769). Bischler-Napieralski cyclisation of the O-methyl ethers of integriamide and isoarnottianamide affords the alkaloids chelirubine (384, R^1R^2=CH$_2$) and chelilutine (384, R^1=R^2=OMe), completing the revision of the structures of these alkaloids. A similar synthesis confirmed that sanguirubine is the 2,3-dimethoxy analogue of chelirubine; sanguilutine is undoubtedly the 2,3-dimethoxy analogue of chelilutine and macarpine is 12-methoxychelirubine (Ishii *et al.*, J.chem.Soc., Perkin Trans., 1984, 2283).

Dihydrosanguilutine, dihydrochelilutine and dihydromacarpine have been synthesised by photochemical cyclisation of the bases (385, R^1=H, R^2=R^3=Me), (385, R^1=H, R^2R^3=CH$_2$) and (385,

(383)

(384)

(385)

(386)

R¹=OMe, R²R³=CH₂) respectively, followed by *N*-methylation.
(Kessar *et al.*, Tetrahedron Letters, 1977, 1459; Takao *et al.*,
Heterocycles, 1981, 16, 221). Dihydrochelirubine and chelirubine have also been prepared by photochemical cyclisation of
the amide (386), which involves the loss of a methoxy group,
followed by reduction and dehydrogenation of the resulting di-
hydro-compound to chelirubine (Ishii *et al.*, Chem.Pharm.Bull.,
1978, 26, 864).

The reaction of esters of methylenedioxyphthalic acid (387)
with the imine (388) affords the ester-lactam (389) and hydrolysis of this to the lactam-acid followed by Arndt-Eistert
homologation and cyclisation yields the ketone (390, R¹R²=O).
Reduction of this to the secondary alcohol (390, R¹=H, R²=OH),
followed by dehydration and aerial oxidation affords oxosangui-
narine (391), from which sanguinarine can be prepared (Shamma
and H. Tomlinson, J.org.Chem., 1978, 43, 2852). A similar
synthesis of oxonitidine and nitidine has been accomplished
(M. Cushman and L. Cheng, J.org.Chem., 1978, 43, 286). Conversion of (389) *via* the acid chloride into the diazoketone, followed by cyclisation in trifluoroacetic acid gives the keto-
lactam (392), which affords (±)-chelidonine (393, R=H) on red-

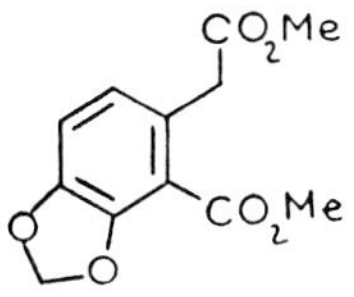

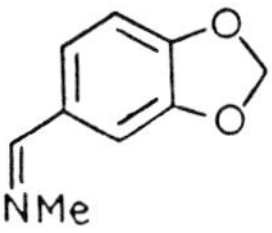

(387)

(388)

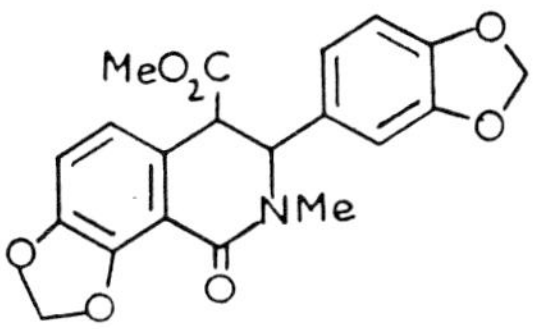

(389)

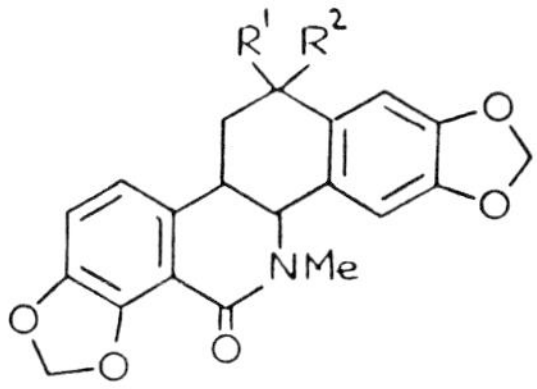

(390)

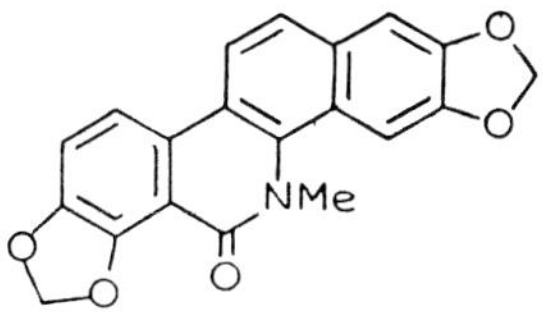

(391)

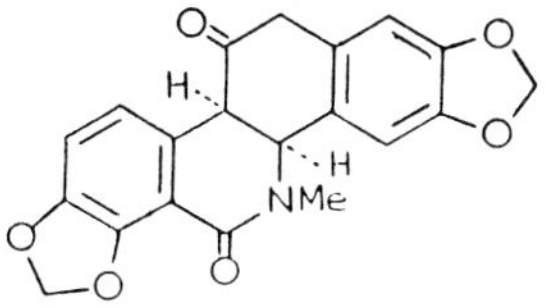

(392)

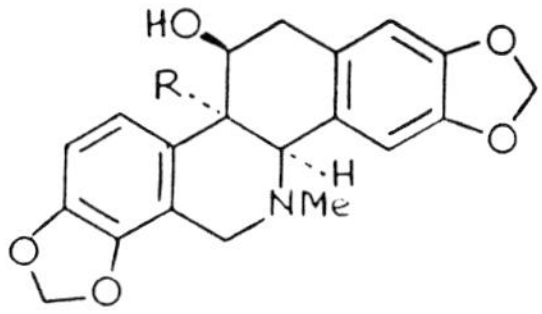

(393)

uction with lithium aluminium hydride (M. Cushman *et al.*, J. org.Chem., 1980, 45, 5067; Tetrahedron Letters, 1980, 21, 3845). A similar synthesis of (±)-corynoline (393, R=Me) has also been achieved. In this the *C*-methylated analogue of the ester (389) is obtained in both the *cis* and the *trans* forms and the *cis* form has been converted into (±)-corynoline by the sequence of reactions described above. When the same re-action sequence is applied to the *trans*-ester, however, the product (397, R^1=H, R^2=OH) is different from isocorynoline and this alkaloid (397, R^1=OH, R^2=H) has been prepared from the *trans* *C*-methylated (389) by conversion of the homologous acid into the ketone (394) and reduction to the alcohol followed by dehydration to the olefin (395). Epoxidation of the olefin gives the epoxide (396) and reduction of this with lithium aluminium hydride gives (397, R^1=OH, R^2=H) (J.R. Falck, S. Manna and C. Mioskowski, J.Amer.chem.Soc., 1083, 105, 2873).

(394)

(395)

(396)

(397)

The amide (398) (prepared by a photocyclisation) has been oxidised with lead (IV) acetate to the phenol acetate (399) which has been converted through the ortho-quinone into the diol (400, R^1=R^2=H). The di-ester of this (400, R^1=MeSO$_2$, R^2=Ac) on treatment with potassium hydroxide in methanol affords (401, R=OMe) which yields (±)-homochelidonine (401, R=H) on hydrogenolysis (Ninomiya, Yamamoto and Naito, J.chem.Soc., Perkin I, 1983, 2171).

As an alternative to the photochemical cyclisation of ring opened berberines, such as (402, R^1=R^2=H) and (402, R^1R^2=O), cyclisations have been achieved by the action of hydrochloric

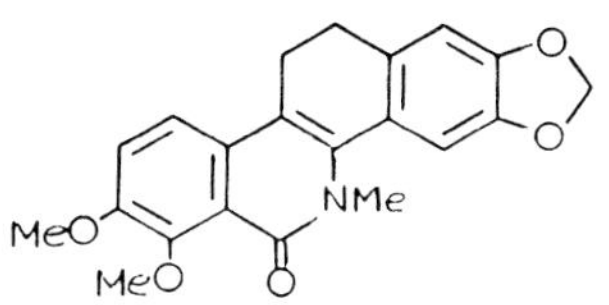

(398)

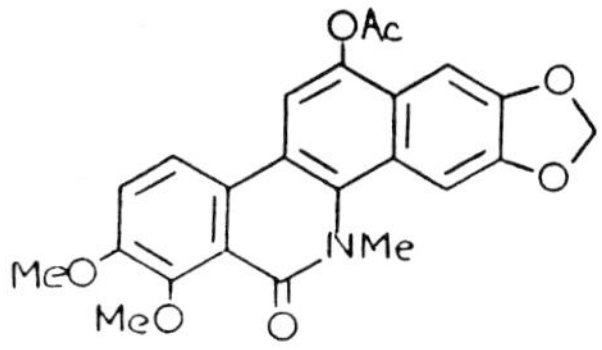

(399)

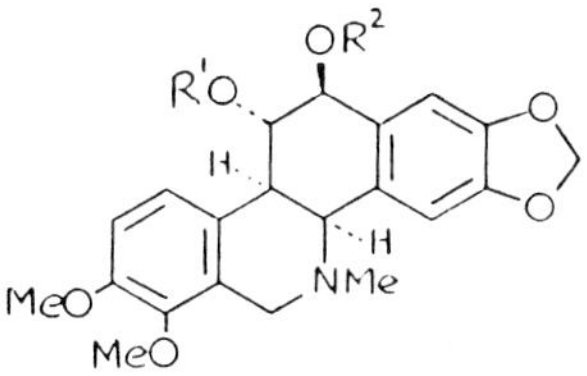

(400)

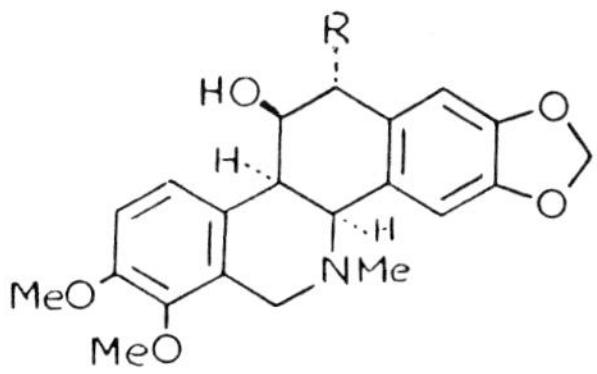

(401)

acid on acetals obtained by the action of thallium III nitrate
in methanol on the olefins. In this way (402, R^1R^2=O) has
been converted into (403), which is cyclised to oxochelerythrine and this is then reduced to the carbinolamine, which gives
chelerythrine in acids. In a similar manner fagaronine and
nitidine have been prepared from analogues of (402, R^1R^2=O)
prepared from the berberine alkaloids dehydrodiscretine and
pseudoberberine respectively (Hanaoka *et al.*, Chem.Comm.,
1984, 25, 5169; Tetrahedron Letters, 1984, 25, 5169).

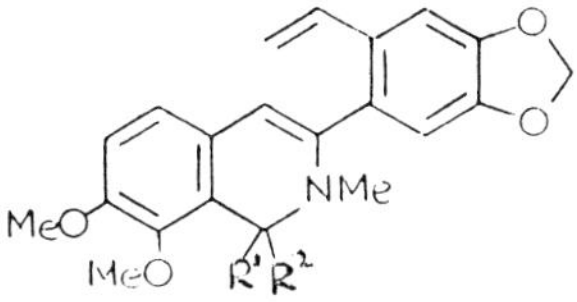

(402)

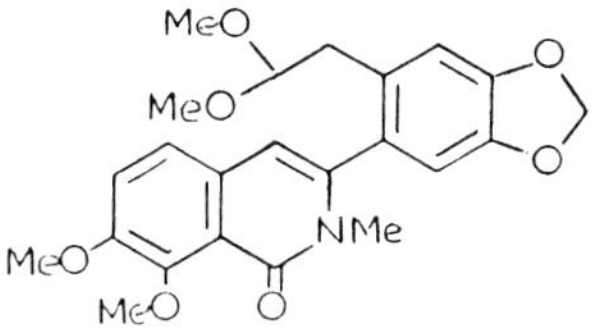

(403)

The photocyclisation of the olefins (402, $R^1=R^2=H$) and (402, $R^1R^2=O$) in the presence of nitrobenzene results in the formation of the adducts (404, $R^1=R^2=H$) and (404, $R^1R^2=O$) of which the latter has been decomposed by heat to oxochelerythrine (M. Onda and H. Yamaguchi, Chem.Pharm.Bull., 1977, 27, 2076). The adduct (404, $R^1=R^2=H$) has been oxidised with dichlorodicyanobenzoquinone to the quinone imine (405) and the amine (406), both of which have been oxidised to the ortho-quinone and this on reduction affords both the *trans*-diol (407) and

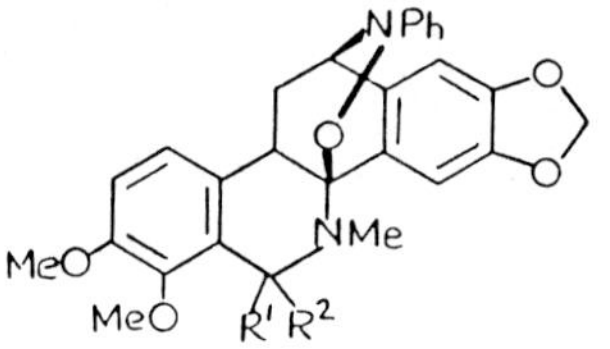

(404)

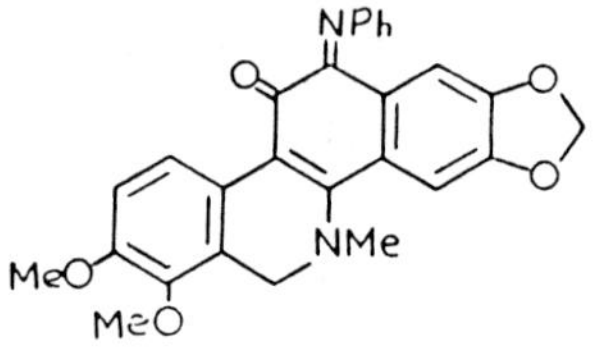

(405)

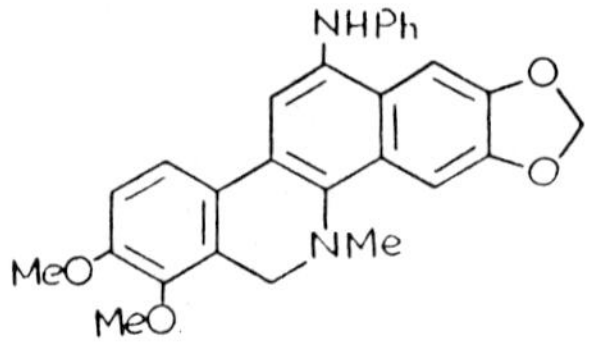

(406)

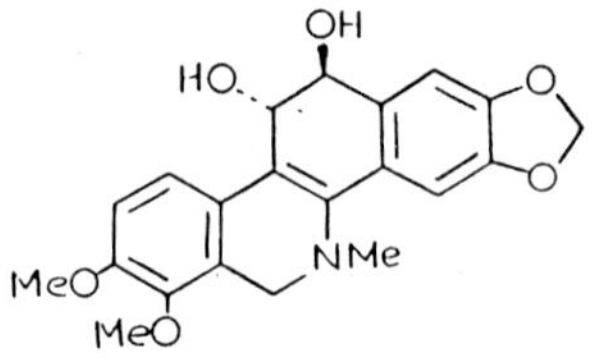

(407)

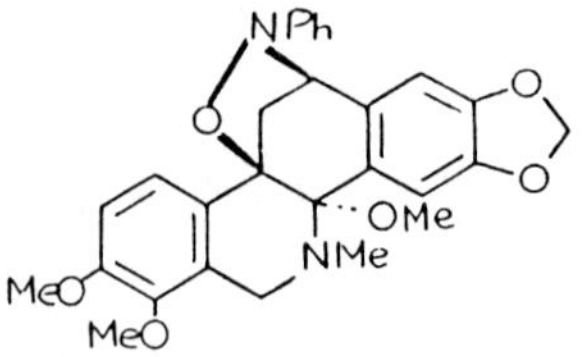

(408)

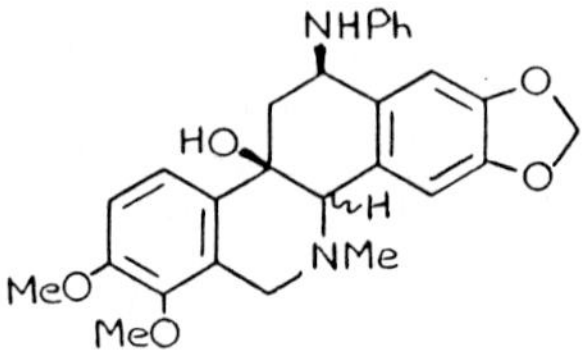

(409)

its *cis*-isomer. Oxidation of the adduct (404, $R^1=R^2=H$) in methanol affords the base (408) which may be reduced to *cis* and *trans* (409). Neither (407) nor (409) has yet been converted into an alkaloid of this group (Yamaguchi, Y. Harigaya and Onda, Chem.Pharm.Bull., 1983, 31, 1601).

The methyl ester of corydalic acid (413) can be isolated from *Corydalis incisa* when the plant is in the vegetative stage and can be classed as a secobenzophenanthridine alkaloid (G. Nonaka, T. Kodera and I. Nishioka, Chem.Pharm.Bull., 1973, 21, 1020). It has been synthesised by the condensation of the acid anhydride (410) with the imine (411) to give an acid that is decarboxylated to a mixture of geometrical isomers, of which (412) is hydrolysed, esterified and reduced to (±)-methyl corydalate (413) (Cushman and W.C. Wong, J.org.Chem., 1984, 49, 1278). It has also been prepared by the reduction, oxidation and esterification of the olefin (414), prepared from the berberine alkaloid corysamine (Hanaoka, S. Yoshida and C. Mukai, Chem.Comm., 1984, 1703).

(410) (411) (412)

(413) (414)

18. *Ipecacuanha Alkaloids*

Three new alkaloids of this group, 9-demethylprotoemetinol (415, R^1=H, R^2=Me), 10-demethylprotoemetinol (415, R^1=Me, R^2 = H) and alancine (416, R=COOH) have been isolated from *Alangium lamarckii*. Both isomers of demethylprotoemetinol have been synthesised from *O*-benzylethers of the related esters by reduction of COOEt to CH_2OH and removal of the benzyl group (Pakrashi *et al.*, Heterocycles, 1982, 19, 230, 2305) and alancine has been prepared by oxidation of a suitably protected derivative of ankorine (416, R=CH_2OH) (S.K. Chattopadhyay *et al.*, Heterocycles, 1984, 22, 1965). Ankorine has been proved to have the structure (416, R=CH_2OH) rather than the previously accepted 11-hydroxy-structure by synthesis of the alkaloid and of all possible stereoisomers of the 11-hydroxy analogue (C. Szantay *et al.*, Ber., 1976, 109, 2420).

(415)   (416)

(417)

Similarly 8-hydroxy rather than 11-hydroxy structures have been demonstrated for alangicine and alangimarckine (417) by synthesis from the ester prepared as an intermediate in the synthesis of ankorine, by conversion into amides of 3-benzyloxy-4-methoxyphenylethylamine and tryptamine respectively, followed by Bischer-Napieralski ring closures (T. Fujii *et al.*, Tetrahedron Letters, 1976, 2553; 1977, 3477; 1978, 3111). The naturally occurring demethyltubulosine has been shown to be the 10-hydroxy compound by synthesis of this and the 9-hydroxy isomer (Fujii *et al.*, Heterocycles, 1980, 14, 299, 971).

Several new syntheses of emetine and its analogues have been reported, most differing only slightly in the preparation of previous intermediates. A more novel approach involved the Michael addition of the anion (418) of 1-methyldimethoxydihydroisoquinoline to the unsaturated ester (419) to gove (420), followed by *C*-ethylation, hydrolysis and decarboxylation and reduction with lithium aluminium hydride (Kametani, S. Surgenor and K. Fukumoto, J.chem.Soc., Perkin I, 1981, 920) to give the alkaloid.

(419)

(418)

(420)

A different approach starts with the synthetic tetrahydroberberine (421, R=CH₂Ph), which is reduced with lithium in liquid ammonia and the product immediately methylated with diazomethane to give the enol ether (422). The reduction presumably involves cleavage of the benzyl ether as an initial process, the anion of the resulting phenol protecting that nucleus from reduction during reduction of the nonphenolic nucleus. Reduction of the methyl ether (421, R=Me) proceeds in both rings to give the tetrahydro-compound (423). Hydrolysis of the enol ether (422) gives the αβ-unsaturated ketone (424), which can be obtained from the tetrahydro compound (423) by a selective aromatisation by *N*-chlorosuccinimide in

methylene chloride, involving allylic chlorination and loss of
hydrogen chloride, which, in the presence of traces of water,
hydrolyses the remaining enol ether. Reduction of the unsaturated ketone (424) proceeds stereospecifically to give (425),
which is converted into an enamine (426) and this gave the
spirodithioketal (427) on treatment with di(toluenesulphonyl)-
propane-1,3-dithiol. The ketal is cleaved to the acid (428)
by potassium hydroxide and the methyl ester is desulphurised
to (429), from which emetine has been prepared by conventional
processes (S. Takano *et al.*, Heterocycles, 1977, 7, 143; J.org.
Chem., 1978, 43, 4169).

Another novel approach starts with norcamphor (430), which
is converted by Baeyer-Villiger oxidation into the lactone
(431, R=H). Stereospecific alkylation of this with sodium
hydride and ethyl iodide gives the equatorial ethyl compound
(431, R=Et), which is converted into the amide (432) on treatment with homoveratrylamine. Oxidation of this alcoholic amide gives the keto-amide which is converted into the dithioke-

(421)

(422)

(423)

(424)

(425)

(426)

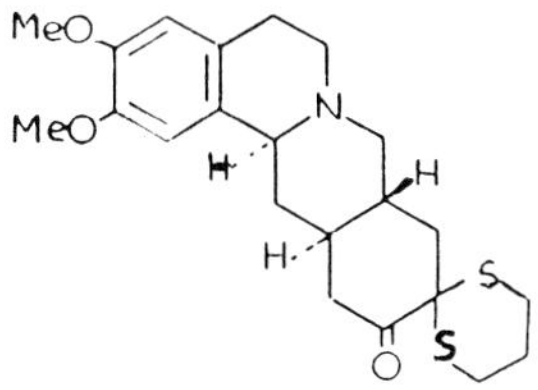

(427)

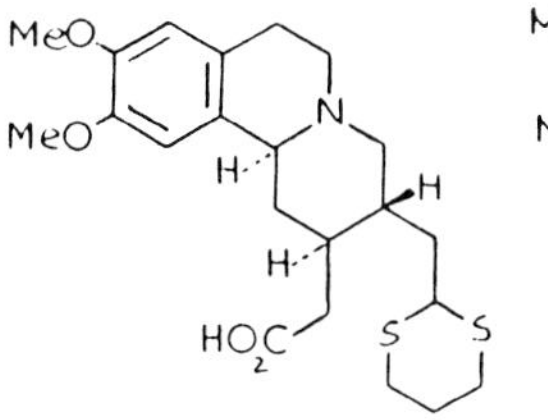

(428)

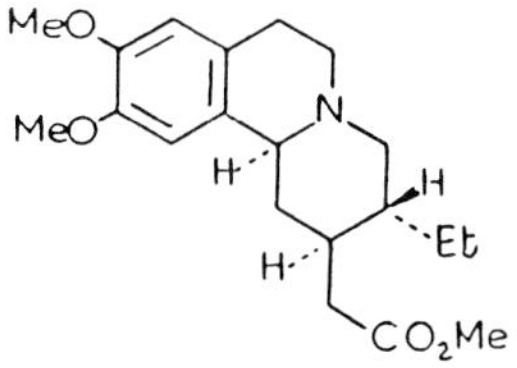

(429)

tal (433) by the processes used to convert (425) into (427) and cleavage of (433) in potassium hydroxide gives the acid (434). Treatment of this with acid effects hydrolysis of the ketal and Pictet-Spengler cyclisation of the tetrahydroisoquinoline ring gives two isomers of the lactam (435) differing in stereochemistry at C-1 in the isoquinoline system. One of these (C-1 αH) gives protoemetinol (415, $R^1=R^2=$Me) on reduction with lithium aluminium hydride (Takano et al., Heterocycles, 1982, 12 (Special Issue), 263).

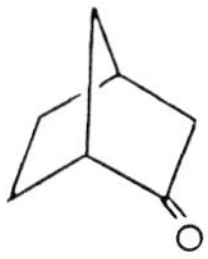

(430)

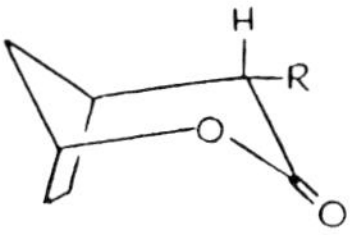

(431)

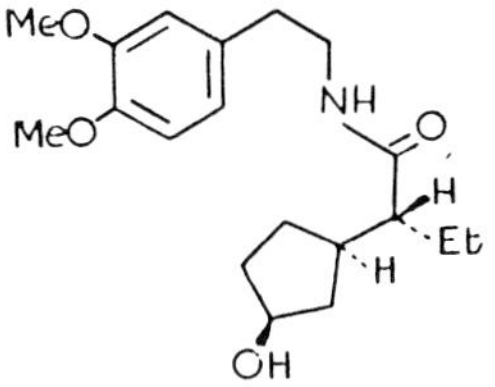

(432)

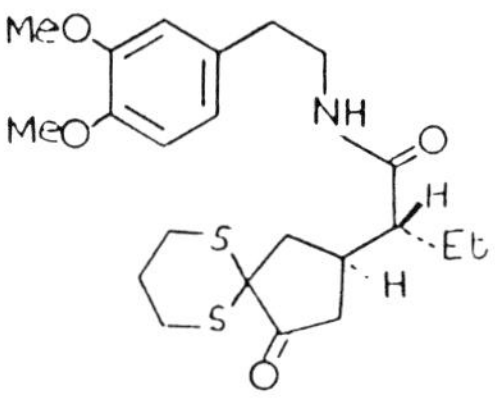

(433)

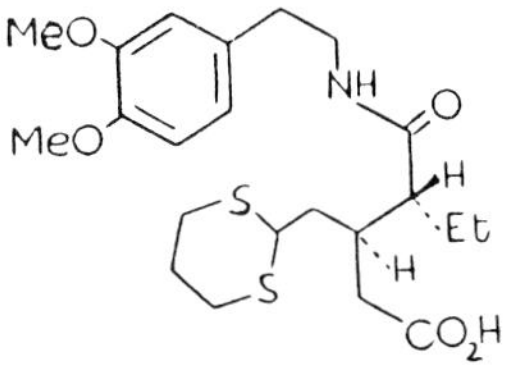

(434)

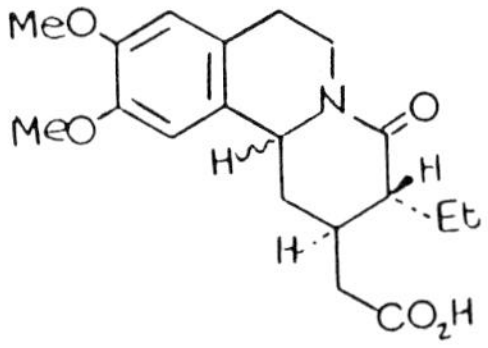

(435)

19. Phenylethylisoquinolines

Two new phenylethylisoquinoline alkaloids, homolaudanosine
(436, R¹=R²=Me) and dysoxyline (436, R¹R²=CH₂), have been isolated from *Dysoxylum lenticellare* (A.J. Aldersanmi, C.J.Kelley
and J.D. Leary, J.nat.Prod., 1983, 46, 127). A new bis-phen-
ylethylisoquinoline, jolantinine, isolated from *Merendera jol-
antae*, has been assigned the structure (437) on spectroscopic
grounds (A.M. Usimanov, M.K. Yusupov and Kh.A. Aslanov, Khim.
Prir, Soedin., 1977, 422), though it is difficult to rational-
ise this structure with known routes of biogenesis of isoquinoline alkaloids and of oxidative coupling of phenols.

Homolaudanosine is oxidised electrochemically to the diben-
zoquinolizinium salt (438) and homoglaucine (439) and its me-
thoperchlorate, under the same conditions, affords the 2'2'-
dimer. (A. Najafi and M. Sainsbury, Heterocycles, 1977, 6,
459; Kupchan *et al.*, J.org.Chem., 1978, 43, 2521). The oxidation of *N*-trifluoroacetylnorhomolaudanosine with vanadium

(436) (437)

(438) (439) (440)

oxyfluoride yields mainly the homoproerythradienone (440) and
not the homomorphinandienone, differing in this respect from
nonphenolic benzylisoquinolines, which give morphinandienones
with this reagent (Kupchan *et al.*, *loc.cit.*)

20. *Colchicine and its Analogues*

The range of known alkaloids of this group has been consider-
ably extended by the isolation and identification of the following:

Colchicine (441, R=Me) and its 2-demethyl-, 2,3-demethyl-
and 10,11-epoxy-derivatives
N-Deacetylcolchicine and its *N*-methyl-, *N*-formyl-, *N*-formyl-
N-methyl-, *N*-acetoacetyl-, *N*-methyl-*N*-2-hydroxybenzyl-
and 2-demethyl-*N*-hydroxyacetyl- derivatives
Colchiceine (441, R=H) and its 2-demethyl-, 3-demethyl-
and *N*-deacetyl- derivatives
β-Lumicolchicine (442) and its 2-demethyl-, 3-demethyl-
and 3-demethyl-*N*-formyl-*N*-deacetyl- derivatives
γ-Lumicolchicine (443) and its 3-demethyl- derivative.

2,3-Demethylcolchicine on treatment with base and di-iodometh-
ane gives cornigerine, which must therefore be the 1-methoxy-
2,3-methylenedioxy-analogue of colchicine and not the 3-
methoxy-1,2-methylenedioxy isomer as originally believed (M.
Roesner, F.-L. Hsu and A. Brossi, J.org.Chem., 1981, 46, 3686).
When colchicine is heated with acetic anhydride it gives
the enol acetate (444, R=Ac), which is partially hydrolysed
by acid-washed alumina to the *N*-acetyl compound (444, R=H).
This, on mild hydrolysis in alkalis, yields the ketone (445),

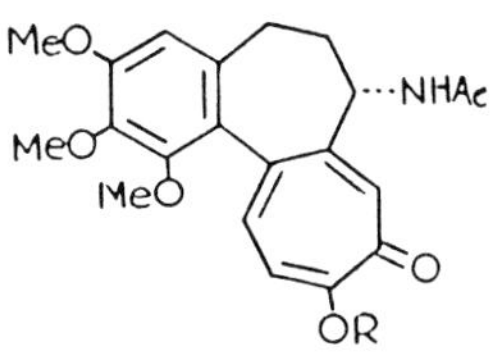

(441) 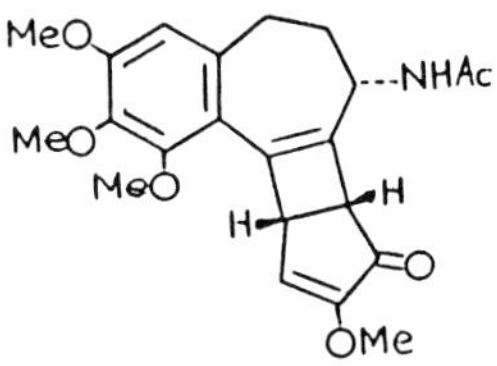   (442) 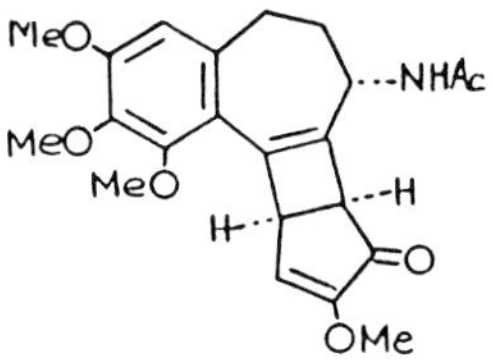   (443)

which can be isomerised to colchicine (441, R=Me). When the
N-acetyl compound (444, R=H) is heated with acetic anhydride
it is isomerised to (446) and when it is photolysed it is converted into the pyrrolenine (447) in the absence of oxygen
and into the ketone (448) in the presence of oxygen (A. Blade-
Font, Tetrahedron Letters, 1977, 2977 and 4097; R. Hunter,
J.J. Bonet and Blade-Font, Affinidad, 1981, 38, 120 and 122).
N-Benzylidene-N-deacetylcolchicine has been converted by potassium hydroxide in methanol into 7-oxodeacetamidocolchiceine
(449, R=H) which is methylated by diazomethane to 7-oxodeacet-
amidocolchicine (449, R=Me) (Blade-Font, Affinidad, 1978, 35,
239).

(444)   (445)   (446)

(447)   (448)   (449)

A high-yield synthesis of deacetamidoisocolchicine (456),
and therefore of colchicine into which it has been converted,
has been accomplished from the methoxyquinone ketal (450).
This dienone is treated with dimethylsulphoxonium methylide
to give the enone (451), which with the appropriate Grignard
reagent gives the tertiary alcohol (452, R=H). Treatment of

the latter with trifluoroacetic acid yields the cyclised and ring-expanded ketone (455, R=H) in a process in which the intermediates (453) and (454) have been identified. Oxidation of the dienone (455) to the tropolone deacetamidoisocolchicine (456) is easily accomplished (D.A. Evans, D.J. Hart and P.M. Koelsch, J.Amer.chem.Soc., 1978, 100, 4593). It may be noted that the intermediate (454) in this process is a compound prepared from the dienone (457), obtained by the oxidation and O-methylation of the diphenol (458) in an earlier synthesis of (456) (E. Kotani, F. Miyazaki and S. Tobinaga, Chem.Comm., 1974, 300).

Variants of this synthesis have been reported starting from the esters (452, R=COOMe) (Evans *et al.*, Pure and Applied Chem., 1979, 51, 1285) and (452, R=COOBu^t) (Evans, S.P. Tanis and Hart, J. Amer.chem.Soc., 1981, 103, 5813).

In the ester series the acid-catalysed cyclisation of the alcohol (452, R=COOMe) gives a mixture of the βγ-unsaturated ester (455, R=COOMe) and its αβ-unsaturated isomer, both of which give the same tropolone on oxidation and this yields (456) on hydrolysis and decarboxylation.

(450)

(451)

(452)

(453)

(454)

(455)

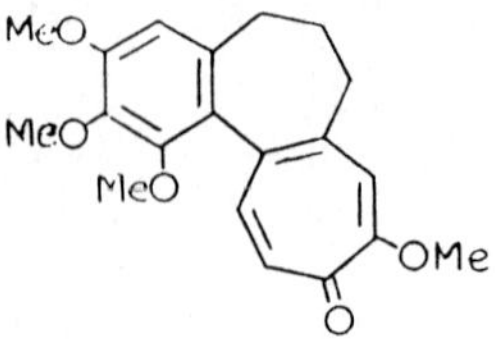

(456) 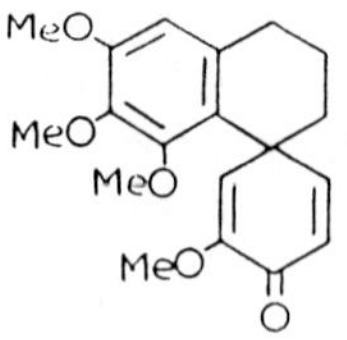 (457) 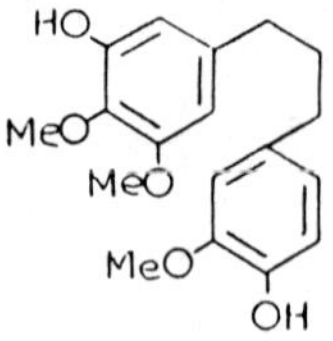 (458)

The coumarin (459), an intermediate in Eschenmoser's synthesis of colchicine (A. Eschenmoser *et al.*, Helv.Chim.Acta, 1961, 44, 540), has been converted into the ketal lactone (461) by the thermal addition of the cyclopentenone ketal (460), a process that presumably involves the generation of a three-carbon dipole, best represented as a delocalised singlet carbene. Hydrolysis of the ketal affords the hydroxy β-keto-acid which loses water and carbon dioxide to give the cyclo-hexatrienone (462). Treatment of this with an excess of hydrazine in ethanol affords the eminoketone (463), hydrolysed by alkali to deacetamidocolchiceine (464, R=H) which yields de-acetamidocolchicine (464, R=Me) and deacetamidoisocolchicine (456) on methylation with diazomethane. This also constitutes a synthesis of colchicine (D.L. Boger and C.E. Brotherton, J.org.Chem., 1985, 50, 3425).

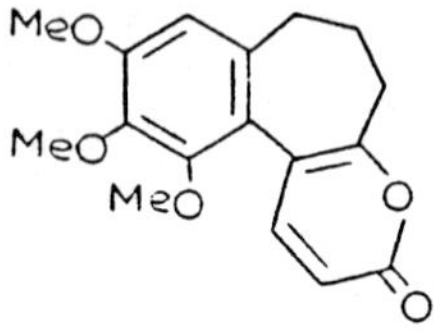

(459) 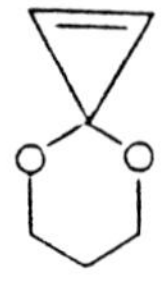 (460) 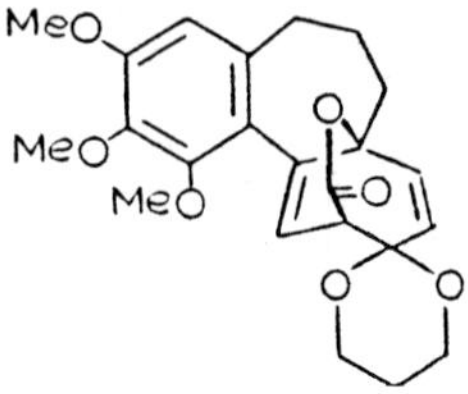 (461)

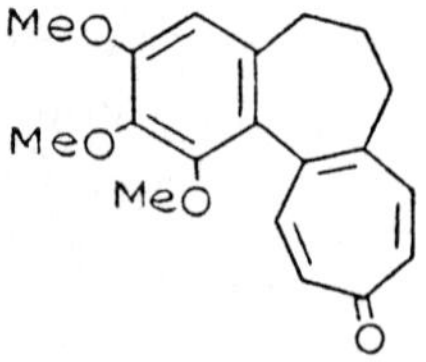

(462) 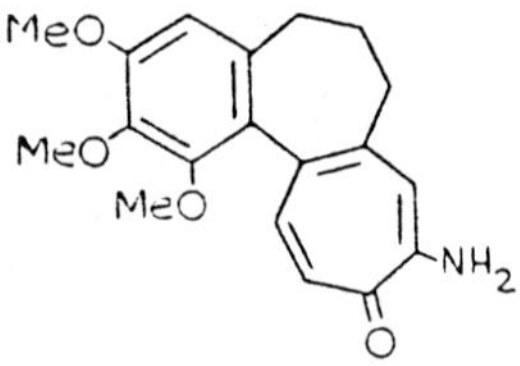 (463) 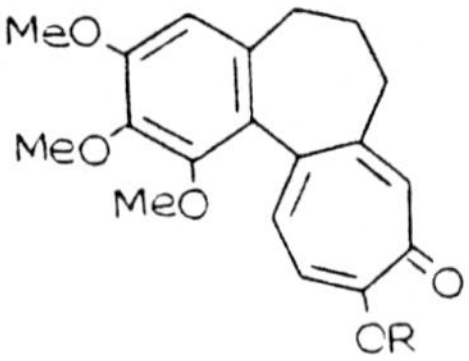 (464)

Chapter 34

DITERPENOID ALKALOIDS

A. R. PINDER

Introduction

General reviews of diterpenoid alkaloids include: S. W.
Pelletier and N. V. Mody in "The Alkaloids. Chemistry and
Physiology," ed. R. H. F. Manske and R. G. A. Rodrigo, Vol.
XVII, Ch. 1, Academic Press, New York, 1979; Vol. XVIII,
Ch. 2, 1981; J. nat. Prod. (Lloydia), 1980, 43, 41; T.
Kametani and H. Fukumoto, Heterocycles, 1977, 8, 540; M. H.
Benn and J. M. Jacyno in "Alkaloids. Chemical and Biological
Perspectives," ed. Pelletier, Vol. 1, Ch. 4 (toxicology and
pharmacology), Wiley, New York, 1984; J. Finer-Moore, E.
Arnold, and J. Clardy; *ibid.*, Vol. 2, Ch. 1 (X-ray diffrac-
tion); Pelletier, Mody *et al.*, Ch. 5 (^{13}C and proton nmr
shift assignments); D. R. Dalton, "The Alkaloids," Ch. 40,
Dekker, New York, 1979. Pelletier and Page have periodically
reviewed these bases in Specialist Periodical Reports (Vols.
7-13, 1977-1983) and in Natural Product Reports 1984, 1, 375
(Royal Society of Chemistry, London). The chromatography of
diterpene alkaloids has been reviewed by A. B. Svendsen and
R. Verpoorte in "Chromatography of Alkaloids, Vol. 23A,
Part A, Ch. 18, Elsevier, Amsterdam, 1983. The use of the
Chromatotron in the separation of these bases has been de-
scribed by Pelletier *et al.* (J. Chromatog., 1985, 322, 223).

There has been an inescapable trend in natural product
chemistry during the last decade. Because of advances and
refinements in physicochemical techniques and their application to organic chemistry it has become possible to arrive
at the structure of a compound with minimal investigation of
its chemistry (in some cases with no chemical studies what-
soever). In no area of organic chemistry is this trend more
noticeable than amongst the diterpene alkaloids. Many new
bases have been isolated, often in very small amounts. This
has precluded much in the way of chemical study, but

techniques such as the various forms of nmr-spectroscopy, mass spectrometry, and X-ray diffraction analysis have permitted detailed structures to be deduced. In view of this situation it has been deemed necessary to adopt a chapter format somewhat different from that used in the 2nd edition. A general account is given of the application of modern physiocochemical methods to structural problems in this area, with examples, followed initially by a discussion of some noteworthy developments in the chemistry of diterpenoid bases, and then by an account of synthetic endeavours. Finally a table is presented which lists new members of the family, with appropriate references.

The alkaloids are usually divided into two groups, containing carbon skeletons with respectively 19 and 20 carbon atoms. Examples from both groups will be discussed.

1. *X-ray Diffraction Analysis*

Single-crystal X-ray diffraction has become the most powerful method available for the elucidation of organic structures, their stereochemistry and absolute configuration. A modern computer-controlled diffractometer allows a structural problem to be solved in a matter of days (and in some cases hours). The technique is ideally suited to complicated structures, and has been used quite extensively in the diterpene alkaloid group. The so-called "direct method" (using a single crystal of the free base) has been applied, for example to the C_{20}-alkaloid veatchine (1) (Pelletier, Mody, and W. H. DeCamp, J. Amer. chem. Soc., 1978, 100, 7976). The "heavy-atom method" (which uses a heavy-metal salt of the alkaloid) was not, in this case, sufficiently informative concerning stereochemistry because formation of the salt involved cleavage of the oxazolidine ring present in the base. This particular X-ray study is of considerable interest because nmr spectral studies reported earlier (Pelletier and Mody, *ibid.*, 1977, 99, 284) indicated that in solution veatchine exists as a pair of C-20 epimers. It is now revealed that this equilibrium persists in the crystalline state, and veatchine exhibits a structure described as being of "epimeric disorder." The 20S:20R ratio is about 3:2. A similar study has been carried out with atisine (as the chloride), isoatisine, and dihydroatisine (Pelletier, Mody, DeCamp, *loc. cit.*). Cuachichicine (2), on the other

(1) Veatchine (20S and 20R)   (2) Cauchichicine

hand, does not show epimeric disorder: its crystal contains
only the epimer shown, analogous in structure to the 20S
epimer of veatchine (Pelletier *et al.*, *ibid.*, 1979, 101,
6741). This rather surprising observation in a base with
almost the same structure as veatchine may be explained by
the proximity of the 14β-hydrogen atom to C-20 (evident
from the X-ray diffraction pattern). Such nearness is known
to cause steric hindrance to ring-closure in atisine; it is
not observed with veatchine. Also, certain 14β-substituted
veatchine-type bases are observed to exist as single C-20
epimers (Pelletier *et al.*, J. org. Chem., 1983, 48, 1787).
Hetisine-13-*O*-acetate (3) is a new base isolated in very
small amounts (such that chemical studies are precluded)
from *Delphinium nuttallianum* Printz. Its structure has been
settled in part by X-ray diffraction analysis of its perchlorate; the absolute configuration remains to be determined
(M. H. Benn *et al.*, Heterocycles, 1986, 24, 1605). Spirasines
V and VI are two new diterpenoid bases from *Spiraea japonica*;

(3) 13-*O*-Acetylhetisine

(4) R₁=CH₃, R₂ = OH Spirasine V
(5) R = OH, R₂ = CH₃ Spirasine VI

application of X-ray diffraction analysis has led to structures (4) and (5), respectively, as relative configurations.
Ring B in (5) has a half-chair conformation, a feature which
explains an unexpected anomaly in its CD curve. The absolute configurations of (4) and (5) have been settled by CD
measurements and correlations with structures of known configuration (X.-t. Liang, J. Clardy *et al.*, Tetrahedron
Letters, 1986, 27, 275). The alkaloid excelsine, originally
believed to be derived from tetrahydrofuran, is in fact an
epoxide (6); this has been revealed by a "heavy-atom" X-ray
analysis of its hydroiodide, which also defined its absolute
stereochemistry. This structural modification accounts for
certain previously unexplained chemical reations (S. Y.
Yunusov *et al.*, Chem. Abs., 1975, 82, 125505t; 1976, 85,
177706r). Full details of an earlier analysis of chasmanine

(6) Excelsine   (7) Chasmanine

(7) have appeared; in this molecule the 1-methoxyl group is α
(Pelletier *et al.*, Acta Cryst., 1977, 33B, 722). Condelphine
has the structure and absolute stereochemistry (8)(*idem*,
ibid., p. 716); the boat shape of ring A is stabilised by
intramolecular hydrogen bonding. A similar state of affairs
exists in jesaconitine, the structure of which (9) has been
corroborated by X-ray analysis of its perchlorate (Pelletier,
DeCamp *et al.*, Cryst. Struct. Comm. 1979, 8, 299). Gadesine,
a new delphinium alkaloid is formulated as (10) on spectral
and chemical evidence; the formulation has been confirmed by
X-ray diffraction analysis (A. G. Gonzalez *et al.*, Tetrahedron Letters, 1979, 79); again ring A has a boat conformation stabilised by two intramolecular hydrogen bonds. The
structure and the stereochemistry of dictysine have been
revised to (11) as a consequence of an X-ray analysis
(Yunusov, unpublished observation; see Pelletier and S. W.
Page, Spec. Periodical Reps., The Alkaloids, 1981, 11, 209).

(8) Condelphine

(9) Jesaconitine

(10) Gadesine

(11) Dictysine

Several X-ray diffraction analyses of degradation products of lycoctonine have been reported; the major outcome is that the methoxyl group at position 1 in the alkaloid is α. This has necessitated stereochemical revision at C-1 of a large group of structurally related bases (O. E. Edwards *et al.*, Acta Cryst., Sect. A, Suppl., 1981, 37, C-211; Pelletier *et al.*, J. Amer. chem. Soc., 1981, 103, 6536), which are now recognised as having a 1α oxygen function. The complete results of a detailed X-ray crystallographic analysis of aconitine (12) have been presented. They reveal that rings A, B and E are chairs, ring C is an envelope, D is a boat and F a half-chair. Hydrogen bonding is important in the determination of ring conformations (P. W. Codding, Acta Cryst., 1982., B38, 2519).

2. *Mass Spectrometry*

Mass spectral studies play an important role in structure determination in this field. One of the major advantages is that only minute amounts of substrate are required for spectral measurement. Under high resolution conditions the compound's molecular weight can be determined with great accuracy; tables are available for translation of this

value into the correct molecular formula of the alkaloid.
Confirmation of the answer is possible *via* a study of the
relative abundances of the M^+, $(M+1)^+$ and $(M+2)^+$ ions.
This type of spectrometry is used to solve structural
problems. For example a base $C_{24}H_{35}NO_4$ has been isolated from *Aconitum karakolicum*, which on alkaline hydrolysis
afforded the known napelline (13). This new compound contains an acetoxy group on chemical and mass spectral evidence. On exposure to acetic anyhdride-pyridine it yields
a diacetate; the original base is therefore a diol, actually
napelline (13) in which one of the three OH groups is
acetylated. One of these groups in the base is easily
oxidised to a ketone with silver oxide, which observation
indicates that it is at position 1. Thus the alkaloid has
an acetoxy group at C-12 and a hydroxyl group at C-15, or
vice versa. The mass spectrum of the base shows a minor
(6%) peak at $(M-Ac)^+$; the diacetate on the other hand shows
a major (36%) corresponding peak. It has been observed that

(12) Aconitine

Napelline (13) $R^1 = R^2 = H$
12-Acetylnapelline (14) $R^1 = H$, $R^2 = Ac$
(15) $R^1 = R^2 = Ac$

in related compounds with 15-acetoxy groups the $(M-Ac)^+$ ion
is a major one. Consequently the new base is a 12-acetoxy
compound of structure (14) and the diacetate is (15) (Yunusov
et al., Chem. Abs., 1977, 86, 152618w). Mass spectrometry is
commonly used when working with very small amounts of material
to establish whether or not a reaction has proceeded according
to expectations: an oxidation of a secondary alcohol group to
a ketone for example can be monitored by observing the extent
of a shift of the parent ion peak M^+ to $(M-2)^+$. The combina-
tion g.c.-m.s. has been applied to cases where extracts of
plant material have yielded mixtures of closely-related

bases, components being in part identified by the m/z values
(G. R. Waller and R. H. Lawrence, Jr., in "Recent Developments in Mass Spectrometry in Biochemistry and Medicine,"
ed. A. Figerio, p. 429, Plenum Press, New York, 1978).
Certain characteristic mass spectral fragmentation patterns
are recognised; for example a new base $C_{26}H_{41}NO_7$ isolated
from *Delphinium ternatum* fragments in a manner peculiar to
lycoctonine-type bases with a 1-methoxyl group. The alkaloid is accordingly assigned a partial structure of this
type (A. S. Marzullaev *et al.*, Chem. Abs., 1979, <u>90</u>, 183153c).

A lengthy table of high resolution mass values and formulae indices has been provided by Pelletier *et al.* (in "Alkaloids: Chemical and Biological Perspectives," Vol. 2,
Chapter 5, Wiley, New York, 1984). The mass spectra of these
bases are, not unexpectedly, complex but there are certain
patterns which are recognisable and are of value. For example
the complete mass spectral analysis of heteratisine (16) has
been reported; this has permitted, by spectral comparisons,
the elucidation of the structures of three closely-related
bases heterophyllidine (17), heterophyllisine (18) and
heterophylline (19) (Pelletier and R. Aneja, Tetrahedron
Letters, 1967, 567). The main fragmentation in (16) and
(17) is loss of the 1-methoxyl group to yield $(M-31)^{\ddot{+}}$,
followed by methane loss to $(M-47)^{\ddot{+}}$. A fragmentation of the
>NEt group to yield $\overset{+}{N}CH_2$ is also characteristic; in some
other bases the main fragmentation centre is the nitrogen
atom (Yunusov *et al.*, Chem. Abs., 1970, <u>73</u>, 15053a). Another

(16) R = CH_3 Heteratisine   (18) R = CH_3 Heterophyllisine
(17) R = H Heterophyllidine   (19) R = H Heterophylline

example of similar fragmentation behaviour with similar structure to be found with the trio delcosine (20), acetyldelcosine
(21) and delsoline (22). Characteristic peaks occur at
$(M-15)^{\ddot{+}}$ (loss of CH_3), $(M-17)^{\ddot{+}}$ (loss of OH), $(M-31)^{\ddot{+}}$ (loss of

OCH_3), $(M-32)^{+\cdot}$ (loss of CH_3OH), and $(M-33)^{+\cdot}$ (loss of CH_3 and H_2O), along with a characteristic peak $(M-189)^{+}$. A pathway for the formation of these fragments has been postulated (G. R. Waller *et al.*, Chem. Abs., 1973, **79**, 137325k).

(20) R = H Delcosine
(21) R = Ac Acetyl-
             delcosine
(22) R = CH_3 Delsoline

C_{19} — diterpenoid alkaloid
           skeleton, with numbering

3. *Nmr Spectroscopy*

This technique is of paramount importance in structural studies in this area, and is used extensively. The chief types are proton and carbon-13 nmr spectroscopy, which will be dealt with separately. A lengthy review chapter on the application of nmr-spectroscopy to C_{19}-diterpenoid alkaloids is noteworthy (Pelletier *et al.*, *op. cit.*; see also *idem*, in "The Alkaloids. Chemistry and Physiology," ed. R. H. F. Manske and R. G. A. Rodrigo, Vol. XVII, Chapter 1, Academic Press, New York, 1979).

(a) Proton magnetic resonance spectroscopy

Pmr-spectroscopy is used routinely in the structural study of diterpene alkaloids. Its value lies largely in its revelation of the nature and position in the molecular frame-work of the twenty or so functional groups commonly encoun-tered. Tables have been compiled of the proton chemical shift range and multiplicity of signals, which are of assistance in settling the functional groups present [Pelletier *et al.*, *op. cit.* ("Alkaloids. Chemical and Bio-logical Perspectives")]. For example, these bases are in most cases *N*-ethyl compounds, but occasionally *N*-methyl. If the former a pmr signal is found in the 0.8-1.10 ppm (δ)

region as a triplet; if the latter no signal appears in this region, but a methyl signal at δ 2.35-2.60 (singlet). If there is an acetoxy group at position 8 the COCH$_3$ protons resonate at δ 1.95-2.05, but when an aromatic ester group is also present at position 14, the former are markedly shielded and resonate at δ 1.25-1.45 (both singlets). Likewise a CH(OH) signal is sensitive to its environment; for example if the proton is 14β it usually resonates at δ 4.0-4.3, but if it is 10β it is somewhat deshielded with a signal at $\sim\delta$ 4.6 ppm. It is apparent from these examples and others that the nature, position, and stereochemistry of functional groups can be largely discerned by a careful analysis of the pmr-spectrum of the base and comparison of it with those of alkaloids of known structure.

(b) Carbon-13 magnetic resonance spectroscopy

This type of spectroscopy is used very extensively in the diterpene alkaloid field, and is as important as X-ray diffraction analysis as a method of structure investigation. The state of the art has been reviewed extensively by Pelletier and co-workers (*opp. cit.*; Heterocycles, 1977, 7, 327). The technique was first applied to the structure determination of two diterpene bases by A. J. Jones and M. H. Benn (Tetrahedron Letters, 1972, 4351; Canad. J. Chem., 1973, 51, 486), who measured the ^{13}C-nmr-spectra of several C$_{19}$-diterpenoid bases and their derivatives. Proton-noise decoupling and additivity relationships permitted assignments of the carbon-13 resonances in these compounds, and a general pattern of ^{13}C chemical shifts in C$_{19}$-diterpene alkaloids emerged. The data were then used to determine the structures of two new bases isolated only in very small amounts. Since then this approach has been extended greatly. A data bank containing ^{13}C and ^{1}H spectral properties along with a catalogue of natural bases has been assembled for use in structural studies (Pelletier, *opp. cit.*). An example of an entry is shown below, reproduced from "Alkaloids. Chemical and Biological Perspectives," Vol. 2, by permission of the author and publisher (John Wiley and Sons, Inc.). Because of observed deviations in ^{13}C-spectra compared with what have become regarded as normal patterns, the structures of some well-known bases in this group have needed revision, and certain "new" bases have been found to be identical with

ACONITINE

$C_{34}H_{47}NO_{11}$; mp 202-205°;
$[\alpha]_D$ + 19° $(CHCl_3)$

*Aconitum callianthum, A. carmichaeli,
A. chinense, A. fauriei, A. flavum,
A. fukatomei, A. grossedentatum, A.
hakusanense, A. ibukiense, A. japoni-
cum, A. karakolicum, A. kusnezoffii,
A. majima, A. mitakense, A. mokchang-
ense, A. nagarum var. heterotrichum f. dielsianum,
A. napellus and ssp., A. pendulum, A. sachalinense,
A. sanyosense, A. senanense, A. soongaricum, A.
stoerckianum, A. subcuneatum, A. tasiromontanum,
A. tianschanicum, A. tortuosum, A. yezoense, A.
zuccarini.*

^{1}H NMR: δ 1.15 (3H, *t*, NCH$_2$-C*H*$_3$), 1.44 (3H, *s*,
OCOC*H*$_3$), 3.29, 3.38, 3.42, and 3.88 (each 3H, *s*,
OC*H*$_3$), 5.04 (1H, *d*, C(14)-β-*H*), and 7.67-8.25
(aromatic protons).

[13]C *Chemical Shift Assignments*

C-1	83.4	C-11	49.8	C-1'	55.7
C-2	36.0	C-12	34.0	C-6'	57.9
C-3	70.4	C-13	74.0	C-16'	60.7
C-4	43.2	C-14	78.9	C-18'	58.9
C-5	46.6	C-15	78.9	C=O	172.2
C-6	82.3	C-16	90.1	CH$_3$	21.3
C-7	44.8	C-17	61.0		
C-8	92.0	C-18	75.6	C=O	165.9
C-9	44.2	C-19	48.8	1	129.6
C-10	40.8	N-CH$_2$	46.9	2	128.6
		CH$_3$	13.3	3	129.8
				4	133.2
				5	129.8
				6	128.6

already known alkaloids. A case in point is cammaconine,
isolated from *Aconitum variegatum* and formulated on the basis
of a chemical correlation, mass spectral analysis, and pmr
studies as (23) (N. M. Mollov *et al.*, Tetrahedron, 1971, 27,
819). A careful examination of its ^{13}C-nmr-spectrum, however,
indicated clearly that the CH$_2$OH group is located at position
4 and the OMe group at position 16, so that cammaconine must

be re-formulated as (24) (Pelletier *et al.*, Heterocycles, 1980, 14, 1751). An extension of this technique has been

(25)

(23) $R^1 = R^3 = Me$, $R^2 = H$
(24) $R^1 = R^2 = Me$, $R^3 = H$ Cammaconine

the application of computer-assisted analysis of ^{13}C-nmr spectra, with prediction of structures. Computer programmes have been developed and also a data-base of the ^{13}C resonances with substructures, which characterize the constitutional and stereochemical environments relating to a large number of diterpenoid alkaloids. Briefly, new ^{13}C-nmr spectra are analysed by comparison with the data-base, which leads to a set of substructures; these are refined using a repetitive interpretation procedure. There results a set of possible structures which are evaluated by spectral prediction, and the spectra are compared with the spectrum of the new alkaloid. The "candidate" spectra are then arranged in order of rank, the top rank member having a structure which can be advanced for the unknown. Such programmes can quickly limit the possible structures for a new alkaloid (J. Finer-Moore *et al.*, J. org. Chem., 1981, 46, 3399; C. Djerassi *et al.*, *ibid.*, pp. 703, 1708; Org. magn. Res., 1981, 15, 375). The procedure has revealed the necessity for structural revision for a number of diterpenoid alkaloids (see Pelletier and Page, Spec. period. Reps., 1981, 11, 209; 1982, 12, 249 for details).

Finally, it is worthy of note that several workers have used two-dimensional nmr-spectroscopy in structural studies. Plotting the ^{13}C-resonances against the [1]H-resonances reveals the connectivities of carbons and their attached hydrogens (see, for example, M. G. Reinecke *et al.*,

Heterocycles, 1986, 24, 49) in a given structure.

Noteworthy Chemical Reactions

A simple, efficient procedure for degrading the oxazolidine
ring in C_{20}-diterpene bases has been discovered (Mody and
Pelletier, Tetrahedron Letters, 1978, 3313). Veatchine (1),
for example, with acetic anhydride-pyridine affords the
diacetate (25), which on refluxing in chloroform yields (26)
and 1,2-diacetoxyethane, by internal S_N2 attack. Atisine,

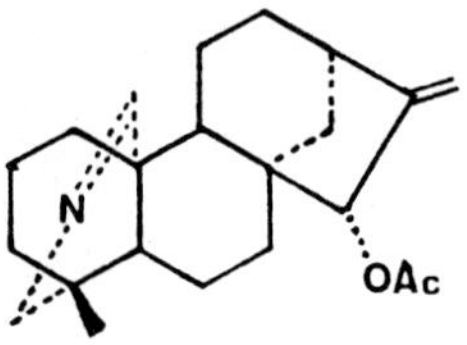

(26) 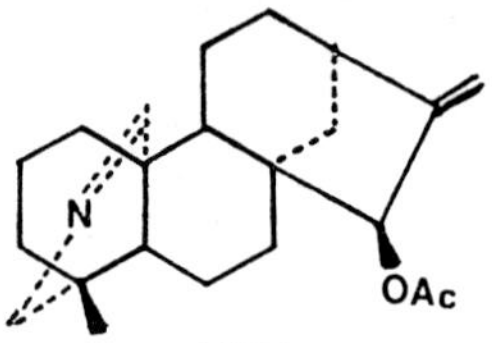   (27)   (28)
                   Lindheimerine   Ovatine

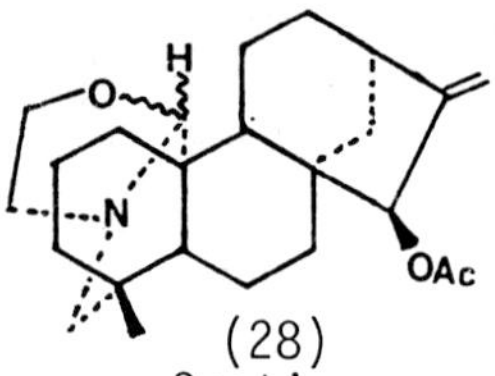

garryfoline, and ovatine, and iso-oxazolidine bases all
behave similarly. Conversely a method for constructing the
oxazolidine ring has been reported (Pelletier, Mody *et al.*,
Synth. Commun., 1979, 9 , 201). Reaction of the appropriate
imine, *e.g.* lindheimerine (27), with ethylene oxide affords
the desired oxazolidine ovatine (28) in almost quantitative
yield. *N*-2-Hydroxyethyl bases such as dihydroatisine (29)
are also convertible into the corresponding iso-oxazolidine
bases, *e.g.* isoatisine (30), by mild oxidation with manganese
dioxide at room temperature; yields are distinctly superior
to those obtained in oxidations effected by osmium tetroxide
or mercury (II) acetate (Pelletier *et al.*, Tetrahedron
Letters, 1978, 5187). Detailed studies using silver oxide

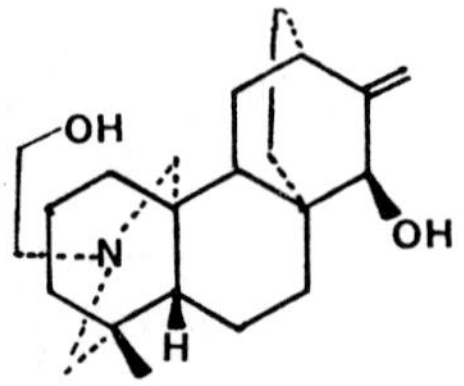

(29)   (30)   (31)
Dihydroatisine   Isoatisine   Veatchinone

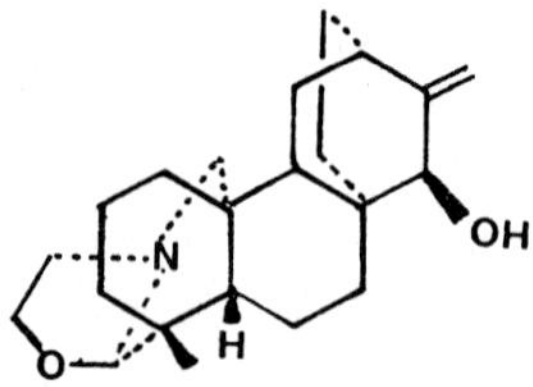

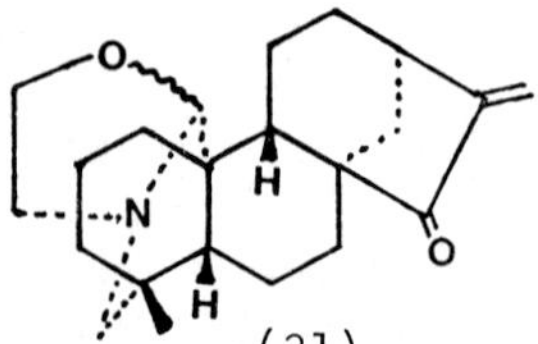

and alkaline potassium ferricyanide as oxidants in this reaction have been made. It has been observed that the former reagent generates only the "iso-compound" [as in (30)], no "normal" oxazolidines [as in (28)] being encountered. The latter oxidant, on the other hand, affords a roughly 1:1 mixture of the two types (Pelletier *et al.*, Heterocycles, 1980, 14, 1155; Tetrahedron Letters, 1980, 21, 3647). The oxazolidine ring (both types) can be reduced selectively to the >NCH$_2$CH$_2$OH base by sodium cyanoborohydride at room temperature and pH 6-7 (Pelletier *et al.*, *ibid.*, 1979, 4939).

The Michael addition of secondary amines to veatchinone (31) occurs only in the presence of alumina of activity III, to afford adducts (32). This behaviour seems to be general for exocyclic αβ-unsaturated ketones, *e.g.* 2-methylene-1-tetralone (Pelletier *et al.*, *ibid.*, 1980, 21, 809).

(32)

Rearrangements

The alkaloid ajaconine (33) undergoes an unusual rearrangement to 7α-hydroxyisoatisine (34) on heating in methanol. The adjoining scheme has been advanced by way of explanation of this behaviour. It is based on a comparison of the rearrangement with the well-known atisine-isoatisine transformation, and the knowledge that ajaconine forms an immonium (iminium) salt with mineral acids rather than an ammonium (≡NH) salt. ^{13}C-nmr-spectral studies on ajaconine reveal that in hydrogen-bonding solvents the ether linkage is cleaved as a result of covalent solvation. It is proposed that with methanol the immonium salt (35) is formed; this suffers a double bond migration to (36) and then cyclisation to (34). This suggestion has been substantiated by using CH$_3$OD as solvent, when a mixture of C-19 deuteriated and C-20 deuteriated ajaconine plus 7α-hydroxyisoatisines labelled at C-19 and C-20 is formed. The cyclisation of the immonium salt (35) to the oxazolidine (37) is much slower

(33) Ajaconine

(35)

(37)

(36)

(34)

than back to ajaconine (33); however the double bond isomer (36) undergoes a ready cyclisation to the product (34) (Pelletier and Mody, J. Amer. chem. Soc., 1979, 101, 492).

The well-known acid-catalysed garryfoline-cuauchichicine rearrangement, peculiar to C_{20} alkaloids with a 15β hydroxyl group, has been re-investigated. It now appears, on ^{13}C-nmr-spectral and deuterium labelling evidence, that the mechanism is one of enol formation followed by *exo*-protonation. Veatchinone (31) is reduced with $NaBD_4$ to an epimeric mixture yielding on acetylation a corresponding mixture of acetates from which (38) is isolable by preparative tlc. Treatment of (38) with acid and re-acetylation affords (39), the ^{13}C-nmr-spectrum of which indicates that it is not deuteriated at C-16. This observation rules out the possibility that the rearrangement occurs *via* a 15→16 hydride shift, as had been suggested earlier (M. F. Barnes and J. MacMillan, J. chem. Soc., C, 1957, 361). Next it was found that isogarryfoline (40) on acid (DCl) treatment yields ketone (41), with deuterium at C-16 and C-17, as shown. The adjoining scheme

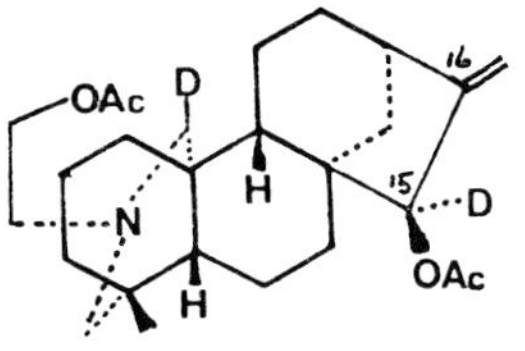

(38) 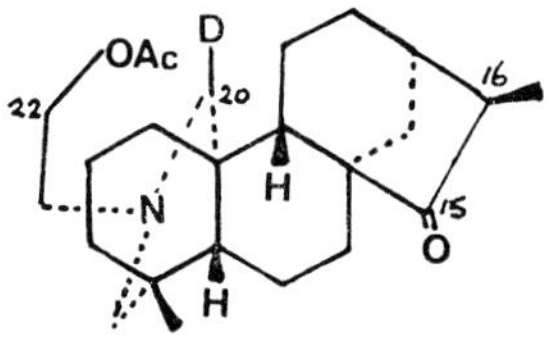   (39)

explains this behaviour, and the approach of D^+ to the less-hindered *exo* side of the molecule as shown accounts for the stereochemistry at C-16 in (41). Finally (41), on treatment with base and then HCl suffers a D→H exchange to (42) during 96 hours. These observations support a pathway of enolisation followed by *exo*-protonation (Pelletier *et al.*, Heterocycles, 1979, 13, 277).

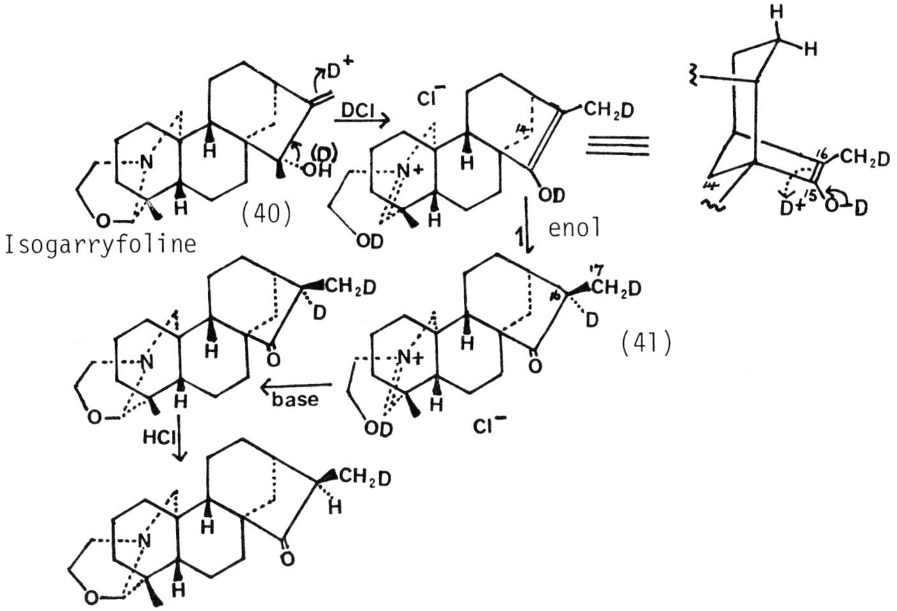

Isogarryfoline

(40)

(41)

(42)

Another well-known rearrangement, that of hetisine (43) by acid catalysis, has been re-investigated in a similar fashion. In fact two products are formed, both adamantane-like in structure, in ratio 19:1. Their structures, settled by nmr-spectral and X-ray analysis, are (44) and (45) respectively. The former was isolated as an alkaloid of *Aconitum heterophyllum* earlier. A pathway for the rearrangement has been suggested (Pelletier *et al.*, Chem. Comm., 1981, 327; Heterocycles, 1983, 20, 413).

(43) (44) (45)

Hetisine

Further study of the rearrangement has revealed that two additional compounds are also formed. They form an inseparable mixture (1:1). On the basis of pmr- and ^{13}C-nmr spectral evidence, confirmed by X-ray diffraction analysis, they are formulated as (46) and (47) (*idem*, Heterocycles, 1983, 20, 143).

(46) $R^1=H, R^2= OH$
(47) $R^1= OH, R^2=H$

An unexpected rearrangement of compound (48), obtainable by oxidation of lycoctonine followed by pinacolic dehydration and Curtius or Hofmann degradation has been observed. Treatment of this primary amine with nitrous acid affords two products (49) and (50); pathways for their formation have been suggested (O. E. Edwards, Canad. J. Chem., 1981, <u>59</u>, 3039).

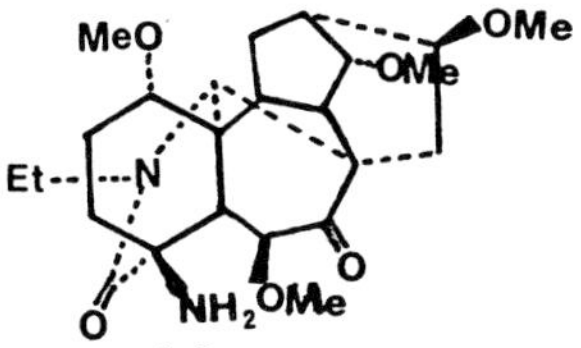

(48) 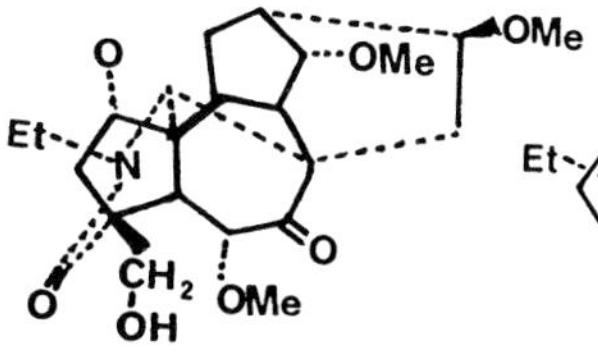   (49) 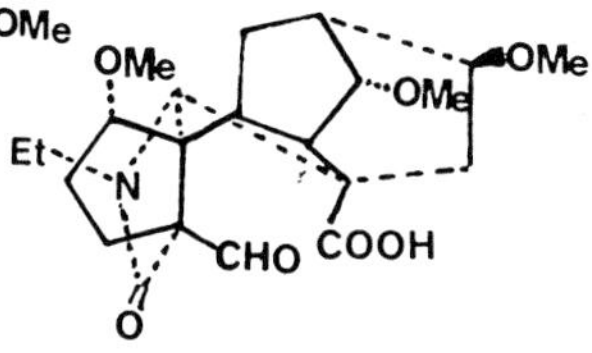   (50)

Synthetic Endeavours

(a) Chasmanine (7)

A total synthesis of the C_{19} alkaloid chasmanine (7) has been described; it represents a landmark in diterpene alkaloid chemistry (K. Wiesner and co-workers, Canad. J. Chem., 1975, <u>53</u>, 2140; 1976, <u>54</u>, 1039; 1977, <u>55</u>, 1091; Heterocycles, 1977, <u>7</u>, 217, and earlier papers cited therein). The first part

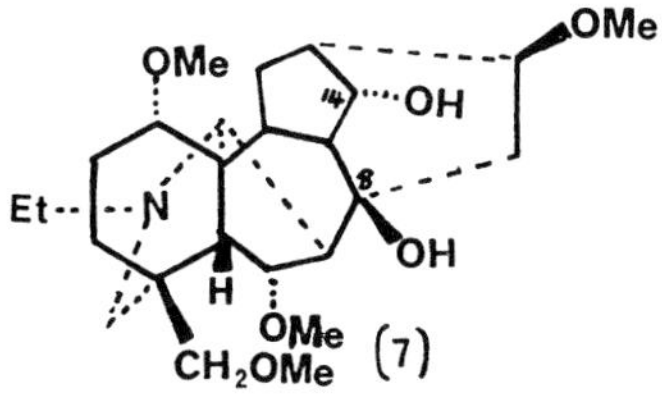

Chasmanine

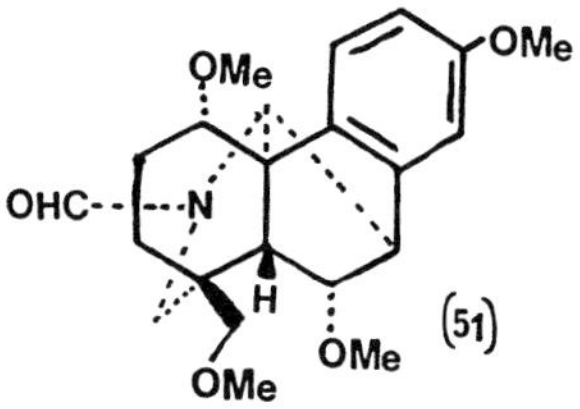

of the necessarily lengthy synthesis is the construction of the intermediate (or synthon) (51) by the pathway outlined in Scheme 1.

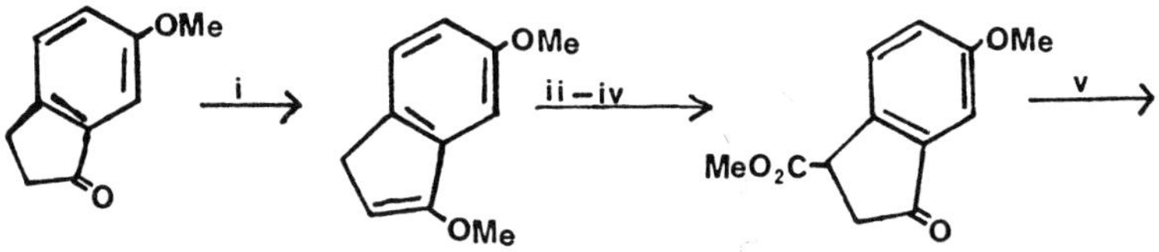

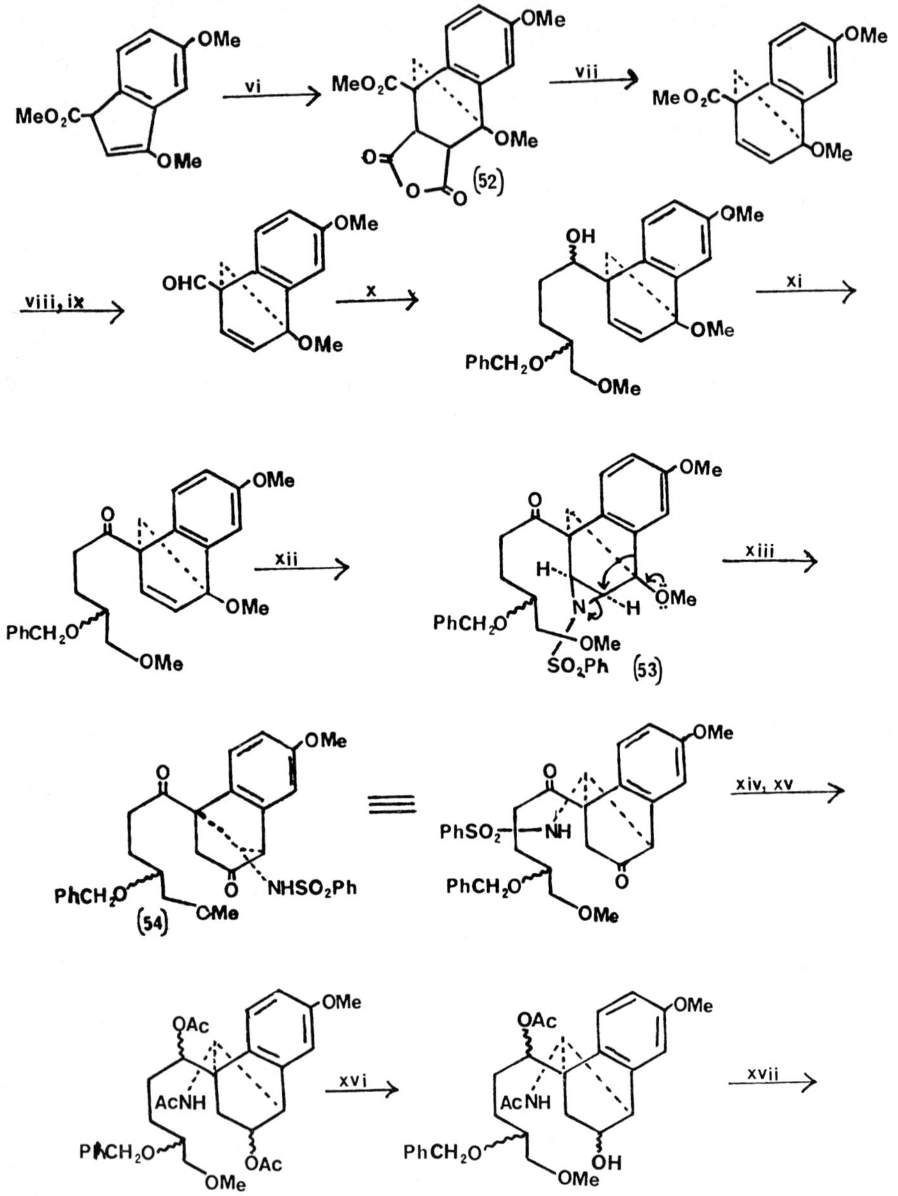
OMe
MeO$_2$C
vi
vii
(52)
viii, ix
OHC
x
OH
PhCH$_2$O
xi
xii
xiii
(53)
SO$_2$Ph
(54)
NHSO$_2$Ph
PhSO$_2$
NH
xiv, xv
OAc
AcNH
xvi
OH
xvii

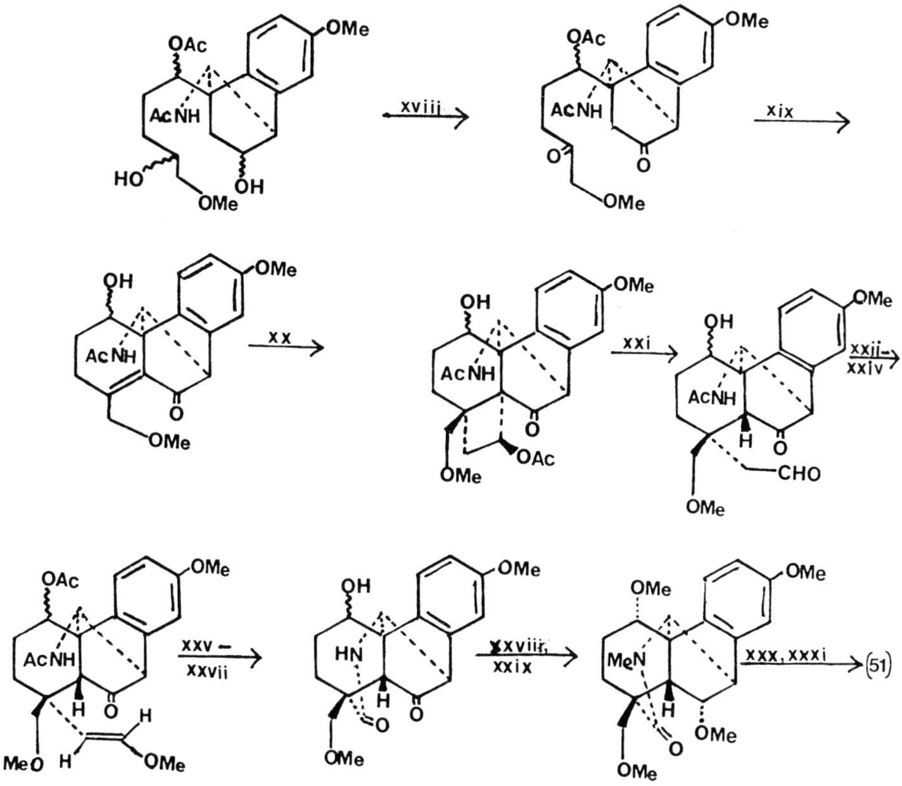

Scheme 1

Reagents: i, HC(OMe)₃, HCl, MeOH, C₆H₆; ii, BuLi; iii, CO₂, -70°; iv, MeOH, HCl; v, HC(OMe)₃, HCl, MeOH, C₆H₆; vi, maleic anhydride, xylene, 175°; vii, (PH₃P)₂ Ni(CO)₂, diglyme, 210° (-CO₂,-CO); viii, LiAlH₄; ix, DCC, DMSO; x, MeOCH₂CHCH₂CH₂-MgBr; xi, py-CrO₃; xii, PhSO₂N₃, HOAc, H⁺, r.t.,: xiii, spontaneous rearrangement; xiv, LiAlH₄, Ac₂O, py; xvi, K₂CO₃, MeOH (partial hydrolysis); xvii, H₂, Pd/C, MeOH; xviii, py-CrO₃; xix, K₂CO₃, MeOH, Δ (-H₂O); xx, CH₂=CHOAc, hν; xxi, KOH, MeOH, r.t.; xxii, HC(OMe)₃, H⁺; xxiii, Ac₂O, py; xxiv, py, xylene, Δ (-MeOH); xxv, NaIO₄-KMnO₄; xxvi, CH₂N₂; xxvii, NaOMe, MeOH, Δ (cylisation); xxviii, LiAl (O-tBu)₃H; xxix, NaH, MeI; xxx, LiAlH₄; xxxi, KMnO₄, HOAc.

Many of the steps used in this synthesis were first tested
using model compounds. The Diels-Alder addition leading to
(52) is based on the known behaviour of indenes in that reaction (E. Wenkert *et al.*, J. org. Chem., 1967, 32, 1126);
the procedure for its decarboxylation/decarbonylation is noteworthy (*cf.* B. M. Trost and F. Chen, Tetrahedron Letters,
1971, 2603). Conversion of the aziridine (53) by spontaneous rearrangement into diketone (54) [see arrows in (53)] is
assisted by the influence of the bridgehead methoxyl group
and the keto-group; the latter hinders the formation of
another, undesirable rearrangement product.
    The conversion of (51) into chasmanine (7), again based on
a number of reactions applied to model compounds, is summarised in Scheme 2.

Scheme 2

(±) - Chasmanine

Reagents: i, Li, NH$_3$; ii, Ac$_2$O, py; iii, HCl, Δ; iv, CH$_2$=C=CH$_2$, hν; v, (CH$_2$OH)$_2$, H$^+$; vi, O$_3$, -78°, then NaBH$_4$; vii, Ac$_2$O, py; viii, HCl; ix, C$_5$H$_5$NHBr$_3$; x, LiBr, Li$_2$CO$_3$, DMF (-HBr); xi, NaOH, MeOH; xii, Ac$_2$O, py; xiii, H$_2$, Rh; xiv, py-CrO$_3$; xv, (CH$_2$OH)$_2$, H$^+$; xvi, KOH, MeOH; xvii, py-CrO$_3$; xvii, NaBH$_4$; xix, NaH, MeI;

xx, H$^+$, H$_2$O; xxi, Br$_2$, Et$_2$O; xxii, [diagram], H$^+$; xxiii, DBN, xylene-DMSO, Δ; xxiv, Hg(OAc)$_2$; xxv, NaBH$_4$; xxvi, H$^+$, H$_2$O; xxvii, H$^+$, H$_2$O.

Improvements in this synthesis have been reported, by way of increased yields and abbreviation (Wiesner *et al.*, Chem. Soc. Rev., 1977, 6, 413; Canad. J. Chem., 1978, 56, 1102, 1451), and another route to the alkaloid has been devised (*idem*, *ibid.*, 1979, 57, 2124; Wiesner, Pure Appl. Chem., 1979, 51, 689). Other synthetic studies in this area have been described by, *inter al.*, W. L. Meyer *et al.*, J. org. Chem., 1977, 42, 2761, 4131; F. Satoh *et al.*, Heterocycles, 1977, 6, 1957; A. K. Banerjee *et al.*, *ibid.*, 1980, 14, 315; Tetrahedron, 1981, 37, 2749; U. R. Ghatak *et al.*, Indian J. Chem., 1980, 19B, 305). A new synthesis of napelline (13) has been reported (Scheme 3); it uses a starting point compound (55) synthesised earlier, and embodies steps and improvements developed in earlier work and using model compounds (Wiesner *et al.*, Canad. J. Chem., 1978, 56, 1102; 1980, 58, 1889).

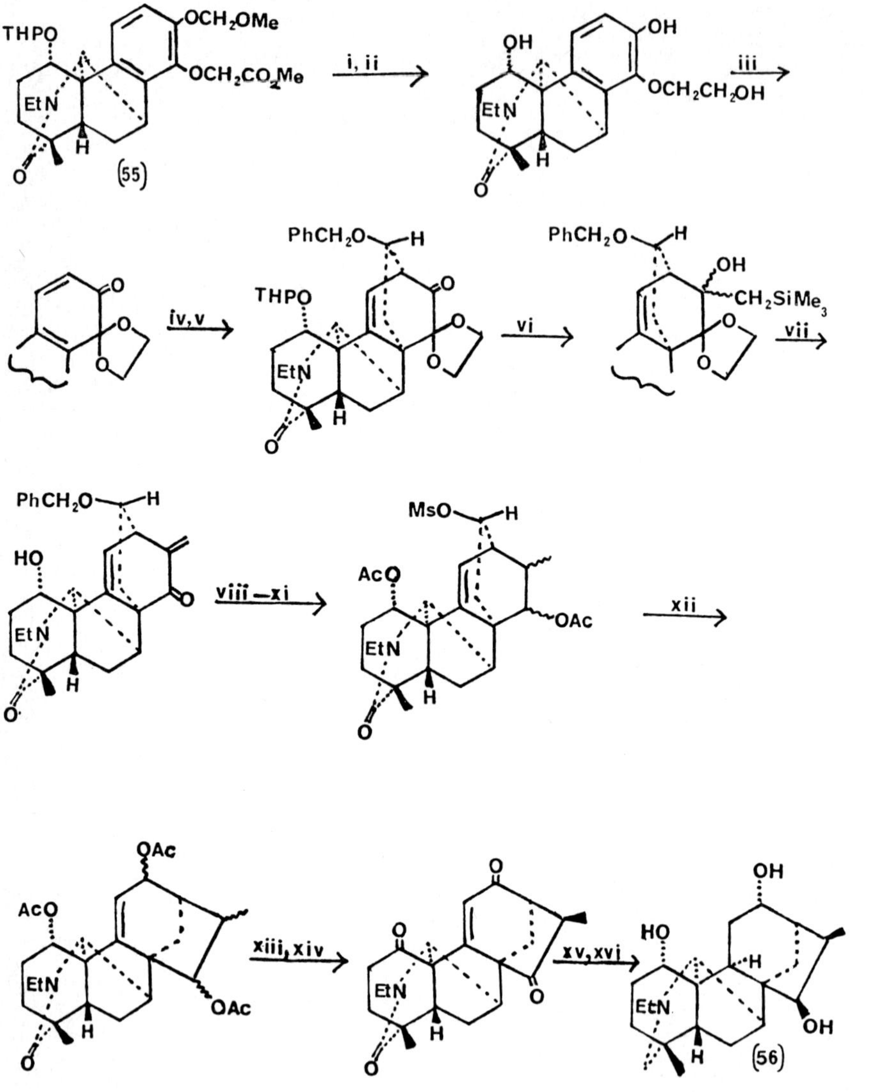

Dihydronapelline

Scheme 3

Reagents: i, $LiBH_4$; ii, H^+, H_2O; iii, $Tl(NO_3)_3$; iv, $PhCH_2OCH=CH_2$; v, DHP, H^+; vi, Me_3SiCH_2MgCl; vii, 70% $HClO_4$, Δ; viii, $LiBH_4$; ix, Ac_2O, py; x, H_2, Pd; xi, MsCl, py; xii, HOAc, Δ; xiii, hydrolysis, xiv, py-CrO_3; xv, H_2; xvi, $LiAlH_4$.

MeO ... OMe ... OH ... (57)

EtN ... OH ... H

Aconosine

OMe ... OMe ... OBz ... OAc ... OH

MeN ... HO ... OMe ... (58)

OMe

Mesaconitine

OMe ... OMe ... OBz ... OMe ... OH

MeN ... OAc ... (59)

OMe ... OMe

Isodelphinine

The final product, (56), is identical with racemic dehydronapelline, which has been converted into napelline (13) earlier (Wiesner *et al.*, Canad. J. Chem., 1974, 52, 2353, 2355).
 Some partial syntheses have been described. For example cammaconine (24) has been converted into aconosine (57) (O. E. Edwards *et al.*, Canad. J. Chem., 1983, 61, 1194), and chasmanine (7) and mesaconitine (58) have both been transformed into isodelphinine (59) (H. Takayama *et al.*, Chem. pharm. Bull. Japan, 1982, 30, 386; T. Mori *et al.*, *ibid.*, 1983, 31, 1422). A very promising synthetic approach to *Daphniphyllum* alkaloids has been reported (J. Orban and J. V. Turner, Tetrahedron Letters, 1983, 24, 2697): the tetracyclic intermediate (62) is reached in four steps as outlined in Scheme 4.

OCH_2OMe $\xrightarrow{i}$ OCH_2OMe $\xrightarrow{ii-iv}$ OCH_2OMe, CO_2H $\xrightarrow{v-vii}$ O, CO_2Me $\equiv$

O, CO_2Me $\xrightarrow{viii}$ $OSiMe_2Bu^t$ (60) → CO_2Me, O, H, $OSiMe_2Bu^t$ $\xrightarrow{ix}$

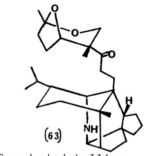

Scheme 4

Reagents: i, Li, NH₃, t-BuOH, THF; ii, BuLi, TMEDA; iii, CO₂, -78°; iv, H⁺; v, K₂CO₃, MeOH; vi, MeI; vii, H⁺, H₂O; viii, Diels-Alder reaction, 140°; ix, CH₂=CHCH₂Br, Bu₄N⁺F⁻, THF; x, O₃, then Zn/HOAc; xi, Bu₄N⁺ F⁻, THF (aldolisation).

Secodaphniphylline

Compound (62) is a logical tetracyclic precursor to seco-daphniphylline (63). The conjugated diene (60) is obtained from 1-acetylcyclopentene, and the structure and stereo-chemistry of intermediate (61) has been corroborated by a single-crystal X-ray diffraction analysis.

An efficient and highly regioselective intramolecular Mannich-type reaction has been developed for the construc-tion of the AEF ring system of aconitine related alkaloids (K. Fukimoto *et al.*, Tetrahedron Letters, 1986, 27, 1167).

New Diterpenoid Alkaloids

In the following table an attempt has been made to list all
diterpenoid alkaloids isolated since 1977. The references
relate in most cases to their isolation, purification, and
structure determination. For work on known alkaloids,
especially structural revision, the review sources listed
at the beginning of the chapter should be consulted. Blank
spaces in the table signify that the appropriate properties
have not been recorded. Only alkaloids for which a reasona-
bly substantiated structure has been advanced are included.
Entries marked with an asterisk may be artefacts. The
literature has been covered up to the end of 1985, and some
1986 references have been included. The alkaloids have been
arranged in order of natural source.

DITERPENOID ALKALOIDS

Alkaloid	Source	M.p. (°C)	$[\alpha]_D$ (°) (Solvent)	Reference
Alkaloids of *Aconitum* spp.				
Puberanine	*A. barbatum* var. *puberulum*		+16.6 (CHCl₃)	1
Puberanidine	A. " var. "		+23.2 (CHCl₃)	1
Puberaconitine	A. " var. "		+34.0 (CHCl₃)	1
Puberaconitidine	A. " var. "		+22.4 (CHCl₃)	1
Senbusine A	*A. carmichaeli*	amorphous		2
Senbusine B	A. "	amorphous		2
Senbusine C (15α-hydroxyneoline)	A. "	206.5-207; 214-216	+11.6 (CHCl₃)	2,3
Hokbusine A	A. "	amorphous	+11.4 (MeOH)	4
Hokbusine B	A. "	183-185		4
Lipohypaconitine	A. "	oil	+13.5 (CHCl₃)	5
Lipodeoxyaconitine	A. "	oil	+12.4 (CHCl₃)	5
Lipomesaconitine	A. "	oil	+13.8 (CHCl₃)	5
Lipoaconitine	A. "		+6.0 (CHCl₃)	5
Columbidine	*A. columbianum*	amorphous	-6.4 (CHCl₃)	6
Crassicauline A	*A. crassicaule*	162.5-164.5	+31.5 (CHCl₃)	7
Crassicauline B	A. "	311-315		7
Crassicaulidine	A. "	206-209		8
Delphinfoline	*A. delphinifolium*	218-220	+30 (EtOH)	9

Alkaloids of *Aconitum* spp. (cont.)

Alkaloid	Source	M.p. (°C)	$[\alpha]_D$ (°) (Solvent)	Reference
14-O-Acetyl-sachaconitine	*A. delphinifolium*	amorphous	+24.8 (CHCl₃)	10
Episcopalisine	*A. episcopale*		−11.7	11
Episcopalisinine	=	152-154	− 3.8	11
Episcopalitine	=	amorphous	− 0.9	11
Episcopalidine	=	210-212		11
Falaconitine	*A. falconieri*	amorphous	+111.5 (EtOH)	12
Mithaconitine	=	amorphous	+94 (EtOH)	12
Falconerine	=		+40.3 (CHCl₃)	13
Falconerine 8-acetate	=	162-163	+13.7 (CHCl₃)	13
N-Deacetyl-ranaconitine	*A. finetianum*	125-127	+43.7	14
N-Deacetyl-finaconitine	=	121-123	+34.9	14
$C_{27}H_{21}N_3O_5$	=	141-142		15
$C_{22}H_{31}NO_3$	=	221-226		15
Finaconitine	=	220-221	+44.7 (EtOH)	16
Flavaconitine	*A. flavum*			17
Liwaconitine	*A. forrestii*	201-202.5	+133.3 (CHCl₃)	18
Forestine	=	amorphous	+21.6 (EtOH)	19
Foresticine	=	79-80	− 1.9 (CHCl₃)	19

Alkaloids of *Aconitum* spp. (cont.)

Alkaloid	Source	M.p. (°C)	$[\alpha]_D$ (°) (Solvent)	Reference
3α, 13-Dihydroxyforesa-conitine	*A. forrestii*	98–103	+19.5 (CHCl$_3$)	20
Foresaconitine (vilmorrianine C)	*A. forrestii* var. *albo-villosum*	153–154	+30.5 (CHCl$_3$)	21,22
Ludaconitine	*A. franchetii*		+28 (EtOH)	23
Gigactonine	*A. gigas*	168–169 (94–95)	+49 (EtOH)	24,25
Yunaconitine	*A. hemsleyanum*	141–143	+37.7 (CHCl$_3$)	22,26
Heterophylloidine (panicutine)	*A. heterophylloides*	159–161	–141.1 (CHCl$_3$)	27,28,29
Ryosenaminol	*A. ibukiense*	287–290	+66.8 (MeOH)	30,31
Ibukinamine	"	243–246	+71.7 (MeOH)	30,31
Ryosenamine	"	213–215	+96.8 (MeOH)	30,31
C$_{20}$H$_{27}$NO$_2$	"	287–291	+68.5° (MeOH)	31
Aljesaconitine A	*A. japonicum*	amorphous	+ 7.5° (EtOH)	32
Aljesaconitine B	"	amorphous	+ 5.8° (EtOH)	32
Sadosine	"	222–224	+53.1 (MeOH)	33
15α-Hydroxyneoline	See Senbusine C			3
C$_{22}$H$_{23}$NO$_2$	*A. jinyangense*	198–200		34
Jynosine	"	254–256 (perchlorate)		34

Alkaloid	Source	M.p. (°C)	$[\alpha]_D$ (°) (Solvent)	Reference
Alkaloids of *Aconitum* spp. (cont.)				
Aconifine	A. *karakolicum*	195–197	+30.6 (CHCl$_3$)	35
12-Acetylnapelline (14)	"	205–206		36,37
Napelline *N*-oxide	"	110–112		37
Karasamine	"			38
1-Benzoylkarasamine	"	206–208		38
2-Isobutyryl-14-hydroxyhetisine	A. *koreanum*	230–231		39
Beiwutine	A. *kusnezoffii*	196–198	+26.9 (CHCl$_3$)	40
Sachaconitine (vilmorrianine D)	A. *miyabei*	129–130	–13.1 (EtOH)	22,41
Isodelphinine	"	167–168	+20.1 (EtOH)	41
Monticamine	A. *monticola*	163–164	+ 3 (MeOH)	42
Monticoline	"	166–167	+15 (MeOH)	42
Songorine *N*-oxide	"	253–255		43
Acomonine	"	208–210		44
Dihydromonticamine	"	156–157		45
1-Deoxydelsoline	"	134–135		45
Aconifine (Wang and Zhu's nagarine)	A. *nagarum*	195–197	+14.8 (MeOH) +30.6 (CHCl$_3$)	8,46

Alkaloids of *Aconitum* spp. (cont.)

Alkaloid	Source	M.p. (°C)	$[\alpha]_D$ (°) (Solvent)	Reference
Nevadenine	*A. nevadense*	resin		47
Nevadensine	=	resin		47
Aconorine	*A. orientale*	amorphous 237 (perchlorate)		48
Paniculatine	*A. paniculatum*	265-268		49
Penduline	*A. pendulum*	166-167		50,51
Polyschistine A	*A. polyschistum*	265-266		52
Polyschistine B	=	182-185		52
Polyschistine C	=	amorphous		52
Ranaconitine	*A. ranunculaefolium*	132-134	+33.2 (CHCl$_3$)	53
Nominine	*A. sanyoense*			54
Sanyonamine	=	276-278		55
Hanamisine	=	124-127		56
14-Dehydro-talatizamine	*A. saposhnikovii*	128-130		57
Vaginadine	*A. scaposum* var. *vaginatum*	147-149	-49.4 (EtOH)	58
Vaginaline	=	209-213 (dec.)	+28.6 (EtOH)	58
Vaginatine	=	86-88	+25.3 (CHCl$_3$)	58
Septentrionine	*A. septentrionale*	123-125	+21.2 (CHCl$_3$)	59
Septentriodine	=	130-135	+56 (EtOH)	59
15-Acetylsongoramine	*A. soongoricum*			60
Dolaconine	*A. stapfianum*	44-46		61
Deoxyjesaconitine	*A. subcuneatum*	174-176	+52 (MeOH)	62,63

Alkaloid	Source	M.p. (°C)	$[\alpha]_D$ (°) (Solvent)	Reference
Alkaloids of *Aconitum* spp. (cont.)				
14-Benzoylneoline	*A. subcuneatum*	amorphous	+ 9.1 (MeOH)	64
Talatisine	*A. talassicum*			65
Umbrosine	*A. umbrosum*	150-151		66
Vilmorrianine A	*A. vilmorrianum*	182-184		22
1-Acetylluciculine	*A. yesoense*	amorphous		67
14-Acetylneoline (bullatine C)	"	198-202	+42.9 (CHCl$_3$)	67
Pyrochasmanine	"	126-129	+243.9	67
Ezochasmanine	"	115-118	+40 (CHCl$_3$)	67,68
Ezochasmaconitine	"	163-165	+26.1 (CHCl$_3$)	67,68
Anisoezochasmaconi- tine	"	136-138.5	+14.1 (CHCl$_3$)	67,68
Dehydroluciduscu- line	*A. yesoense* var. *macroyesoense*	186-189	+ 2.6 (EtOH)	69
N-De-ethyldehydro- lucidusculine	"	amorphous	- 9.6 (EtOH)	69
Alkaloids of *Anopterus* spp.				
Anopterimine	*A. macleayanus*	235-238	+106 (CHCl$_3$)	70
Anopterimine N-oxide	"	233-235	+95 (CHCl$_3$)	70
Hydroxyanopterimine	"	247-249	-14 (MeOH)	70

Alkaloid	Source	M.p. (°C)	$[\alpha]_D$ (°) (Solvent)	Reference
Alkaloids of *Anopterus* spp. (cont.)				
Dihydroxyanopteri- mine	*A. macleayanus*	242-244	-9 (MeOH)	70
Alkaloids of *Daphniphyllum* spp.				
Daphnigracine	*D. gracile*	liquid		71,72
Daphnigraciline	"	76-78		71,72
Hydroxydaphgraci- line	"	liquid		72
Epioxodaphni- graciline	"	102-104		71
Oxodaphnigracine	"	116-117		71
Oxodaphnigraciline	"	107-109		71
Daphgracine	"	liquid		72
Daphgraciline	"	liquid		72
Deoxy- yuzurimine	*D. humile*	132		73
Isodaphnilactone- B	"	liquid		73

Alkaloids of *Delphinium* spp.

Alkaloid	Source	M.p. (°C)	$[\alpha]_D$ (°) (Solvent)	Reference
Ajacusine	*D. ajacis*	158–161	+65.2 (EtOH)	74
Ajadine	"	134–136		74
Ambiguine	"	106–108	+38 (CHCl₃)	75
14-Acetylbrowniine	"	129–130	+27.8 (CHCl₃)	76,77
Dihydroajaconine	"	99–100		74
Bicoloridine	*D. bicolor*	amorphous	+10 (CHCl₃)	78,79,80,81
Bicolorine	"	190–191	+16 (CHCl₃)	78,79,81
6-0-Acetylbi-colorine	"	165–167	+19 (CHCl₃)	81
14-Benzoylbrowni-ine	*D. bitermatum*	114–116	+53	82
14-Benzoylilien-sine	"	148–150	+63.8 (CHCl₃)	82
14-Dehydroilien-sine	"	212.5–213.5	+25.2 (CHCl₃)	82
Delbiterine	"	137–138		82
Iliensine (delcosine)	"	203–204	+57 (CHCl₃)	83
Bonvalone	*D. bonvalotii*	235–236	-89.3 (CHCl₃)	84
Bonvalol	"	165–166	-26.3 (CHCl₃)	84
Bonvalotine	"	218–220	-35.7 (CHCl₃)	84
Delbrunine	*D. brunonianum*	178	0 (EtOH)	85
Delbruline	"	129–131	0 (CHCl₃)	85

Alkaloids of *Delphinium* spp. (cont.)

Alkaloid	Source	M.p. (°C)	$[\alpha]_D$ (°) (Solvent)	Reference
Delbrusine	*D. brunonianum*	141	+16.8 (CHCl₃)	85
Brunonine	=	208-209	+174 (EtOH)	86
Cardiopetamine	*D. cardiopetalum*	302-305 (dec.)	+65 (EtOH)	87
15-Acetyl-cardiopetamine	=	236-237	+12 (EtOH)	87
14-Benzoyl-gadesine	=	resin		88
14-Benzoyldi-hydrogadesine	=	199-202		88
Cardiopetaline	=	179-181	-16 (EtOH)	89
Cardiopetalidine	=	223-227	+1.1 (EtOH)	89
13-Acetyl-hetisinone	=	219-220		90
Delcaroline	*D. carolinianum*	amorphous 160-162 (perchlorate)	+49.8 (MeOH)	91
Delavaine A	*D. delavayi* var. *pogonanthum*	amorphous	+39.4 (CHCl₃)	92
Delavaine B	=	amorphous	+31.7 (CHCl₃)	92
14-Acetyl-delectine	*D. dictyocarpum*	118-120	+42 (CHCl₃)	93
N-Acetyldelectine	=	116-118	+30 (CHCl₃)	94
Delectinine	=	167-169	+42 (CHCl₃)	93
Dictionine	=	246-248		93
Delectine	=	107-109		95

Alkaloids of *Delphinium* spp. (cont.)

Alkaloid	Source	M.p. (°C)	$[\alpha]_D$ (°) (Solvent)	Reference
Acetonyldictysine*	*D. dictyocarpum*	151-153		96,97,98
Acetonyldehydro-dictysine*	=	143-145		96,97,98
14-Benzoyldictyo-carpine	=	143-145		98
Dictyocarpine	=	214.5-216.5	-14.7 (CHCl$_3$)	99
			-12.8 (MeOH)	
Dictysine (11)	=	184-186		96,97,100
Geyerine	*D. geyeri*	amorphous	+9.6 (EtOH)	101
Geyeridine	=	gum		101
Geyerinine	=	gum		101
Glaudelsine	*D. glaucescens*	amorphous	+36.1 (CHCl$_3$)	102
Glaucenine	=	amorphous	-45 (CHCl$_3$)	102
Glaucerine	=	amorphous	-48.5 (CHCl$_3$)	102
Glaucephine	=	amorphous	-33.6 (CHCl$_3$)	102
Glaucedine	=	117-120	+39.1 (MeOH)	102
Dictyocarpinine	=	205-206.5	-5 (MeOH)	102
Dehydrodelconine	*D. iliense*	141-143	-64 (MeOH)	103
Ilidine	=	141-143		103
Delcoridine	=			104
Hetisine 13-O-acetate (3)	*D. nuttallianum*	amorphous		105
18-Methoxy-gadesine	*D. orientale*	180-184		106

Alkaloid	Source	M.p. (°C)	[α]$_D$ (°) (Solvent)	Reference
Alkaloids of *Delphinium* spp. (cont.)				
18-Hydroxy-14-O-methylgadesine	*D. orientale*	110-114		107
Gadesine (10)	*D. pentagynum*	174-177	+76 (EtOH)	108
Pentagyline	"	198-200		109
Gadenine	"	147-150		109
Gadeline	"	resin		88
Dihydrogadesine	"	136-138	+54 (EtOH)	88
14-Acetylgadesine	"			88
14-Acetyldihydro-gadesine	"	resin		88
Pentagynine	"	198-201	+72 (EtOH)	110
Dihydropentagynine	"	150-154	+43 (EtOH)	110
Pentagydine	"	130-131		110
Delphidine	*D. staphisagria*	98-100	+16.6 (EtOH)	111
Delphirine	"	95-100	+3.8 (EtOH)	112
Delstaphisine	"	182-184	-11.0 (EtOH)	113
Delstaphisagrine	"	amorphous	+3.8 (EtOH)	113
Delstaphisagnine	"	amorphous	+20.0 (EtOH)	113
Staphisagnine	"	resin	-104.5 (benzene)	114
Staphisagrine	"	229-231	-105.6 (benzene)	114
Staphigine	"	225-227	-116 (benzene)	115
Staphirine	"	222-225	-126 (benzene)	115
Tatsiensine	*D. tatsienense*	amorphous	+17.4	116

Alkaloid	Source	M.p. (°C)	[α]$_D$ (°) (Solvent)	Reference
Alkaloids of Delphinium spp. (cont.)				
Deacetylambiguine	D. tatsienense	amorphous	+36.6	116
Tatsinine	=	163–164	+9 (EtOH)	117
Deltatsine	=	amorphous	+28.6 (EtOH)	118
Tricornine	D. tricorne	187–189	+47.3 (EtOH)	119
Delvestine	D. vestitum	185–187	+18.6 (CHCl$_3$)	120
Delvestidine	=	amorphous	+22.1 (CHCl$_3$)	120
Virescenine	D. virescens	68–70	+16.9 (EtOH)	121
14-Acetylvir-escenine	=	157–159	+31.8 (CHCl$_3$)	121
Alkaloids of Garrya spp.				
Ovatine (28)	G. ovata	113–114	−79.4 (CHCl$_3$)	122,123
Lindheimerine (27)	=	amorphous	−113.8 (CHCl$_3$)	122,128
Alkaloids of Spiraea spp.				
Spiredine	S. japonica	163		124
Spirasine V (4)	=	177–179	−47 (CHCl$_3$)	125
Spirasine VI (5)	=	202–203	−107 (CHCl$_3$)	125
Spirasine I	=	244–246	−131 (CHCl$_3$)	126
Spirasine II	=	208–209		126

Alkaloid	Source	M.p. (°C)	$[\alpha]_D$ (°C) (Solvent)	Reference
Alkaloids of *Spireaea* spp. (cont.)				
Spirasine VII	*S. japonica*	191-193	-78 (CHCl$_3$)	126
Spirasine VIII	"	207-209	-57 (CHCl$_3$)	126

References

1. *Y. De-quan* and *B. C. Das*, Planta Med., 1983, 49, 85.
2. *H. Hikino et al.*, J. nat. Prod., 1982, 45, 128.
3. *H. Takayama et al.*, Chem. pharm. Bull. Japan, 1981, 29, 3078.
4. *H. Hikino et al.*, J.nat. Prod., 1983, 46, 178.
5. *I. Kitagawa et al.*, J. pharm. Soc. Japan, 1984, 104, 848; Chem. pharm. Bull. Japan 1982, 30, 758.
6. *S. W. Pelletier et al.*, Heterocycles, 1985, 23, 331.
7. *F. P. Wang* and *Q.-C. Fang*, Planta Med., 1981, 42, 375.
8. *F. Wang* and *Q. Fang*, Planta Med., 1983, 47, 39.
9. *M. H. Benn et al.*, Tetrahedron Letters, 1981, 22, 483.
10. *M. H. Benn et al.*, Phytochem., 1986, 25, 973.
11. *F. Wang* and *Q. Fang*, Chem. Abs., 1982, 97, 212631j; 1984, 100, 64981m.
12. *S. W. Pelletier et al.*, Chem. Comm. 1977, 12; Phytochem., 1977, 16, 623.
13. *S. W. Pelletier et al.*, Heterocycles, 1986, 24, 1061.
14. *S. Jiang et al.*, Chem. Abs., 1982, 97, 107027u; 1984, 100, 20505e.
15. *B.-R. Chen et al.*, Chem. Abs., 1981, 95, 175625w.
16. *S. Jiang et al.*, Chem. Abs., 1982, 96, 118986m; 1982, 97, 20736a.
17. *Y. Liu* and *G. Chang*, Chem. Abs., 1982, 97, 141679t.
18. *C. Wang et al.*, Planta Med., 1983, 48, 55.
19. *S. W. Pelletier et al.*, J. nat. Prod., 1984, 47, 474.
20. *E. Breitmaier et al.*, Annalen, 1985, 1297.
21. *W.-s. Chen* and *E. Breitmaier*, Ber., 1981, 114, 394.
22. *C.-R. Yang et al.*, Chem. Abs., 1981, 95, 58094m.
23. *D. Chen* and *W. Song*, Chem. Abs., 1982, 97, 88655d.
24. *S. Sakai et al.*, Heterocycles, 1977, 8, 207.
25. *S. Sakai et al.*, Chem. Abs., 1979, 90, 152417n.
26. *S.-Y. Chen*, Chem. Abs., 1979, 91, 20833f.
27. *A. Katz* and *E. Staehelin*, Helv., 1982, 65, 286.
28. *S. W. Pelletier et al.*, Heterocycles, 1986, 24, 1275.
29. *S. W. Pelletier et al.*, Tetrahedron Letters, 1981, 22, 313.
30. *S. Sakai et al.*, Chem. pharm. Bull. Japan, 1983, 31, 3338.
31. *S. Sakai et al.*, J. pharm. Soc. Japan, 1984, 104, 222.
32. *T. Amiya et al.*, Chem. pharm. Bull. Japan, 1985, 33, 4717.
33. *T. Okamoto et al.*, Chem. pharm. Bull Japan, 1983, 31, 360.

388

34. *D. Chen* and *W. Song*, Chem. Abs., 1982, 96, 65677c.
35. *M. S. Yunusov et al.*, Chem. Abs., 1981, 94, 153427k.
36. *M. S. Yunusov et al.*, Chem. Abs., 1977, 86, 152618w.
37. *M. S. Yunusov et al.*, Chem. Abs., 1979, 90, 39084p.
38. *M. S. Yunusov et al.*, Chem. Abs., 1983, 98, 104284k.
39. *M. G. Reinecke et al.*, Heterocycles, 1986, 24, 49.
40. *Y.-G. Wang et al.*, Chem. Abs., 1981, 94, 117772k.
41. *S. W. Pelletier et al.*, Tetrahedron Letters, 1977, 4027.
42. *M. S. Yunusov et al.*, Chem. Abs., 1982, 96, 123043f.
43. *M. S. Yunusov et al.*, Chem. Abs., 1978, 89, 43860k.
44. *S. Y. Yunusov et al.*, Chem. Abs., 1975, 82, 54165u.
45. *M. S. Yunusov et al.*, Chem. Abs., 1982, 97, 212774h.
46. *S. W. Pelletier et al.*, Hetercycles, 1982, 17, 91; 1982, 19, 1523.
47. *A. G. Gonzalez et al.*, Heterocycles, 1985, 23, 2979.
48. *S. Y. Yunusov et al.*, Chem. Abs., 1976, 84, 150807s.
49. *A. Katz* and *E. Staehelin*, Tetrahedron Letters, 1982, 23, 1155.
50. *Y. Zhu* and *R. Zhu*, Heterocycles, 1982, 17, 607.
51. *L. Liu et al.*, Chem. Abs., 1983, 99, 67495c.
52. *Y. Fujimoto et al.*, Heterocycles, 1985, 23, 803.
53. *N. Mollov et al.*, Chem. Abs., 1964, 61, 12324g; *S. W. Pelletier et al.*, Tetrahedron Letters, 1978, 5045.
54. *E. Ochiai et al.*, J. pharm. Soc. Japan, 1956, 76, 1414; Chem. Abs., 1957, 51, 6661i.
55. *S. Sakai et al.*, Chem. pharm. Bull. Japan, 1982, 30, 4576.
56. *T. Okamoto et al.*, Chem. pharm. Bull. Japan, 1983, 31, 1431.
57. *M. S. Yunusov et al.*, Chem. Abs., 1982, 97, 141697x.
58. *Q. P. Jiang* and *W. L. Sung*, Heterocycles, 1986, 24, 877.
59. *S. W. Pelletier et al.*, Heterocycles, 1979, 12, 377; J. Am. chem. Soc., 1981, 103, 6536; *M. Shamma et al.*, J. nat. Prod., 1979, 42, 615.
60. *M. S. Yunusov et al.*, Chem. Abs., 1981, 94, 117791r.
61. *S. Luo* and *W. Chen*, Chem. Abs., 1982, 97, 107079n.
62. *H. Bando et al.*, Heterocycles, 1981, 16, 1723.
63. *T. Mori et al.*, Chem. pharm. Bull. Japan, 1983, 31, 2884.
64. *T. Amiya et al.*, Chem. pharm. Bull. Japan, 1985, 33, 3658.

65. *Z. Karimov* and *M. G. Zhamierashvili*, Chem. Abs., 1982, 96, 20311n.
66. *V. A. Tel' nov et al.*, Chem. Abs., 1977, 86, 167854u.
67. *H. Takayama et al.*, Chem. Abs., 1982, 97, 36082v.
68. *S. Sakai et al.*, Heterocycles, 1981, 15, 403.
69. *T. Amiya et al.*, Heterocycles, 1985, 23, 2473.
70. *J. A. Lamberton et al.*, Austral. J. Chem., 1976, 29, 1319.
71. *J. A. Lamberton et al.*, Chem. Letters, 1975, 923.
72. *J. A. Lamberton et al.*, Chem. Letters, 1980, 393.
73. *S. Yamamura* and *Y. Terada*, Chem. Letters, 1976, 1381.
74. *S. W. Pelletier et al.*, Heterocycles, 1978, 9, 463.
75. *S. W. Pelletier et al.*, Heterocycles, 1978, 9, 1241.
76. *M. S. Yunusov et al.*, Chem. Abs., 1978, 89, 43862n.
77. *S. W. Pelletier et al.*, Heterocycles, 1977, 7, 327.
78. *S. W. Pelletier et al.*, Tetrahedron Letters, 1976, 3025.
79. *M. H. Benn et al.*, Tetrahedron Letters, 1980, 21, 127.
80. *P. W. Codding* and *K. A. Kerr*, Acta Cryst., 1981, 37B, 379.
81. *M. H. Benn et al.*, Phytochem., 1986, 25, 1511.
82. *M. S. Yunusov et al.*, Chem. Abs., 1978, 89, 39366m.
83. *S. Y. Yunusov et al.*, Chem. Abs., 1975, 83, 79431m; 1976, 84, 150805q.
84. *Q. P. Jiang* and *W. L. Sung*, Heterocycles, 1984, 22, 2429.
85. *W. Deng* and *W. L. Sung*, Heterocycles, 1986, 24, 873.
86. *W. Deng* and *W. L. Sung*, Heterocycles, 1986, 24, 869.
87. *A. G. Gonzalez et al.*, Tetrahedron Letters, 1983, 24, 3765.
88. *A. G. Gonzalez et al.*, Heterocycles, 1986, 24, 1513.
89. *A. G. Gonzalez et al.*, Tetrahedron Letters, 1980, 21, 1155.
90. *A. G. Gonzalez et al.*, Chem. Abs., 1982, 97, 20730u.
91. *S. W. Pelletier et al.*, Heterocycles, 1981, 16, 747.
92. *S. W. Pelletier et al.*, Heterocycles, 1986, 24, 1853.
93. *M. S. Yunusov et al.*, Chem. Abs., 1978, 88, 170367j; 1978, 89, 103711z.
94. *S. Y. Yunusov et al.*, Chem. Abs., 1977, 87, 65340b.
95. *S. Y. Yunusov et al.*, Chem. Abs., 1976, 84, 105862v.
96. *M. S. Yunusov et al.*, Chem. Abs., 1982, 97, 56076z.
97. *B. Tashkhodzhaev*, Chem. Abs., 1982, 97, 163295s.
98. *M. S. Yunusov et al.*, Chem. Abs., 1982, 96, 100872m.

99. *S. W. Pelletier* and *K. I. Varughese*, J. nat. Prod., 1984, <u>47</u>, 643.
100. *M. S. Yunusov et al.*, Chem. Abs., 1980, <u>93</u>, 46911w.
101. *F. R. Stermitz et al.*, J. org. Chem., 1986, <u>51</u>, 390.
102. *S. W. Pelletier et al.*, J. org. Chem., 1981, <u>46</u>, 3284.
103. *M. S. Yunusov et al.*, Chem. Abs., 1978, <u>89</u>, 43859s.
104. *M. S. Yunusov et al.*, Chem. Abs., 1981, <u>94</u>, 171022d.
105. *M. H. Benn et al.*, Heterocycles, 1986, <u>24</u>, 1605.
106. *A. G. Gonzalez et al.*, Heterocycles, 1983, <u>20</u>, 409.
107. *A. G. Gonzalez et al.*, Tetrahedron Letters, 1981, <u>22</u>, 4843,
108. *A. G. Gonzalez et al.*, Tetrahedron Letters, 1979, 79.
109. *A. G. Gonzalez et al.*, Heterocycles, 1984, <u>22</u>, 17.
110. *A. G. Gonzalez et al.*, Phytochem., 1982, <u>21</u>, 1781; Tetrahedron Letters, 1983, <u>24</u>, 959.
111. *S. W. Pelletier et al.*, Phytochem., 1977, <u>16</u>, 404.
112. *S. W. Pelletier* and *J. Bhattacharyya*, Tetrahedron Letters, 1976, 4679.
113. *S. W. Pelletier* and *M. M. Badawi*, Heterocycles, 1985, <u>23</u>, 2873.
114. *S. W. Pelletier et al.*, Tetrahedron Letters, 1976, 1749.
115. *S. W. Pelletier et al.*, J. org. Chem., 1976, <u>41</u>, 3042.
116. *S. W. Pelletier et al.*, Heterocycles, 1983, <u>20</u>, 1347.
117. *S. W. Pelletier et al.*, Tetrahedron Letters, 1984, <u>25</u>, 1211.
118. *S. W. Pelletier et al.*, Heterocycles, 1984, <u>22</u>, 2037.
119. *S. W. Pelletier* and *J. Bhattacharyya*, Tetrahedron Letters, 1977, 2735; Phytochem., 1977, <u>16</u>, 1464; Heterocycles, 1977, <u>7</u>, 327; J. Am. chem. Soc., 1981, <u>103</u>, 6536.
120. *S. W. Pelletier et al.*, Heterocycles, 1985, <u>23</u>, 2483.
121. *S. W. Pelletier et al.*, Heterocycles, 1979, <u>12</u>, 779.
122. *S. W. Pelletier et al.*, Heterocycles, 1978, <u>9</u>, 1409.
123. *S. W. Pelletier et al.*, J. org. Chem., 1981, <u>46</u>, 1840.
124. *V. D. Gorbunov et al.*, Chem. Abs., 1976, <u>85</u>, 59577b.
125. *S. Fang et al.*, Tetrahedron Letters, 1986, <u>27</u>, 275.
126. *F. Sun et al.*, Heterocycles, 1986, <u>24</u>, 2105.

Acknowledgement

The author is indebted to Professor S. W. Pelletier, University of Georgia, for his helpful co-operation in the preparation of this chapter.

Chapter 35

STEROIDAL ALKALOIDS

A. R. PINDER

Introduction

Several texts and reviews covering various aspects of steroidal alkaloids are available. Amongst these are "Chemistry
of the Alkaloids," ed. S. W. Pelletier, Chapter 19 (by Y.
Sato), van Nostrand Reinhold, New York, 1970; D. R. Dalton,
"The Alkaloids," Part 7, Chapters 37 and 41, Marcel Dekker,
New York, 1979; "The Alkaloids: Chemistry and Physiology,"
ed. R. G. A. Rodrigo, Vol. 19, Chapter 2 (by H. Ripperger
and K. Schreiber), Academic Press, New York, 1981. Periodic
reviews are to be found in Specialist Periodical Reports,
Vols. 8-13, and Natural Product Reports, Vol. 1, 1984 (Royal
Society of Chemistry, London). Numerous recently-isolated
steroidal bases are listed in J. S. Glasby, "Encyclopaedia
of the Alkaloids," Vol. 4, Plenum Press, New York and London,
1983. The mass spectra of steroidal alkaloids have been
reviewed (H. Budzikiewicz, Mass Spec. Rev., 1982, 1, 125,
Org. Mass Spec., 1982, 17, 107). Isoprenoid alkaloids (J.
G. Riddick, Encycl. Plant Physiol., New Series, 1980, 8
[Secondary Plant Products], p. 167), the photochemistry of
alkaloids (S. P. Singh *et al*., Chem. Rev., 1980, 80, 269),
and their liquid chromatography (E. Heftmann, J. liq.
Chromatog., 1979, 2, 1137) have been reviewed. Enzymic
transformations of steroidal bases have been discussed
(H.L. Holland, in "The Alkaloids," ed. R. H. F. Manske and
R. G. A. Rodrigo, Vol. 18, Chapter 5, Academic Press, New
York, 1981), as has the circular dichroism of their *N*-
salicylidene derivatives (H. E. Smith *et al*., J. org. Chem.,
1976, 41, 704; 1982, 47, 2525).

Because of extensive developments in steroidal alkaloid
chemistry since the publication of the 2nd edition it is
necessary to adopt a classification somewhat different from
that used earlier. The alkaloids are discussed in nine

sections, seven of which are based on their biological origin, the eighth covers miscellaneous alkaloids, and the last is concerned with biosynthesis.

1. Apocynaceae Alkaloids

(a) 3-Aminosteroids

The mechanism of the acid-catalysed "backbone" rearrangement of holamine (1) to isoholamine (2) has been explored by ^{2}H- and ^{13}C-nmr spectroscopy on the product resulting when D_2SO_4 is used as catalyst (F. Frappier *et al.*, J. org. Chem., 1981, 46, 4314). It is

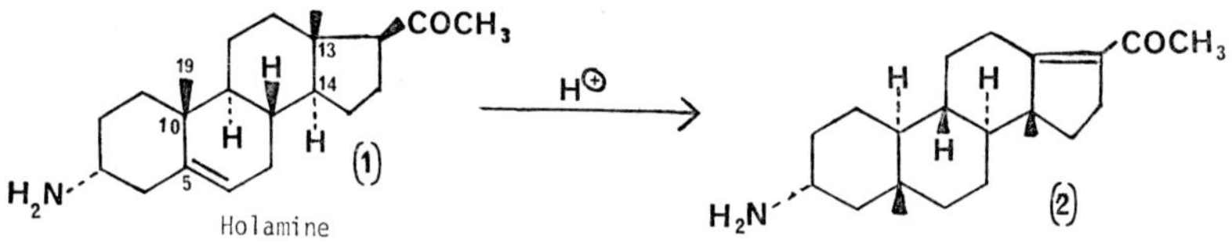

Holamine

Ischolamine

suggested that initially a carbocation is generated at C-5, followed by migration of the C-10 methyl group from C-10 to C-5. The charge then migrates from C-10 to C-14 along the backbone, proceeding either *via* 1,2 hydride ion shifts or by protonation-deprotonation. Finally there is an energetically-favoured C-13 to C-14 methyl shift.

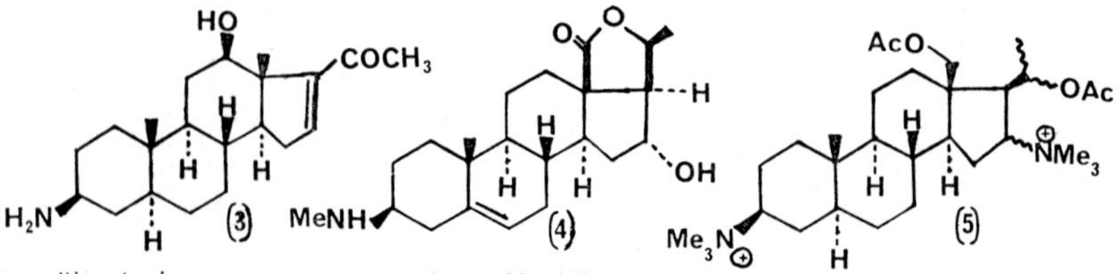

Kisantamine

Paravallaridine

Kisantamine, occurring in leaves of *Holarrhena congolensis*, is formulated as (3) from spectroscopic studies on its *N*-acetyl derivative (H. Dadoun and A. Cavé, Plant Med. Phytother., 1978, 12, 225). Paravallaridine (4) has been converted into several bisquaternary ammonium salts (5), all of which are curare-like in physiological activity (J. LeMen *et al.*, Eur. J. Med. Chem-Chim. Ther., 1982, 17, 43).

(b)   20-Aminosteroids

A stereospecific synthesis of funtuphyllamine A (7) has been reported (G. Demailly and G. Solladié, Tetrahedron Letters, 1975, 2471): diborane reduction of the chiral iminopregnane (6), followed by hydrogenolysis, yields the alkaloidal base (7) sterospecifically.

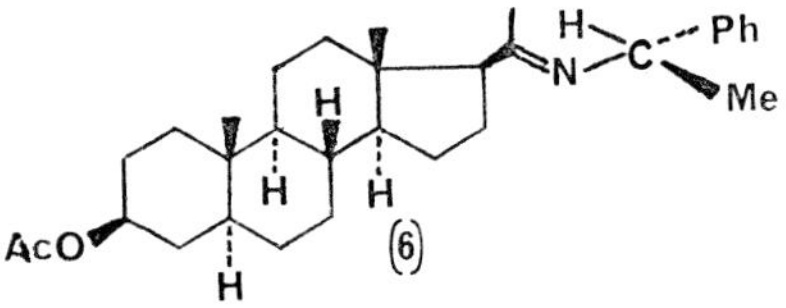

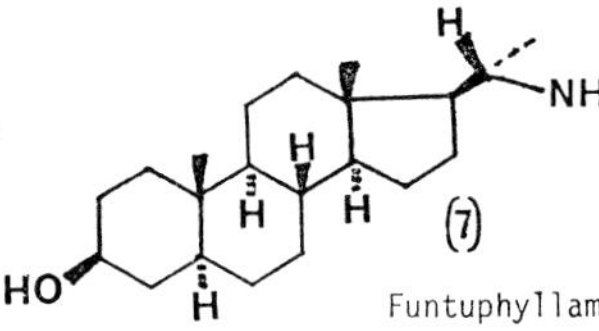

Funtuphyllamine A

A new alkaloid holacetine (8) occurs in the root bark of *Holarrhena antidysenterica*; its structure has been settled by chemical correlation with the known funtumafrine C (R. N. Rej *et al.*, Phytochem., 1976, 15, 1173).

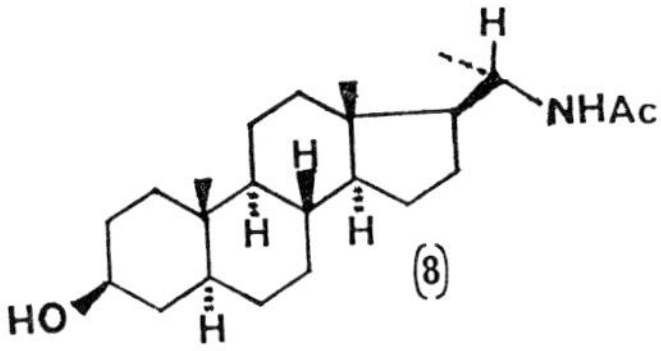

Holacetine

(c) 3,20-Diaminosteroids

H. antidysenterica seeds have yielded a new alkaloid holarricine (9), formulated after detailed spectroscopic and
chemical studies. It has been correlated with holarrhimine
(10) by Clemmensen reduction, and its two carbonyl functions
were located by analysis of its mass spectral fragmentation
pattern (S. Siddiqui and B. S. Siddiqui, Chem. Abs., 1983,
98, 50327).

(9) Holarricine (R=O)
(10) Holarrhimine (R=H_2)

(11) Irehdiamine F

Irehdiamine F has been isolated from the roots of *Vahadenia
laurentii;* its structure (11) has been settled by X-ray
diffraction analysis of its hydrochloride (J. Lamotte *et al.*,
Acta Cryst., 1977, B33, 2392). 20-Epi-irehdiamine I (12),
found in seeds of *Funtumia elastica*, has been formulated on
the basis of its nmr and mass spectra; its structure has
been corroborated by a synthesis from progesterone by an
unexceptional pathway (M. D. L. Tolela and P. Foche, Planta
Med., 1979, 35, 48).

(12)
20-Epi-irehdiamine I

(d) 3-Aminoconanines (conessanes)

A partial synthesis of dihydroholarrhenine has been described. 3β-Acetoxy-5α-pregnan-12,20-dione ethylene ketal
(13) was oximated and then reduced to a pair of epimeric
20-amines, with concomitant hydrolysis of the acetoxy group.
The 20α amine, separable by fractional crystallization, was
converted into the urethane (14), reduction of which afforded
(15), oxidisable to the corresponding 3-ketone.   Reductive
amination of the latter yielded epimeric 3-dimethylamino-
compounds (16).   Hofmann-Löffler cyclization of the 3β
epimer *via* its N-chloride furnished 12-oxodihydroconessine
(17), catalytic hydrogenation of which gave dihydroholarrhenine (18) (G. van de Woude and L. van Hove, Bull. Soc.
chim. Belge, 1975, 84, 911).

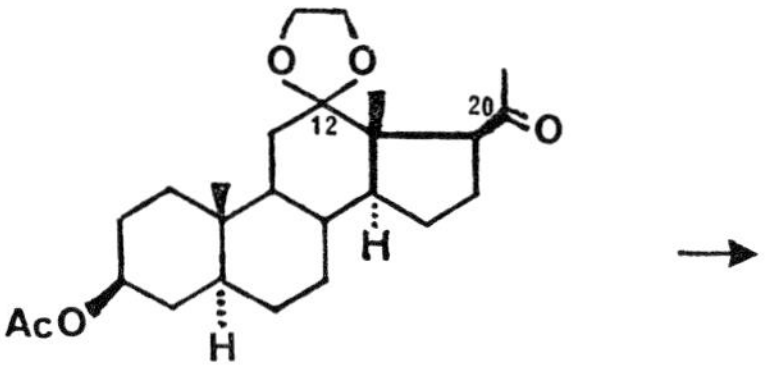

(13)

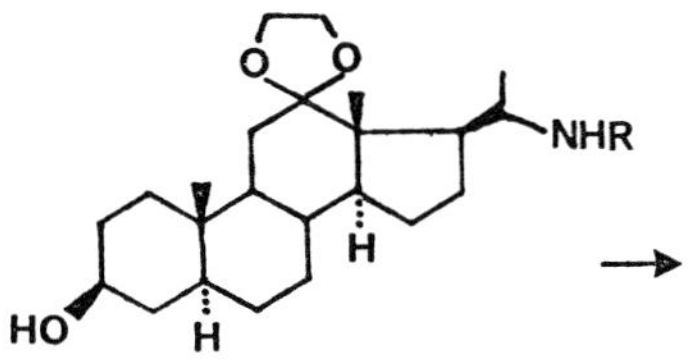

(14) R = CO₂Et

(15) R = Me

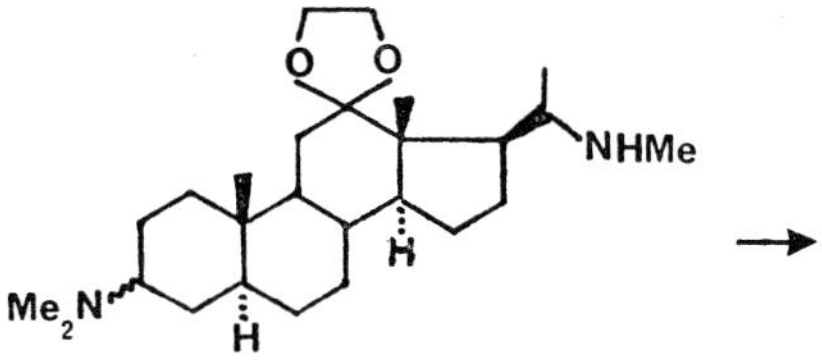

(16)

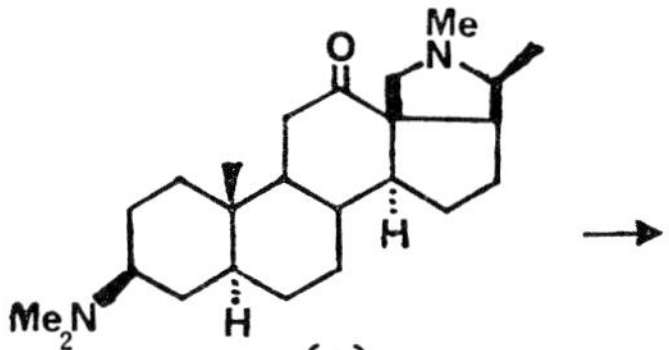

(17)

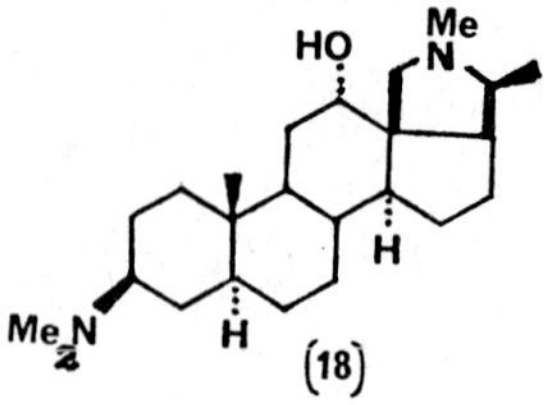

(18)

Dihydroholarrhenine

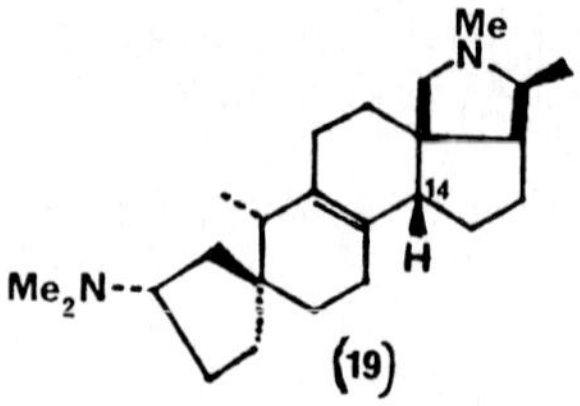

(19)

Neoconessine

The mechanism of the "backbone" rearrangement of conessine into isoconessine and into neoconessine has been investigated by ^{13}C- and ^{3}H-nmr spectroscopy. The rearrangements were carried out with catalysis by D_2SO_4 and $HTSO_4$, and the spectra of the products were analysed carefully and positions of incorporation discerned (Frappier *et al.*, J. Org. Chem., 1982, <u>47</u>, 3783). Neoconessine (19) has been assigned 14β stereochemistry on the basis of this study.

Holarrhesine an ester-base found in the bark of *H. flori-bunda*, is formulated as (20) chiefly on the basis of spectroscopic analysis and because of its facile hydrolysis to the known holarrheline (21) (G. A. Hoyer *et al.*, Planta Med., 1978, <u>34</u>, 47).

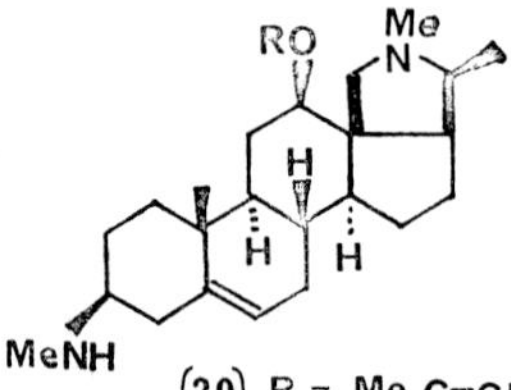

(20) R = $Me_2C=CHCH_2CO$ Holarrhesine

(21) R = H Holarrheline

The phytochemistry and pharmacology of *H. antidysenterica* alkaloids have been reviewed (J. P. Gupta *et al.*, Chem. Abs., 1982, 96, 57612).

(e) Miscellaneous alkaloids

Several antibiotic aza-D-homosteroids of general structure (22) have been isolated from *Geotrichum flavo-brunneum* (K. H. Michel *et al.*, J. Antibiotics, 1975, 28, 102).

R_1	R_2	R_3	R_4
Me	Me	⟋OH ⋯H	H_2
H	H	⟋OH ⋯H	H_2
H	H	⟋OAc ⋯H	H_2
H	H	O	H_2
H	H	⟋OH ⋯H	⟋OH ˋH
H	H	⟋OH ⋯H	O

2. Salamandra Alkaloids

The synthesis of *Salamandra* alkaloids has been reviewed (K. Oka and S. Hara, Chem. Abs., 1979, 90, 187183; Oka, *ibid.*, 1930, 93, 46951). The structure of cycloneosamandione has been revised to (23), with a "normal" A/B ring junction. The new formulation has been confirmed by a total synthesis, starting from oxime (24) which on Beckmann rearrangement afforded lactam (25); this was converted in four steps into

(23) (Oka and Hara, J. Amer. chem. Soc., 1977, 99, 3859).
This structural amendment requires a similar revision of the
structure of cycloneosamandaridine but, in fact, the carbinolamine (26), synthesised from lactam (25), proved not to be
identical with the natural base (Oka and Hara, *loc.cit.*).
The structure of the latter has consequently been revised to
(27) (*idem*, J. org. Chem., 1978, 43, 4408).

(23) Cycloneosamandione   (24)   (25)

(26)   (27) Cycloneosamandaridine ?

A synthesis of samandaridine has been reported, from 17β-
hydroxy-2-hydroxymethylene-5β-androstane-3-one (28) as
follows (Y. Shimizu, Tetrahedron Letters, 1972, 2919), J.
org. Chem., 1976, 41, 1930). The final product had already
been transformed into samandarine, samandarone, and
samandaridine (29).

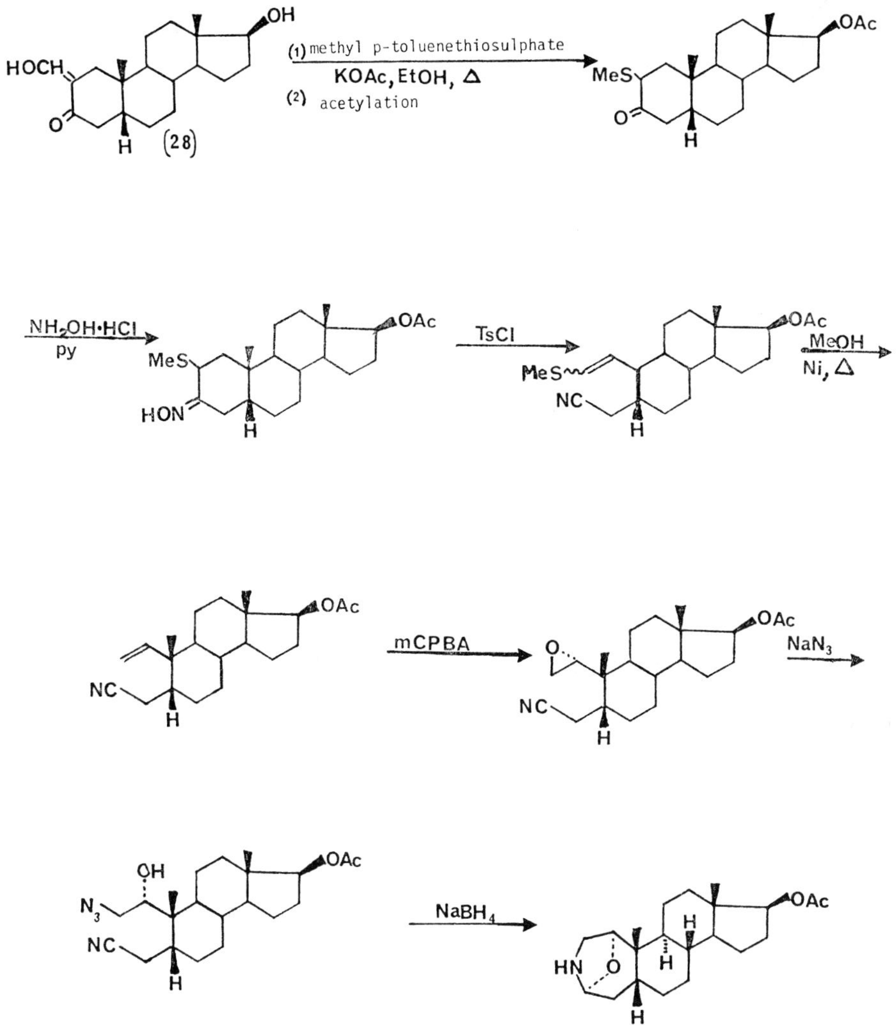
OH
HOCH
O
H
(28)
(1) methyl p-toluenethiosulphate
KOAc, EtOH, Δ
(2) acetylation
OAc
MeS
O
H
NH2OH·HCl
py
OAc
MeS
HON
H
TsCl
OAc
MeS
NC
H
MeOH
Ni, Δ
OAc
NC
H
mCPBA
OAc
O
NC
H
NaN3
OH
OAc
N3
NC
H
NaBH4
OAc
H
HN
O
H
H

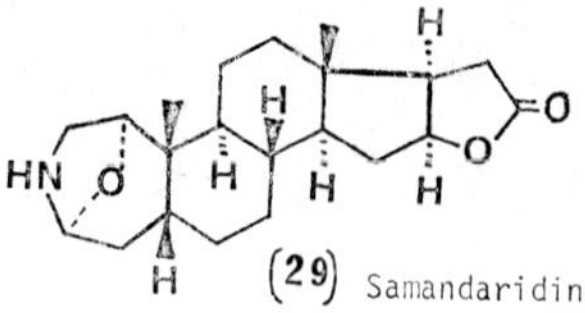

(29) Samandaridine

3. Buxus Alkaloids

Buxus alkaloids based on 19-nor-B-homopregna-9a(10), 9(11)-diene have been reviewed (W. Turowska-Jones and U. Wrzeciono, Chem. Abs., 1976, 85, 21695). Recent work on *Buxus* bases generally has also been reviewed [J. Tomko and Z. Voticky, IUPAC Int. Symp. chem. nat. Prod., 1978, 4, Part 1, 260).

Amongst newly-isolated *Buxus* bases are N_3-acetylcycloproto-buxine C (30) from *B. sempervirens* (S. Y. Yunusov, Khim. prirod. Soed., 1975, 176), buxaminol B (31) and cyclobullatine A (32), the last two both from *B. sempervirens* var. *bullata* Kirchn.(Voticky *et al.*, Coll. Czech. Chem. Comm., 1975, 40, 3055). The structures are based on spectroscopic properties and correlations with known bases.

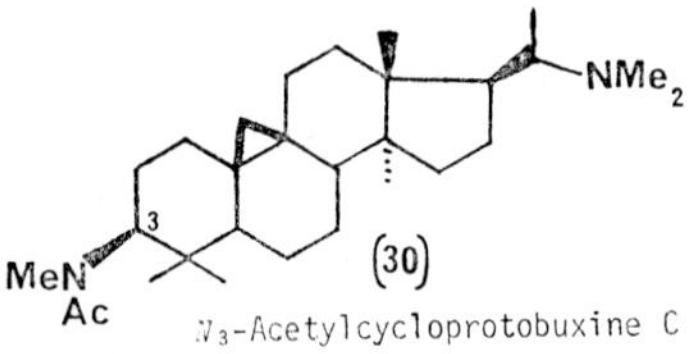

(30) N_3-Acetylcycloprotobuxine C

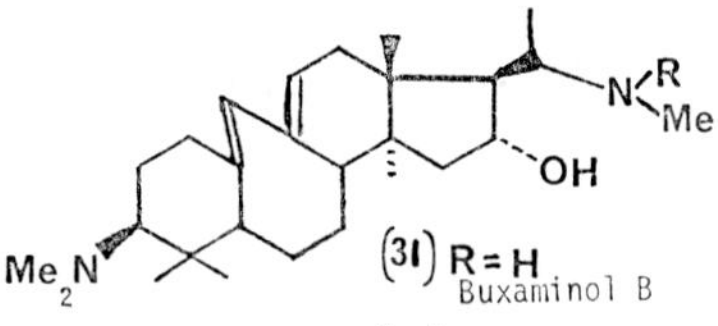

(31) R = H Buxaminol B

(32) R = Me Cyclobullatine A

ℓ-Cycloprotobuxine C has been assigned the unusual 20R structure (33) as a result of mass and nmr spectral comparisons with cycloprotobuxine C, of established structure. The α-configuration for the 3-methylamino group follows from a comparison of hydrolysis rates of 3-*N*-acyl compounds (Yunusov, *et al.,* Chem. Abs., 1975, <u>83</u>,128654; 1977, <u>86</u>, 121587).

ℓ-Cycloprotobuxine C

Cyclobuxamine-H

That cyclobuxamine H (34) has a 4α-methyl group is confirmed by a demonstration that in a Ruschig degradation epimerization of a 4β-methyl group may occur simultaneously (Voticky and V. Paulik, Coll. Czech. chem. Comm., 1977, <u>42</u>, 541).

Formal total syntheses of buxandonine (35), cycloprotobuxine F (36) and cycloprotobuxine A (37) have been achieved (Scheme 1) (C. Singh and S. Dev, Tetrahedron, 1977, <u>33</u>, 1053). The same starting material has also been used in syntheses of cyclobuxophyllinine M, cyclobuxophylline K, and buxanine M (Dev *et al.*, *ibid.*, 1981, <u>37</u>, 2935).

Buxozine C is a new alkaloid of *B. sempervirens* it is formulated as the novel structure (38) largely on spectral evidence, but also because it suffers hydrogenolysis to the known cyclovirobuxine C (39) (Voticky *et al.*, Coll. Czech. chem. Comm., 1977, <u>42</u>, 2549; Phytochem., 1977, <u>16</u>, 1860).

Cyclobuxoviricine is a new base from the leaves of *B. papilosa*. Its formulation as (40) is based on spectral studies (A.-ur-Rahman *et al.*, Phytochem., 1985, <u>24</u>, 3082).

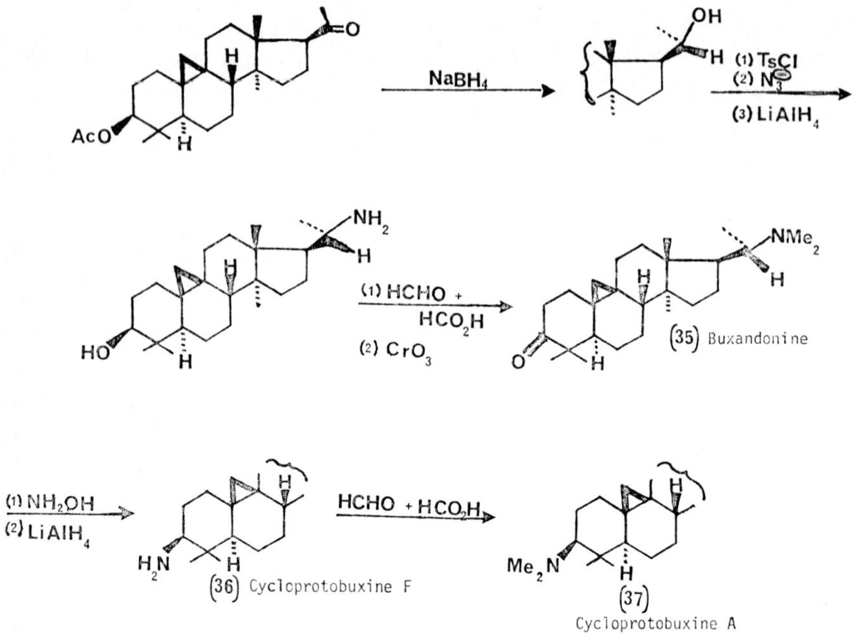
NaBH4
(1) TsCl
(2) N3
(3) LiAlH4
(1) HCHO + HCO2H
(2) CrO3
(35) Buxandonine
(1) NH2OH
(2) LiAlH4
HCHO + HCO2H
(36) Cycloprotobuxine F
(37) Cycloprotobuxine A

Scheme 1

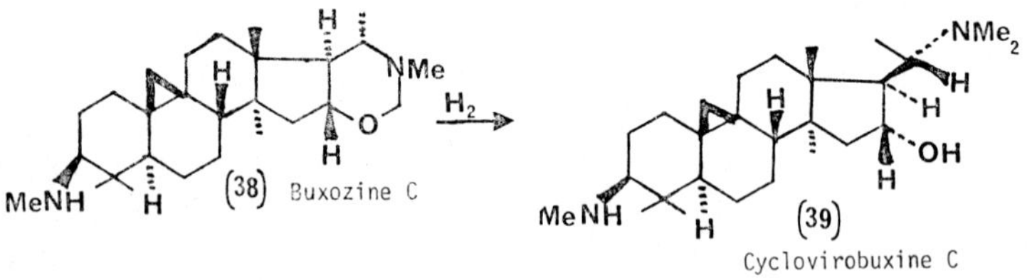
H2
(38) Buxozine C
(39) Cyclovirobuxine C

(40)
Cyclobuxoviricine

(41) RR' = O
Buxaquamarine

(42) R = NHMe, R' = H
Papilinine

Buxaquamarine, from the same source, represents an unusual, not previously encountered structural type, formulated as (41) largely on spectral (chiefly mass) evidence (*idem*, Heterocycles, 1985, 23, 1951). Papilinine (42), a third such base, should clearly be derivable from buxaquamarine by reductive amination (*idem*, Z. Naturforsch., 1985, 40B, 565).

Numerous new alkaloids continue to be isolated from *Buxus* spp.; for details see, *inter al.*, Voticky *et al.*, Coll. Czech. chem. Comm., 1981, 46, 1425; A.-ur-Rahman *et al.*, Heterocycles, 1983, 20, 69; 1985, 23, 1961; Phytochem., 1985, 24, 3082; Z. Naturforsch., 1985, 40B, 565, 567.

4. Pachysandra Alkaloids

Full details relating to structural studies on spiropachysine (43), a major alkaloid of *Pachysandra terminalis* Sieb. and Zucc. leaves, have been published (T. Kikuchi *et al.*, Chem. pharm. Bull. Japan, 1975, 23, 416). The sterochemistry at the spiro-position 3 follows from circular dichroism comparisons with structurally related compounds.

(43) Spiropachysine

5. Solanum Alkaloids

The mass spectra of Solanum bases have been reviewed (J. Tamas and M. Mak, Chem. Abs., 1977, 87, 102514; 1983, 98, 198590). Complete assignments for ^{13}C-nmr resonances of a number of Solanum alkaloids have been published (G. J. Bird et al., Austral. J. Chem., 1979, 32, 783, 797). H.p.l.c. has been used to advantage in the separation of Solanum bases (I. R. Hunter et al., J. Chromatog., 1976, 119, 223).

(a) Spirosolane Alkaloids

Solasodine (44) yields two N,O-diformyl derivatives. The difference between these isomers has been shown by spectroscopic study to be one of stereochemistry at C-22 rather than of restricted rotation about the C=N partial double bond of the amide group, or sterically hindered inversion of the nitrogen atom, proposed earlier. The two isomers have 22R, 25R (major) and 22S, 25R (minor) configurations. When either isomer is heated ring F is cleaved; when the ring is re-formed the former isomer is kinetically favoured, whilst the latter is thermodynamically more stable. Analysis of their pmr- and ^{13}C-nmr-spectra reveals that in both ring F is a twist-boat, the two being formulated as (45) and (46) (W. Gaffield et al., Austral. J. Chem., 1983, 36, 325). The alkaloid also forms two N-methyl-O-acetyl derivatives, and two *N*-formyl-*O*-acetyl derivatives; presumably a similar explanation applies (G. Kusano *et al.*, Heterocycles, 1975, 3, 697).

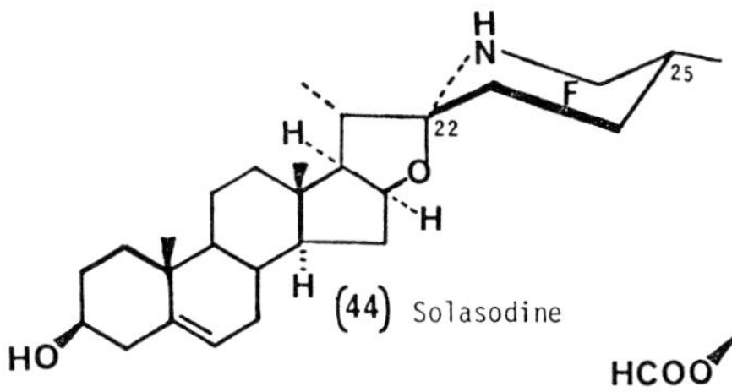

(44) Solasodine

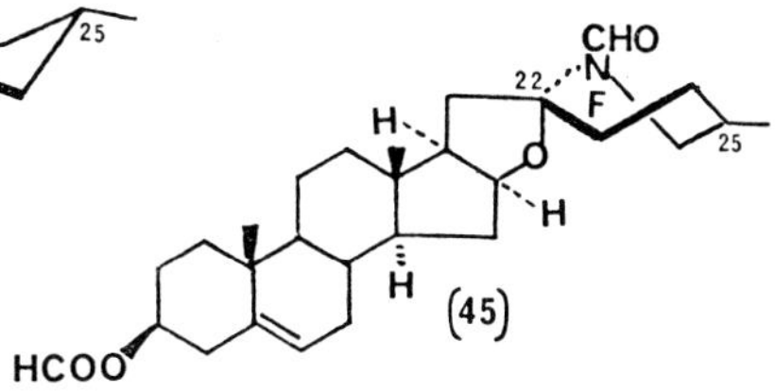

(45)

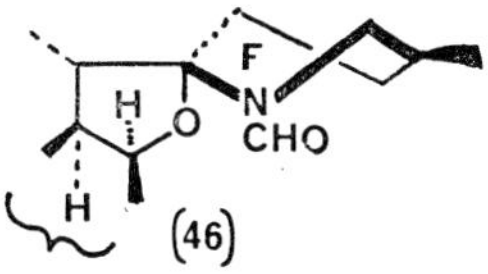

(46)

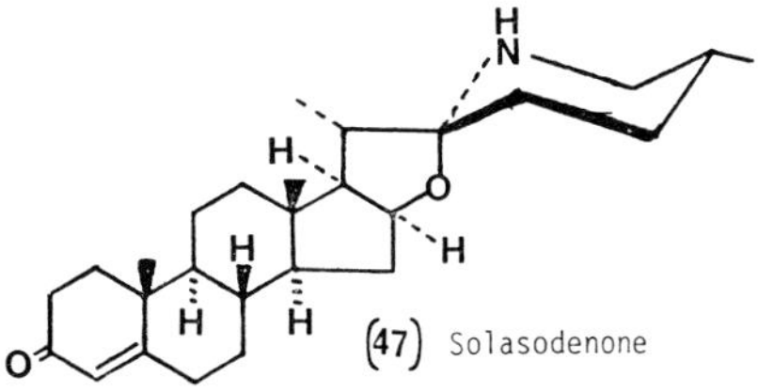

(47) Solasodenone

Solasodenone, a new base from *S. hainanense*, is formulated as (47) on spectral evidence. Its absolute configuration follows from o.r.d. studies and molecular rotation differences (G. Adam *et al.*, Phytochem., 1978, 17, 1070). This formulation is confirmed by the observed Oppenauer oxidation of solasodine (44) to the alkaloid. Solaparnaine occurs in the green berries of *S. asperum* Vahl. An analysis of its mass and ^{13}C-nmr-spectra points to structure (48) (J. Bhattacharyya, Heterocycles, 1958, 23, 3111). Solandunalidine, from *S. dunalianum*, is a diacidic base (49), formulated on extensive spectral evidence and by a link-up with tomatidine (Bird *et al.*, Tetrahedron Letters, 1978, 159; Austral. J. Chem., 1979, 32, 611).

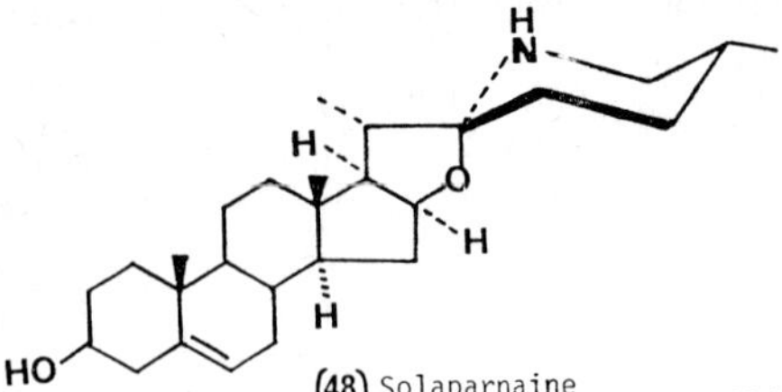

(48) Solaparnaine

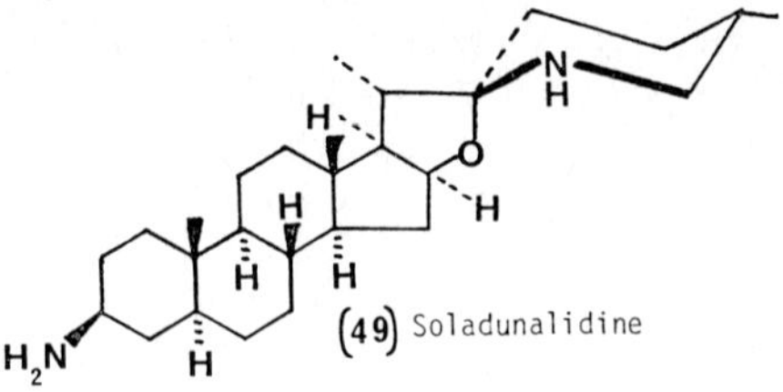

(49) Soladunalidine

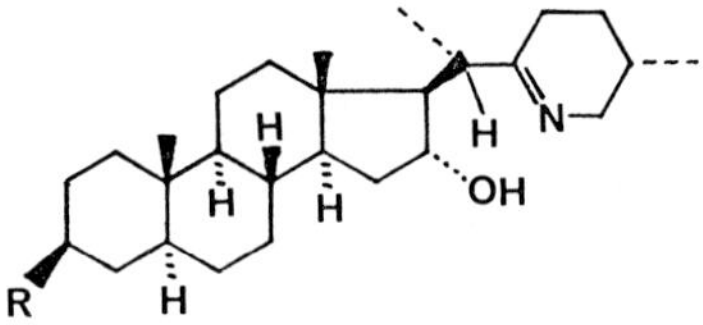

(50) R = OH 25-Isosolafloridine

(51) R = NH₂ Solacallinidine

(b) Alkaloids related to 20-piperidyl-5α-pregnane

S. callium contains two glycoalkaloids which on hydrolysis afford 25-isosolafloridine (50) and solacallinidine (51), respectively. These have been investigated spectrally and the hydrochloride of the former subjected to an X-ray diffraction analysis (Bird et al., Tetrahedron Letters, 1976, 3653; Acta Cryst., 1977, 33B, 3782). A chemical correlation between the pair has been established (idem, Austral. J. Chem., 1979, 32, 597).

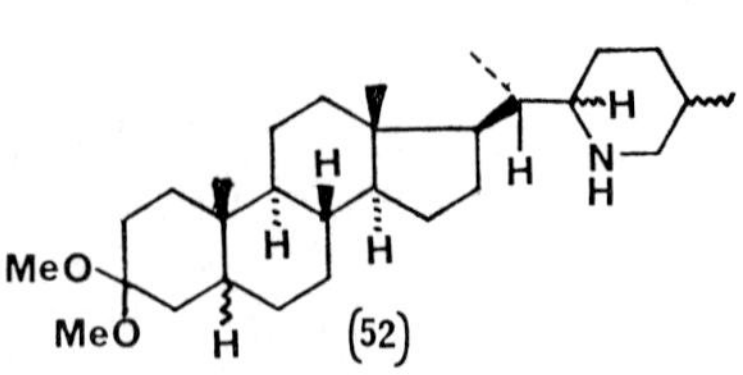

Solaquidine

(52)

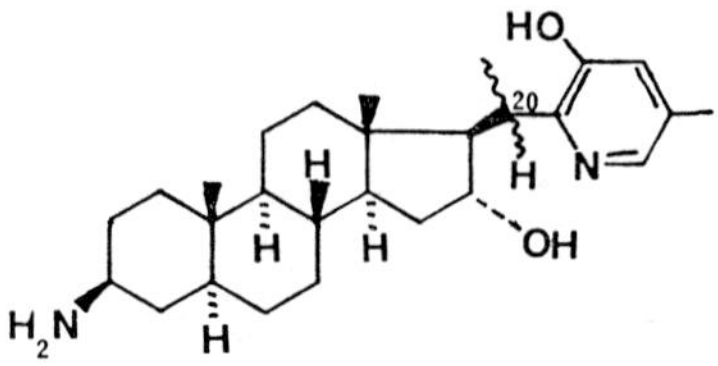

(53) S at C-20 Solaseaforthine

(54) R at C-20 Isosolaseaforthine

Solaquidine, a constituent of *S. pseudoquina*, has been examined in spectroscopic detail and is formulated as the ketal (52). Since methanol was used for extraction of the base it may be an artefact (A. Usubilliga *et al.*, Phytochem., 1977, 16, 1861). Two pyridine-derived alkaloids solasea-forthine and isosolaseaforthine, have been assigned structures (53) and (54), largely on spectroscopic and c.d. evidence. They are diastereomers differing in configuration at C-20 (S and R) (Pakrashi *et al.*, Tetrahedron Letters, 1978, 3871). Solafilidine (55) and its desacetyl derivative (56) have been isolated as major alkaloids of the dried fruits of *S. eucadorensis* (P. Martinod *et al.*, Chem. Abs., 1979, 90, 148439).

(55) R = Ac Solafilidine

(56) R = H

(57) Solaverbascine

Solaverbascine, occurring in *S. verbascifolium* leaves, has been assigned structure (57) from spectral study and because it is formed by reductive ring-cleavage of solaso-dine (44) (G. Adam *et al.*, Phytochem., 1980, 19, 1002); conversely it affords solasodine on manganese dioxide oxidation (Adam and H. T. Huong, Tetrahedron Letters, 1980, 21, 1931; J. prakt. Chem., 1981, 323, 839). The latter reaction occurs *via* 22-N double bond formation followed by spontaneous cyclization.

(c) Alkaloids related to 22-pyrrolidyl-5α-pregnane

Tomatillidine has been re-formulated as (58) as a result of certain new observations: (i) a reference compound with a 2-piperidein-3-one unit underwent a ring contraction during silica gel chromatography to a 2-acyl-2-pyrrolideine;

(ii) borohydride reduction of 5,6-dihydrotomatillidine
followed by acetylation afforded three triacetyl derivatives which were separated and examined by decoupled pmr-
spectroscopy and shown to have the gross structure (59);
a piperidine-derived structure was to be expected on the
basis of the original formulation, (iii) a synthesis of
tomatillidine from solasodine (44) has been achieved; the
last step involves a ring contraction promoted by silica
gel, exactly as observed under (i) above, and (iv),
spectral data obtained from the alkaloid agree much more
satisfactorily with the new formulation (Kusano *et al.*,
Chem. pharm. Bull. Japan, 1976, 24, 661).

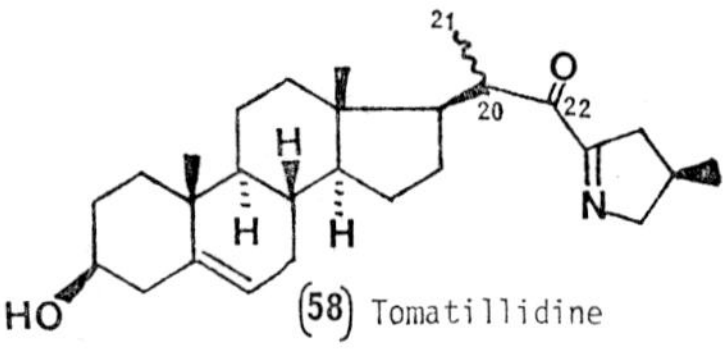

(58) Tomatillidine 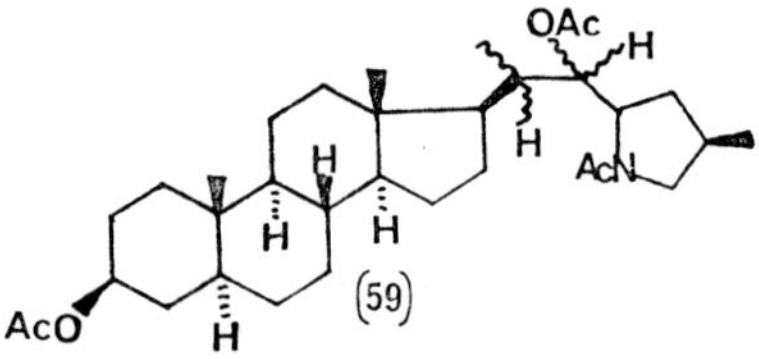 (59)

As a consequence of an X-ray diffraction analysis
solamaladine is re-formulated as (60) rather than as an
isomeric piperideinone structure as suggested earlier,
(Usubillaga *et al*, Acta Cryst., 1982, 38B, 966).

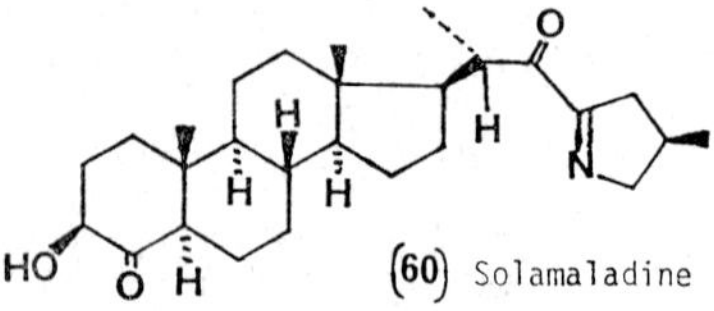

(60) Solamaladine

(d) Alkaloids containing an indolizidine unit

Three isomeric bases *solanogantine*, solanogantamine and isosolanogantamine have been found in *S. giganteum* leaves They are formulated as (61), (62) and (63) respectively, on spectral evidence from the bases and their derivatives (Pakrashi *et al.*, Tetrahedron Letters, 1977, 645, 814; J. Indian chem. Soc., 1978, 55, 1109).

(61) Solanogantine

(62) R = β-NH$_2$ Solanogantamine

(63) R = α-NH$_2$ Isosolanogantamine

Solanopubamine, from the aerial parts of *S. pubescens*, has been investigated spectroscopically. It is isomeric with solanogantamine (62) and since its structure and stereochemistry have been settled (64) it follows that the C-25 methyl group in solanogantamine must be β (Kaneko *et al.*, Phytochem., 1985, 24, 1369). The indolizidine group in these bases was detected by the so-called Bohlmann ir-band at ∿ 2750 cm^{-1}.

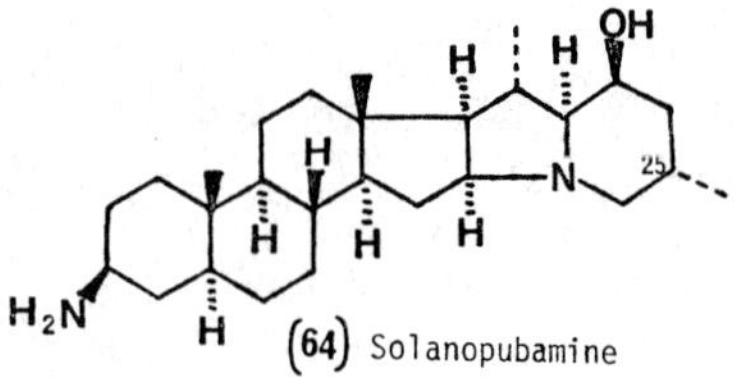

412

(64) Solanopubamine

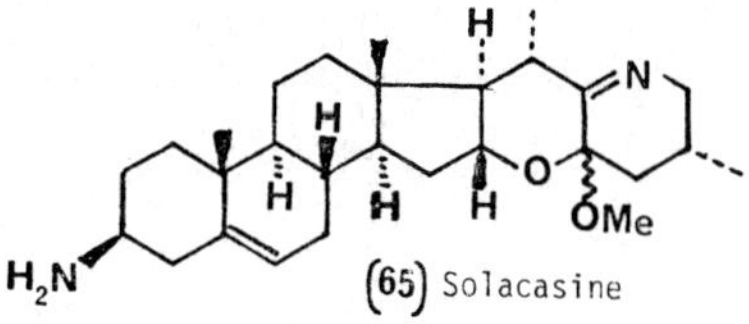

(65) Solacasine

(e) Miscellaneous alkaloids

An antimicrobial base solacasine has been found in the
flowers of *S. pseudocapsicum*. In its mass spectrum it
showed peaks at m/z 56 and 82, associated with 3β-amino-
steroids. An azomethine linkage is responsible for its ir-
absorption at 1660 cm.$^{-1}$; on reduction (NaBH$_4$) a primary-
secondary diamine structure results, convertible into a
trimethyl derivative on Eschweiler-Clark methylation.
Structure (65) has been advanced and confirmed by the formation of dihydrosolacasine by methanolysis of solano-
capsine (L. A. Mitscher *et al.*, Experientia, 1976, 32, 415).
Solanoforthine, a similarly-structured base from *S. sea-
forthianum* is formulated as (66) largely on spectral evidence
(Pakrashi *et al.*, Tetrahedron, 1977, 33, 1371).

(66) Solanoforthine

6. Veratrum and Fritillaria Alkaloids

The pharmacological properties of *Veratrum* alkaloids have been reviewed (H. P. Buech, Chem. Abs., 1976, 85, 87115). A review of investigations conducted by J. P. Kutney's group has been published (Kutney, Bio-org. Chem., 1977, 6, 371). Reviews on *Veratrum* ester bases (N. V. Bondarenko, Chem. Abs., 1978, 89, 110064) and on the synthesis of C-nor-D-homo structured bases (E. Brown and M. Ragault, Tetrahedron, 1979, 35, 911) have appeared. Field desorption mass spectrometry has been used extensively in structural studies on glycoalkaloids in this area (Kawasaki *et al.*, Phytochem., 1982, 21, 187).

(a) Piperidylpregnane and indolizidine alkaloids

Several alkaloids of this group are related structurally to members of the *Solanum* family (see p.406). For example muldamine, a previously known alkaloid of *V. californicum*, has now been reformulated as (67), and teinemine (deacetyl-muldamine) is now (68) (R. F. Keeler *et al.*, Phytochem., 1982, 21, 2397). The new assignments are the result of spectral investigation, and, in the latter case, of X-ray diffraction analysis.

(67) R = Ac Muldamine
(68) R = H   Teinemine

Veralodinine
(69) G = β-D-glucosyl

Veralodinine, found in *V. lobelianum*, is formulated as (69) on chemical and spectroscopic grounds (Yunusov *et al.*, Chem. Abs., 1975, 83, 79473). Further chemical studies have confirmed the structure (70) advanced earlier for veracintine (Voticky, J. Tomko *et al.*, Coll. Czech. chem. Comm., 1976, 41, 2964). Its β-D-glucoside (71) is gluco-veracintine, occurring in *V. album*, subsp. *lobelianum* (Tomko *et al.*, Chem. Abs., 1978, 89, 39410).

(70) R =H   Veracintine

(71) R = β-D-glucosyl   Glucoveracintine

Hapepunine (72), occurring in *Fritillaria camtschatcensis*, is the first natural 16β-hydroxy-22,26-epiminocholestane derivative encountered (Mitsuhashi *et al.*, Tetrahedron Letters, 1978, 2099). Anrakorinine, from the same source, affords a tosylate which on reduction (LiAlH₄) yields hapepunine (72). In its pmr-spectrum it lacks a methyl singlet at δ 0.96 ppm (present in hapepunine), but has an AB quartet (2H) at 3.62 and 3.88 ppm. Consequently it is formulated as 18-hydroxyhapepunine (73) (*idem*, Phytochem., 1981, 20, 157). Several indolizidine bases, also related to *Solanum* Alkaloids, have been encountered in the *Fritillaria* group. Camtschatcanidine, for example, from *F. camtschatcensis*, is (74), on the basis of spectral comparison with solanidine: on reduction (LiAlH₄) of its O-tosyl-derivative solanidine (75) is formed (Mitsuhashi *et al.*,

Phytochem., 1981, 20, 327). Stenantine and stenantidine, occurring in the aerial parts of *Rhinopetalum stenantherum*, are glycosides of solanidine (75), and have been subjected to partial and total hydrolysis, with product identification, and to spectral study, and are formulated as a consequence as (76) and (77) respectively (Yunusov *et al.*, Chem. Abs., 1982, 96, 20313).

(72) R = CH₃ Hapepunine

(73) R = CH₂OH Anrakorinine

(74) R = H, R' = OH Camtschatcanidine

(75) R = R' = H Solanidine

(76) R = β-D-Glucopyranosyl-(1→6)-[α-L-rhamnopyranosyl-(1→4)]-β-D-glucopyranosyl

Stenantine

(77) R = β-D- glucopyranosyl-(1→6)-β-D-glucopyranosyl

Stenantidine

(b) C-Nor-D-homosteroidal alkaloids

(i) Alkamines
The total synthesis of verticine (79) is set out below (Kutney *et al.*, J. Amer. chem. Soc., 1977, 99, 963, 964). The starting diacetoxyketone (78) is available by breakdown of hecogenin.

The structure of imperialine has been revised to (80), with a *cis* D/E ring junction, on physical and chemical evidence. In particular, an X-ray diffraction analysis of

(79) Verticine

the methobromide reveals the structure and stereochemistry depicted (S. Ito *et al.*, Tetrahedron Letters, 1976, 3161). Likewise, the stereochemistry of veramarine must be modified to (81), with a 16β-hydroxyl group, on the basis of a similar analysis of its 3-*O*-acetate (Tomko and F. Pavelcik, *ibid.*, 1979, 887; Acta Cryst., 1979, 35B, 1790).

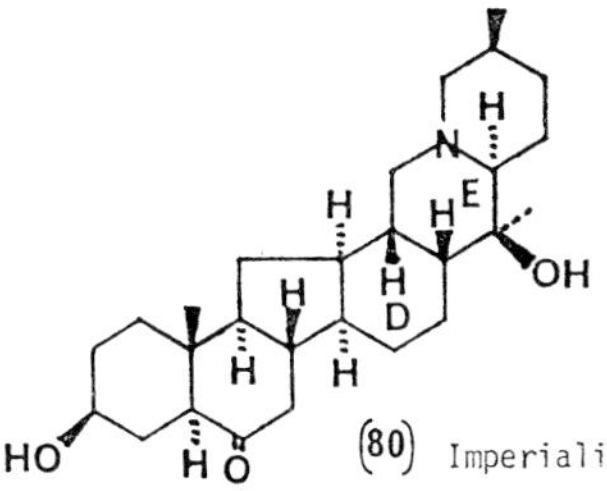

(80) Imperialine

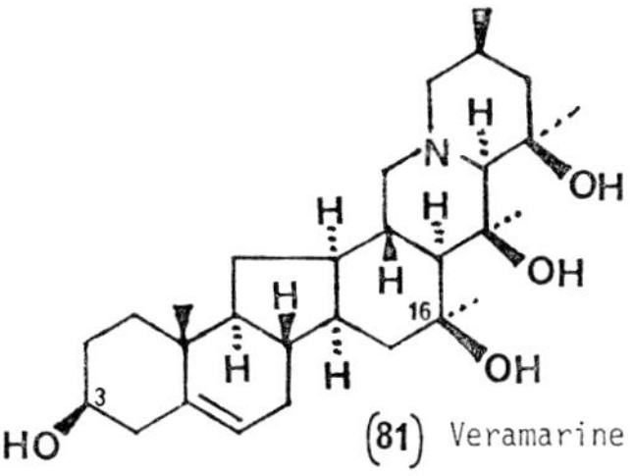

(81) Veramarine

Shinonomenine, found in *V. grandiflorum* seedlings is formulated as (82) on the basis of an X-ray study on its hydroiodide (Mitsuhashi *et al.*, Tetrahedron Letters, 1978, 4801).

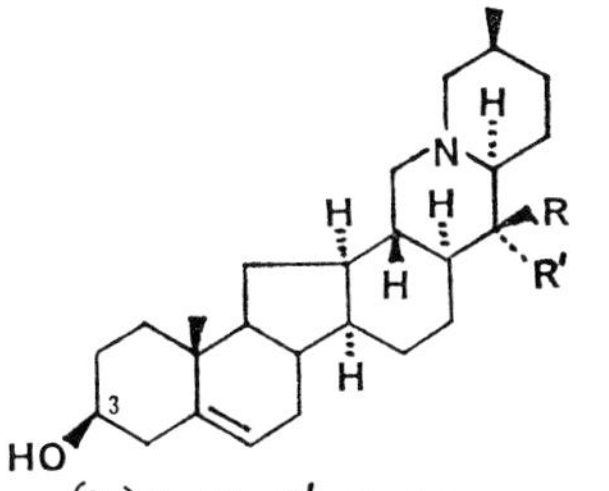

(82) R = Me, R'=H Shinonomenine
(83) R =OH, R'= Me Veraflorizine

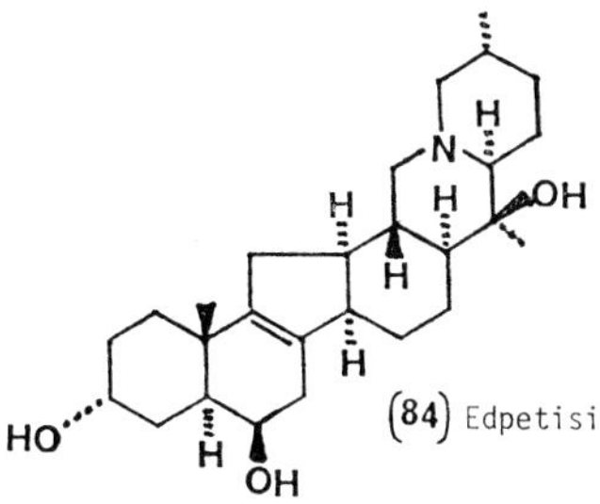

(84) Edpetisidine

The structure of veraflorizine (83), from the same source, has been deduced from spectral measurements on it and its 3-*O*-acetyl derivative, which has also been partially synthesised from verticinone (fritillarine) (Mitsuhashi *et al.*, *loc. cit.*). Edpetisidinine, a new base from *Petilium eduardi* is formulated as (84) on spectroscopic evidence (Yunusov *et al.*, Chem. Abs., 1979, 90, 23341).

Two new bases isolated from *F. delavayi* are delavine (85) and delavinone (86), formulated from physical measurements, with confirmation for the latter by X-ray diffraction analysis (Kaneko *et al.*, Chem. pharm. Bull. Japan, 1985, 33, 2614).

(85) R = OH, H, R'=H, R"= Me
Delavine

(86) R =O, R'=H, R"=Me
Delavinone

(87)
Procevine

Procevine, also from *V. grandiflorum* seedlings, has a novel structure, based on spectroscopic analysis and biogenetic considerations. Its formulation (87) has been confirmed by a synthetic link-up with isorubijervine (Mitsuhashi *et al.*, Tetrahedron Letters, 1978, 4801). The base, also known as pseudosolanidine, had been encountered earlier as a synthetic product (S. W. Pelletier and W. A. Jacobs, J. Amer. chem. Soc., 1953, 75, 4442).

(ii) Ester-alkaloids

The structure and stereochemistry of veratridine (88) have been confirmed by X-ray diffraction analysis of its hydro-perchlorate (P. W. Codding, J. Amer. chem. Soc., 1983, 105, 3172). *Veratrum lobelianum* has yielded two new germine derivatives, germinaline (89) and an un-named base 15-ℓ-2-methylbutanoylgermine (90). The former was formu-lated on the basis of spectroscopic and chemical compari-sons with the known germitetrine (91), which on partial hydrolysis afforded germinaline. The latter was correlated with neogermitrine (92) (Yunusov *et al.*, Chem. Abs., 1983, 99, 71046, 71047).

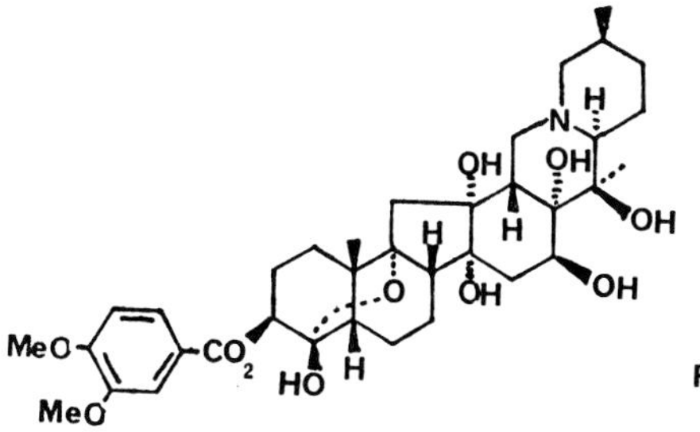

(88) Veratridine

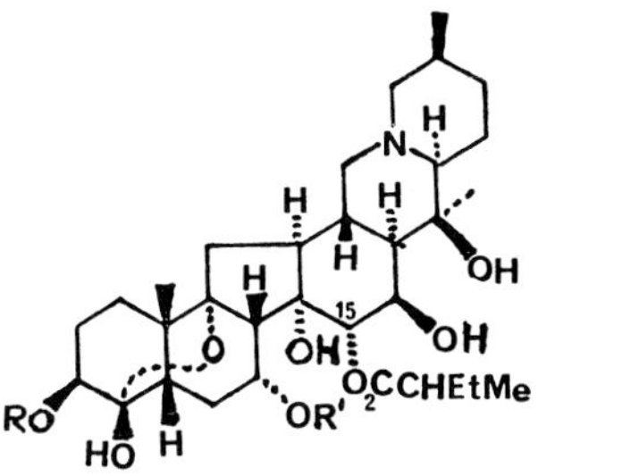

(89) R = MeCH(OAc)CMe(OH)CO, R' = H
Germinaline

(90) R = R' = H

(91) R = MeCH(OAc)CMe(OH)CO, R' = Ac
Germitetrine

(92) R = R' = Ac Neogermitrine

Full details of the total syntheses of verarine and 5α,6-dihydroveratramine are now available (Kutney *et al.*, Canad. J. Chem., 1975, <u>53</u>, 1775, 1796). This is a wide-spectrum synthesis, since the latter has been converted into several other *Veratrum* bases. Germinalinine, a new alkaloid from *V. lobelianum*, is derived from germine and is formulated as (93) on chemical and spectroscopic evidence. It yields a diacetate identical with the known germbudine triacetate, yields germbudine (94) on methanolysis, and its pmr-spectrum is closely similar to that of germbudine (Yunusov *et al.*, Chem. Abs., 1976, <u>84</u>, 59816).

(93) R = Ac, R′ = H Germinalinine

(94) R = R′ = H Germbudine

7. Asclepiadaceae Alkaloids

Stephanthranilines C and A are new bases from *Stephanotis japonica*; they are steroidal esters of *N*-methylanthranilic acid. The former, on spectral evidence, is assigned structure and sterochemistry represented by (95) (S. Terada *et al.*, Tetrahedron Letters, 1978, 1995). The latter on total hydrolysis affords the known steroid sarcostin (96), and a careful analysis of its pmr- and ^{13}C-nmr-spectra, with appropriate comparisons with model esters derived

(95) R = [structure with NHMe / CO]

Stephanthraniline C

(96) R = R' = H

(97) R = [structure with NHMe / CO], R' = Ac

Stephanthraniline A

from sarcostin, pointed to structure (97). Mild basic hydrolysis of the alkaloid (removal of acetyl group) is accompanied by migration of the remaining ester group to give (96) (R=H, R'=o-MeNHC$_6$H$_4$CO)(Terada *et al.*, Chem. pharm. Bull. Japan, 1977, 25, 2802). Two minor bases from the same source, stephanthraniline B and dihydrogagaminine, both afford dihydrosarcostin on hydrolysis. They are consequently assigned structures (98) and (99) respectively (Terada and Mitsuhashi, *ibid.*, 1979, 27, 2304).

Tomentomine, a new alkaloid of *Marsdenia tomentosa*, on basic hydrolysis affords the known aglycone tomentogenin (100). Analysis of its spectra points to structure (101) for the base (H. Seto *et al.*, *ibid.*, 1977, 25, 876).

8. Miscellaneous Steroidal Alkaloids

The skin of the Colombian frog *Phyllobates terribilis* contains many alkaloids, including batrachotoxin (102), homobatrachotoxin (103), 4β-hydroxybatrachotoxin (104), and 4β-hydroxyhomobatrachotoxin (105). Their ^{13}C-nmr-spectra have been analysed carefully and assignments reported (T. Tokuyama and J. W. Daly, Tetrahedron, 1983, 39, 41).

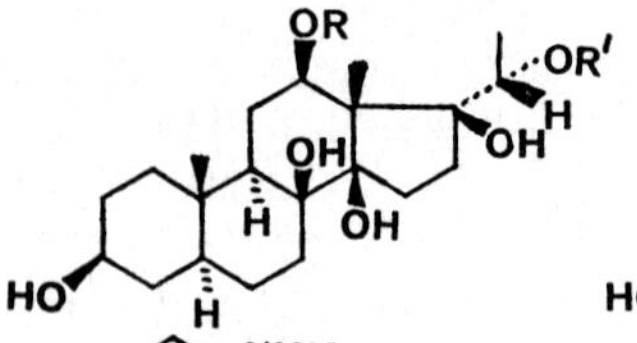

(98) R = [structure] NHMe, R' = Ac
CO

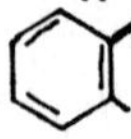

Stephanthraniline B

(99) R = PhCH≐CHCO, R' = nicotinoyl
Dihydrogagaminine

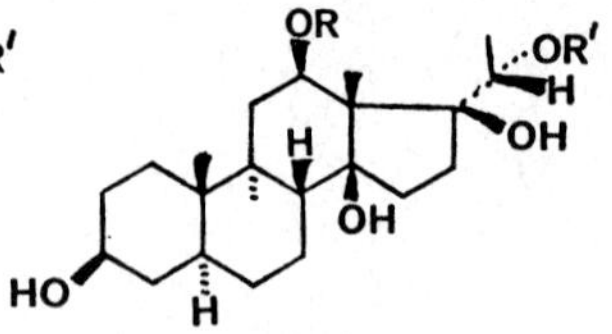

(100) R = R' = H

(101) R = PhCH≐CHCO, R' = nicotinoyl
Tomentomine

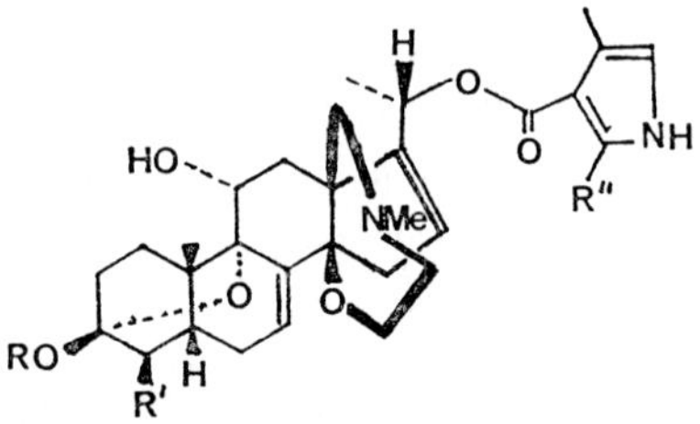

(102) R = R' = H, R" = Me
(103) R = R' = H, R" = Et
(104) R = H, R' = OH, R" = Me
(105) R = H, R' = OH, R" = Et

Two alkaloids have been isolated from sponges of *Plakina* spp. They are the plakinamines A (106) and B (107), their structures being established mainly by comparison of their ^{13}C-nmr-spectra with those of model compounds synthesized from ergosterol. These bases, both of which have anti-microbial properties, are the first steroidal alkaloids to be encountered in a marine organism (R. M. Rosser and D. J. Faulkner, J. org. Chem., 1984, 49, 5157).

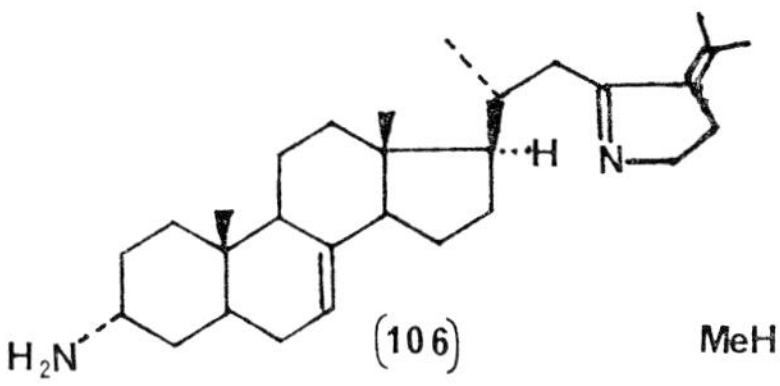

Plakinamine A

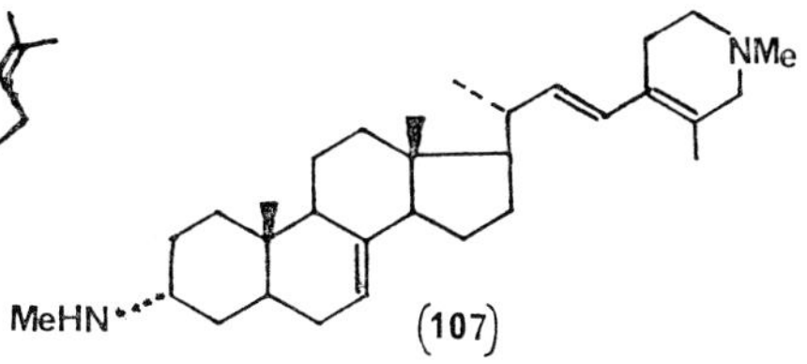

Plakinamine B

9. Biosynthesis of Steroidal Alkaloids

It appears that the biosynthesis of tomatidine (108) and related alkaloids occurs *via* amination of 26-hydroxylated steroids such as (109), followed by formation of the tetrahydrofuran ring (F. Ronchetti *et al.*, Phytochem., 1975, 14, 2423).

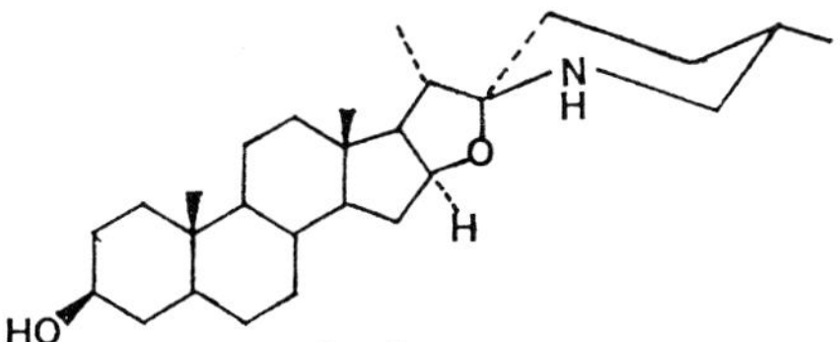

Tomatidine

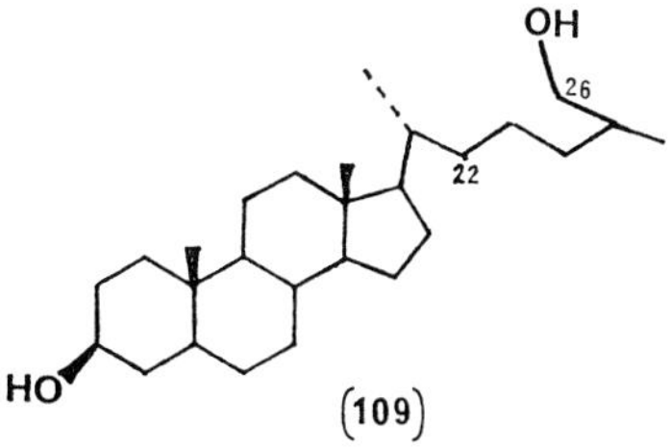

Verazine (110) has been shown to be an early intermediate in the biosynthesis of *V. grandiflorum* alkaloids, for example solanidine (75); arginine is a primary source of the nitrogen (Kaneko *et al.*, *ibid.*, p. 1295; 1976, 15, 1391). Hydroxylation of one of the terminal methyl groups of cholesterol is an early step in steroidal biosynthesis; it is followed by functionalization at C-22, then nitrogen-ring formation. Formation of the tetrahydrofuran ring (where applicable) is a much later process (Ronchetti *et al.*, *loc. cit.*; Chem. Comm., 1977, 286; R. Tschesche *et al.*, Phytochem., 1976, 15, 1387; 1978, 17, 251). (25R)-26-Aminocholesterol (111) has been found to be a significant precursor of solasodine (44) in *S. laciniatum* (Tschesche and H. R. Brennecke, *ibid.*, 1980, 19, 1449). The diol (112) on the other hand is poorly utilized by the plant; this appears to confirm that replacement of the 26-hydroxyl group occurs before oxygenation at C-16.

(110)
Verazine

(111) R = H

(112) R = OH

TABLE

Group	Alkaloid	M.p (°C)	$[\alpha]_D$(°) (solvent)
Apocynaceae	Kisantamine		
	Holacetine	258	+6.9 (EtOH)
	Holarricine		
	Irehdiamine F		
	20-Epi-irehdiamine I	151-153	-36 ($CHCl_3$)
	Holarrhesine		
Buxus	$\underline{N}_3$-Acetyl-cycloproto-buxine C		
	Buxaminol B	225	+20 (MeOH)
	Cyclobullatine A	275	-99 (EtOH)
	ℓ-Cycloproto-buxine C		
	Cyclobuxamine H	209-211 (decomp.)	+30 ($CHCl_3$)
	Buxozine C	137	+65 ($CHCl_3$)
	Cyclobuxoviri-cine	amorphous	-54 ($CHCl_3$)
	Buxaquamarine	gum	+24 ($CHCl_3$)
	Papilinine	gum	+29.4 ($CHCl_3$)
Pachysandra	Spiropachysine	290-292	+35 ($CHCl_3$)
Solanum	Solasodenone	178	+28 ($CHCl_3$)
	Solaparnaine	228-230	-77.8 (MeOH)
	Solandunalidine	145-153	+1.3 ($CHCl_3$)
	25-Isosolaflori-dine	164.5-166.5	+44.8 ($CHCl_3$)
	Solacallinidine	175-178	+51.3 ($CHCl_3$)
	Solaquidine	278-281	
	Solaseaforthine	172-178	+22 (MeOH)
	Isosolaseafor-thine	172-180	+26 (MeOH)
	Solafilidine		
	Desacetylsola-filidine		
	Solaverbascine	263-265	-67.9 ($CHCl_3$)

	Solamaladine	178-180	
	Solanogantine	syrup	
	Solanogantamine	180	+35 ($CHCl_3$)
	Isosolanogant-amine	252-254	+31 ($CHCl_3$)
	Solanopubamine	263	+30.5 (MeOH)
	Solacasine	215-220	+29 (MeOH)
	Solanoforthine	208-210	-26.6 ($CHCl_3$)
Veratum and Fritillaria	Teinemine	205-207	-38.7 ($CHCl_3$)
	Veralodinine		
	Glucoveracintine		
	Hapepunine	196.5-198.5	-72.6
	Camtschatcani-dine	261-265	-19.4 (MeOH)
	Stenantine		
	Stenantidine		
	Shinonomenine	95-96	-90.7 ($CHCl_3$)
	Veraflorizine	175-176	-91 ($CHCl_3$)
	Edpetisidinine		
	Delavine	182-183	-20 ($CHCl_3$)
	Delavinone	182-184	-54 ($CHCl_3$)
	Procevine (pseudosolani-dine)	235-237	-12.2 ($CHCl_3$)
	Germinaline		
	15-$\underline{l}$-2-Methyl-butanoyl-germine		
	Germinalinine		
	Germbudine		
Asclepiadaceae	Stephanthrani-line C	amorphous	4.2 ($CHCl_3$)
	Stephanthrani-line A	170-173	+17.9 ($CHCl_3$)
	Stephanthrani-line B	165-168	-24.6 ($CHCl_3$)
	Dihydrogagamine	amorphous	+105 ($CHCl_3$)
	Tomentomine	155-157	+137 ($CHCl_3$)
Miscellaneous	Batrachotoxin		
	Homobatracho-toxin		
	4β-Hydroxybatracho-toxin		

4β-Hydroxyhomoba-
trachotoxin

Plakinamine A	129-130 (decomp.)	+16 (CHCl$_3$)
Plakinamine B*	180-200	+29 (MeOH)

*dihydrochloride

Guide to the Index

This index is constructed in a similar manner to the volume indexes of the first edition of the Chemistry of Carbon Compounds. However, to make the index easier to use, more descriptive entries have been made for the commonly occurring individual, and groups of chemicals.

The indexes cover primarily the chemical compounds mentioned in the text, and also include reactions and techniques, where named, and some sources of chemical compounds such as plant and animal species, oils, etc.

Chemical compounds have been indexed alphabetically under the names used by authors, editing being restricted to ensuring uniformity of entries under the same heading. In view of the alternative nomenclature that can often be used, a limited amount of cross-referencing has been done where it is considered to be helpful, but attention is particularly drawn to Convention 2 below.

For this and the succeeding volumes, the indexing conventions listed below have been adopted.

1. *Alphabetisation*

(a) The following prefixes have not been counted for alphabetising:

n-	*o-*	*as-*	*meso-*	D	*C*
sec-	*m-*	*sym-*	*cis-*	DL	*O-*
tert-	*p-*	*gem-*	*trans-*	L	*N-*
	vic-				*S-*
		lin-			*Bz-*
					Py-

Some prefixes and numbering have been omitted in the index, where they do not usefully contribute to the reference.

(b) The following prefixes have been alphabetised:

Allo	Epi	Neo
Anti	Hetero	Nor
Cyclo	Homo	Pseudo
	Iso	

(c) A letter by letter alphabetical sequence is followed for entries, firstly for the main entry, followed by the descriptive entry. The only exception to this sequence is the placing of plural entries in front of the corresponding individual entries to prevent these being overlooked by a strict alphabetical sequence which could lead to a considerable separation of plural from individual entries. Thus "butanes" will come before *n*-butane, "butenes" before 1-butene, and 2-butene, etc.

2. *Cross references*

In view of the many alternative trivial and systematic names for chemical compounds, the indexes should be searched under any alternative names which may be indicated in the main body of the text. Only a limited amount of cross-referencing has been carried out, where it is considered that it would be helpful to the user.

3. *Esters*

In the case of lower alcohols esters are indexed only under the acid, e.g. propionic methyl ester, not methyl propionate. Ethyl is normally omitted e.g. acetic ester.

4. *Derivatives*

Simple derivatives are not normally indexed if they follow in the same short section of the text.

5. *Collective and plural entries*

In place of "– derivatives" or "– compounds" the plural entry has normally been used. Plural entries have occasionally been used where compiunds of the same name but differing numbering appear in the same section of the text.

6. *Main entries*

The main entry of the more common individual compounds is indicated by heavy type. Multiple entries, such as headings and sub-headings over several pages are shown by "–", e.g., 67–74, 137–139, etc.

INDEX